Nanotechnology In Construction

Nanotechnology In Construction

Edited by

P.J.M. Bartos, J.J. Hughes, P. Trtik and W. Zhu
University of Paisley, Paisley, Scotland

RS•C

advancing the chemical sciences

The proceedings of the 1ˢᵗ International Symposium on Nanotechnology in Construction held on 23–25 June 2003, at the University of Paisley, Paisley, Scotland

Special Publication No. 292

ISBN 0-85404-623-2

A catalogue record for this book is available from the British Library

Published by The Royal Society of Chemistry,
Thomas Graham House, Science Park, Milton Road,
Cambridge CB4 0WF, UK

Registered Charity Number 207890

For further information see our web site at www.rsc.org

Printed by Athenaeum Press Ltd, Gateshead, Tyne and Wear, UK

Preface

Nano-scale science emerged almost a century ago when molecular and then atomic- size objects were first observed and identified. Since then, *nanoscience* has developed greatly, particularly in the fields of physics, chemistry, medicine and fundamental materials science.

Nanotechnology emerged much later, with a rapid advance since the early 1990s, when demand for characterisation, manipulation, production or even assembly of objects at the nano-scale increased *and* the leading edge of technology advanced sufficiently to provide the first generation of effective tools for such applications.

Unlike nanoscience, *nanotechnology is an enabling technology*, which was particularly driven by the advancing micro-electronics industry, where continued miniaturisation was commercially highly desirable. The exploitation of nanotechnology progressed most rapidly in areas where there were immediate applications, leading to products of very high value or products which were mass-produced, where there was a strong market pull. In such areas the high cost of nanotechnology-based Research and Development facilities were recoverable over a commercially acceptable short to medium term. Over the last few years, promoters of nanotechnology succeeded in raising steeply the public profile of nanotechnology. Significant exploitation of new opportunities for investment commenced, assisted by a decline in growth of IT based business.

The construction industry differs substantially from other industries, especially in the unique nature of its product and in historically very low Research and Development investment, which often relies on adaptation and exploitation of advances from other scientific and technical fields. Construction is a very significant global economic area. However, largely because of its special characteristics, very little is known about the existing and likely future impact of nanotechnology. Where many conferences and similar events were already held in other, non-construction related, scientific and technical sectors, networking on an international scale advanced rapidly. It was time to organise an international forum on the specific topic of *construction and nanotechnology*, with a very wide scope, covering the whole spectrum of Research and Development and commercial construction and construction-related activities. The Advanced Concrete and Masonry Centre of the University of Paisley took on the organisation of the first such event in 2000, leading to the 1st International Symposium for Nanotechnology in Construction being held in Paisley in June 2003.

It was not a coincidence that the organisation of the 1st Symposium on Nanotechnology in Construction (NICOM 1) was carried out by the Advanced Concrete and Masonry Centre (ACMC) and the attached Scottish Centre for Nanotechnology in Construction Materials in Paisley, Scotland. The ACMC already had a track record of active research, which depended on the very early introduction and exploitation of nanotechnology in the construction materials field and commenced well before the recent rapid rise in promotion of nanotechnology overall. Research into advanced (simultaneously stronger and tougher) composites for construction applications, based on reinforcement of brittle matrices (e.g. ceramic, carbon, but particularly cementitious) by multifilament reinforcing elements, had been carried out at the University of Paisley since the late 1970s. Further development and exploitation of this promising novel approach relied critically on characterisation and subsequent design/engineering of bond within and outside the fibre bundles. Progress slowed down in the absence of a technology to enable measurements and characterisation of key properties at the sub-micron scale. A sharp technology-watch was therefore maintained and when the first generation, then leading edge, nanotechnology-based instrumentation was developed for other applications (thin films/coatings, micro-electronics etc.), it was adapted for the new purpose and installed at the ACM Centre in Paisley in the early 1990s. The ACMC facilities, which enabled a gradual move from micro to nano-scale load application and indentation, were continually improved as technology advanced. In conjunction with the application of other advanced techniques, such as Focused Ion Beam micro-fabrication, which facilitated hitherto un-achievable shaping of diamond indenters, research progressed and Proof of the Concept for design and performance of the new generation of advanced composites was obtained in the late 1990s. This provided the required evidence for funding of a second, currently state-of-the-art, nano-indentation facility housed in a new specialist laboratory built away from the busy main Campus of the University of Paisley. A Scottish Centre for Nanotechnology in Construction Materials (SCNCM) was established as an integral part of the Advanced Concrete and Masonry Centre as part of project NANOCOM (2000-2003). The work of the SCNCM has now widened, as immediate applications for nano-scale based characterisation in the two other groups (Concrete Technology; Heritage Masonry) became apparent and are being presently pursued.

The experience, track record and current research of the team at Paisley enabled it to establish important collaborations in the broad field of Nanotechnology and Construction in the UK, and worldwide. A 'horizontal' RILEM Technical Committee TC-NCM on Nanotechnology of Construction Materials with a wide remit, chaired by Professor PJM Bartos, immediately attracted full membership of experts world-wide. It held its inaugural meeting in Madrid in September 2002.

The significance of nanotechnology related Research and Development has become recognised worldwide and government funding authorities and private/corporate investors have made substantial investments. This has included the European Commission, which provided financial support for the 1st International Symposium on Nanotechnology in Construction through project NANOCONEX (Growth, 2002-2003). The European Commission later adopted nanotechnology as one of the key topics within its 6th Framework (2002-2006) of support for research. At the Symposium, Mr Hervé Péro outlined the future activities of the EU in which nanotechnology and construction will be involved.

Members of the RILEM TC-NCM and key partners in project NANOCONEX formed the Scientific Committee for the Symposium, chaired by Professor PJM Bartos. Their assistance in selection and reviewing of contributions and in chairing of Sessions at the Symposium is gratefully acknowledged. In addition, the US National Science Foundation, represented by Dr Ken Chong, supported the event by providing a review of developments in the USA and sponsoring participation at the Symposium of a group of leading US researchers in nanotechnology and construction.

There were many organisations and companies which supported the Symposium and whose contributions were much appreciated. However, the event, which attracted 130 delegates from 30 countries worldwide, would not have taken place without a number of colleagues at the University of Paisley who went well beyond their call of duty to ensure that the event was a success. It is impracticable to name all, however special thanks go to

Mrs Margaret Nocher, Administrator of the ACM Centre, for her efficient management of the very complex organisation during the preparation and running of the Symposium. This was significantly assisted by Mrs Lorraine Dymond and Mr Graham Brooks, whose untiring efforts helped to guarantee the financial stability of the event.

Dr Pavel Trtik, in early stages, and subsequently Dr John J Hughes, chaired the Organising Committee which co-ordinated all activities.

These proceedings, published by the Royal Society for Chemistry, contain only contributions which were selected and presented at the Symposium. The topics and emphasis of the contributions reflected the fragmented and unbalanced nature of current nano-related research in this very wide area. They also reflected the diversity of prospects for its further development, commercial exploitation and even societal impacts within different sectors of construction. It has been a considerable challenge for the Editors to provide a rational sequence in structure of the proceedings, if not a complete 'balance'. A number of contributions were genuinely multi-disciplinary in their content: an expected and desirable feature, which nevertheless complicated the editing of this publication. A few contributions, particularly from the United States, were presented but could not be provided as full papers. In such cases, the Editors decided to include extended abstracts in the Proceedings, with references, to maintain coverage of research in progress.

The "NICOM 1" Symposium was a path-finding event, the first of its kind, promoting awareness and then integration of Research and Development across traditional scientific and technical boundaries. Delegates at the Symposium unanimously endorsed this aspect as the best way forward, which maximises realistic potential for exploitation of nanotechnology in the whole of the construction domain. Learning about related/relevant research in other disciplines and being able to meet investigators from areas of research not normally encountered in specialised networks and through discussion of their work was much appreciated by all. It is expected that 'sectorial' events focusing on specific aspects, sections or products related to construction and nanotechnology will be held in the near future. Another wide-coverage, integrating and multidisciplinary 2nd International Symposium on Nanotechnology and Construction is then expected to follow in 2-3 years' time.

Professor Peter J.M. Bartos

Dr. John J. Hughes

Dr. Pavel Trtik

Dr. Wenzhong Zhu

Paisley, September 2003

Contents

Modelling

Materials and Products

Applications

Organising Committee

Symposium Chair
  Prof Peter M.J. Bartos (Advanced Concrete & Masonry Centre (ACM), University of Paisley)

Chair of the Organising Committee
  Dr John J. Hughes (ACM Centre, University of Paisley)

Administrator
  Mrs Margaret Nocher (ACM Centre, University of Paisley)

  Mr Graham Brooks (Innovation & Research Office, University of Paisley)
  Mrs Lorraine Dymond (Innovation & Research Office, University of Paisley)
  Mr John Gibbs (ACM Centre, University of Paisley)
  Dr Wenzhong Zhu (ACM Centre, University of Paisley)

Scientific Committee

Chairman
>Prof Peter M.J. Bartos (ACM Centre, University of Paisley)

Secretary
>Dr Pavel Trtik (ETH, Zurich)

Members
>Dr James Beaudoin (NRC, Canada)
>Prof Arnon Bentur (Technion, Israel)
>Prof W. Brameshuber (RWTH Aachen, Germany)
>Dr Ravindra Gettu (UPC, Spain)
>Dr David Ho (NUS, Singapore)
>Prof Toshiharu Kishi (University of Tokyo, Japan)
>Prof Karl Kromp (University of Vienna, Austria)
>Dr Jacques Lukasik (Lafarge, France)
>Dr Martin Meyer (FIEM, Finland)
>Prof Jan van Mier (ETH, Switzerland)
>Dr Bernhard Middendorf (University of Kassel, Germany)
>Prof Frank Placido (Thin Film Centre, University of Paisley, UK)
>Dr Antonio Porro (Labein, Spain)
>Prof Marco di Prisco (Politecnico di Milano, Italy)
>Prof Hans-Wolf Reinhardt (University of Stuttgart, Germany)
>Dr Ottilia Saxl (Institute of Nanotechnology, UK)
>Prof Karen Scrivener (EPFL, Switzerland)
>Prof Surendra Shah (ACBM, USA)
>Dr Jan Skalny (Material Service Life, USA)
>Dr Åke Skarendahl (BIC, Sweden)
>Prof Henrik Stang (DTU, Denmark)
>Dr Johan Vyncke (BBRI, Belgium)
>Prof Anthony Walton (University of Edinburgh, UK)

Sponsors

The University of Paisley provided essential support in infrastructure and administration for the Symposium. The following sponsors are thanked for their additional support:

Scottish Enterprise Renfrewshire
MTS
The European Commission
Lafarge
Micro Materials
FEI Company
Akzo Nobel- Eka Chemicals
Renfrewshire Council
The Engineering and Physical Sciences Research Council
Institute of Nanotechnology
The Institute of Materials
The Institution of Civil Engineers
Concrete Society
British Cement Association
The Institution of Structural Engineers

Part 1: Nanotechnology in Construction in the 21st Century

FROM NANOTECHNOLOGY TO NEW PRODUCTION SYSTEMS: THE EU PERSPECTIVE

Hervé Péro

DG Research - EUROPEAN COMMISSION – 200 rue de la Loi, Brussels

1 INTRODUCTION

It is my pleasure to be able to give a key note address on such an exciting subject, not only because I am an engineer by training and have worked several years in industry - therefore it reminds me of very good times - but also because research on nanotechnology and its applications represent a key factor for the development of high added value products and will surely provide the basis for a competitive and sustainable development of European industry.

2 NANOTECHNOLOGY

Nanotechnology is a relatively young field of science and technology, with an enormous market potential and societal and economic impacts, and for all industrial sectors. Nanotechnology is truly multidisciplinary. Research at the nano-scale frontier is unified by the need to develop knowledge, tools, techniques and expertise on atomic and molecular interactions for applications in real products. Nanotechnology covers a wide range of research and innovation aspects, for example: magnetic random access memories; simplification and use of biological molecular functionalities; nano-wires, nano-crystals, carbon nano-tubes, and quantum effects; industrial production of nano-coatings; epitaxial self-assembly, etc. Nanocomposites for example, which are hybrids of greatly differing components – often comprising an inorganic and an organic component – are probably among the most promising new materials. Their applications range from mechanically reinforced lightweight components to components for batteries, sensors, adhesives, packaging materials, pigments, building and construction materials and artificial body parts.

The development of a strong European position in this field, and the establishment of a European nanotechnology industry, requires a concerted approach at the European level in order to:
- Merge and facilitate complementary and unique competencies.
- Define strategic plans and positioning (roadmaps).
- Share large investments and/or common use of research facilities .
- Set up common R&D open platforms.
- Initiate cores for EU collaborations.

- Increase attractiveness of research groups for e.g. junior researchers.

The European Union via the Framework Programmes funds only part of the research in Europe (its contribution corresponds to roughly 6% of overall European investment in research). The current Framework Programme (period 2003-2006) is open to international co-operation virtually with all Countries in the world. Despite its reduced quantity, by quality the Union's research plays a key role for European integration and acts as catalyst of much larger impact. In orientative terms, the total European investment in the nano-research is being estimated around 700M Euro per year.

3   RESEARCH AND THE CONSTRUCTION SECTOR

The **Construction sector**, with annual turnover of almost 1000 Billion €, total directly employed workforce of more than 11 million people, and another 15 million indirect employment is Europe's major industrial sector contributing with about 10% to the GDP. It is of enormous importance for European social and economic cohesion, considering also the facts that it is largely dominated by **small and medium size enterprises and it motivates the economic activity of all other sectors by consuming products and services and providing space and infrastructure**. Europe is world leader with 30% of the overall market but construction is mainly local business as less than 4% of the market is international.

Facing challenges of competitiveness and the needs for modernisation, the commitment of the sector in research is a key priority. The construction sector drastically needs research activities, for the competitiveness of its large projects of course, for ensuring safety of infrastructures such as bridges or tunnels, for maintaining the world cultural heritage, but also to support modernisation of the numerous SMEs (97% of enterprises).

However, the fact that this industry is very fragmented makes the changes happen very slowly. Tight regulation is also one of the characteristics of this sector, which needs to be taken into account in forecasting any technical progress. Long life span, high costs, and particular business model are also obstacles for drastic innovation. However, the signs of changes are there; we see more and more public-private partnerships, service contracts, performance based approaches, products/services, etc.

The Construction sector has continuously participated in European Research Programmes for more than 15 years and has clearly demonstrated its interests in different fields of research and innovation:

- Higher performance and intelligent materials, including for repair and rehabilitation of existing structures.
- Innovative systems that optimise the "design-production-service-end of life" value-chain through the development of new tools based on information technologies.
- New technologies for processing of multicultural applications/products.
- New production methods that drastically reduce the amount of water, energy and waste, as well as environment technologies linked with recycling or recovery of products.

Figure 1   *Fire in tunnels (network FIT, project UPTUN)*

Figure 2   *Lightweight membrane construction (TENSINET)*

Innovation in construction is largely "invisible" for the general public. It involves very often the use of new technologies and/or materials for the design, construction or maintenance of well-known products. As generic and high added value products, many elements of the built environment are at the basis of the competitiveness of European industry as a whole, as well as the basis of many clean and safe technologies for a better world.

Figure 3   *Construction: Nonwovens for insulation of interiors.*

Figure 4   *Construction: Invisible safety = textile-strengthened reinforced roads.*

Under the FP5 Growth, (Competitive and Sustainable Growth Programme), construction related projects were mainly supported by the Key Action 1 "Innovative Products, Processes and Organisation". At the same time, long-term research relevant to new and improved materials was treated under the Generic technologies: "Materials and their technologies for production and transformation".

The GROWTH programme, for the period 1999-2002, funded about 93 projects directly linked to the construction sector corresponding to a substantial amount of funding equivalent to 6-7% of the total. The construction sector has also substantially participated in other Community Research programmes (IST and EESD).

For what concerns FP6, I am pleased that this sector does not only recognise the need to integrate, and better structure its research efforts but that in fact, this movement has already taken place. This is evident from the analysis of the Expression of Interests, in which the construction industry participated actively.

The construction sector participated also well in the first call for NMP. A total of 18 proposals directly linked to the sector were submitted under different research priorities, over a total of 406 proposals. For different reasons none of these proposals succeeded to pass to the second stage. I would like to reassure you that we are considering both the interest and the particularities of the sector in research and together with the programme committee and the advisory group we plan to address these issues in the next NMP call.

An analysis of the received proposals for the first cal NMP-1, closed in March 2003, shows that the required multi-disciplinary approach to "problem-solving" proposals was not always satisfactory: *proposals were very often limited to opening traditional partnerships to new disciplines, instead of proposing a real multi-disciplinary treatment of problems.* In addition a real need for integration of research capacities was generally not demonstrated, in particular for the Networks of Excellence: *such networks should be proposed when this is clearly beneficial for the efficiency of research activities in comparison with a dispersed situation at national / regional level (subsidiarity principle).*

Research is needed for the European Construction Industry but it should help shifting existing paths towards sustainable development.

Indeed the "business as usual" approach and even the efforts towards minimisation of impacts on environment are dramatically leading in the medium to long term to a huge crisis at world level. The only solution is to turn towards sustainable development, which means ensuring the same - if not better - conditions to the next generations as we have today, without negative impact on our economic activity. Sustainable development may be a utopia... but what a stimulating goal for researchers and industrialists!

At the same time that **Sustainable Development** is the cornerstone of any EC action plan, strengthening **industrial competitiveness** in Europe constitutes the key to achieving the strategic goals of sustainable industrial development. It can be stated that the Lisbon Strategy, the various EC communications of industrial policy and on innovation, and the European Strategy for Sustainable Development pave the way Europe would like to take: the way towards a world-leading manufacturing sector that addresses the needs of a sustainable society!

Enormous efforts remain: consumption of resources needs to be drastically reduced, in buildings, during their construction and during materials processing. Also, if energy consumption of buildings could be reduced by 30% (which is easily achievable), the Kyoto objective of reducing CO_2 emissions at the level of 8% could be achieved.

4   FROM NANOTECHNOLOGY TO NEW CONSTRUCTION PROCESSES

Nanosciences and nanotechnologies represent a new revolutionary approach in the way of thinking and producing, as they somehow revert to the traditional scientific approach and production processes, from "big" to "small".

The new goal being achieved runs from "small" to "big", building by adding together atom by atom and molecule by molecule. Conceptually, this is easy: it is just an imitation of nature. In practice, it is a huge revolution that is allowing us to create better performing products and processes, within an ideal context of sustainable development.

Industry will be one of the major beneficiaries of nanotechnology. Now it is approaching it in a rather "top-down" direction, whilst mostly universities are engaged in exploring "bottom-up", self-organising, self-assembling routes. Both ways need research. The more we improve knowledge and command at the nanoscale, the quicker we will be able to use it in industrial production.

Nanotechnology opens the way towards new production routes, towards new, more efficient, better performing and intelligent materials, towards new design of structures and related monitoring and maintenance systems. Nanoparticles and nanocomposite materials are already earning a lot of money. Over the next five years, for example, nanotechnology-based paints, pigments, coatings and detergents will be doing good business. In the construction industry, nanotechnology has the potential to improve construction materials, including steel, polymers and concrete. Concrete in particular, which is a complicated, nanoscale structure of hydrates of cement, additives and aggregates is an excellent candidate for nanotechnological manipulation and control of properties. Programming the time-release of chemical admixtures in concrete can provide maximum effectiveness at the construction site, while reinforcing cement binders with nanodiameter fibres and rods can result in higher performance of cementitious materials in general by impeding crack formation and growth.

Nanotechnology can provide tools for understanding basic phenomena and be able to respond to today's challenges. For example, the owner specifications for the Öresund link between Denmark and Sweden required 120 years of service life for the concrete with only maintenance work - no major repair. This is perfectly feasible; recent work has shown that the Byzantine Hagia Sofia in Istanbul has such long life thanks to the self repairing properties of the mortar used almost 15 centuries ago. This opportunity should not be missed by the Construction sector.

To pass from a "macro-centric" to a "nano-centric" system, a vast multidisciplinary knowledge, a strong ability for integration and an attitude to complexity are needed. Researchers need the courage to abandon rigid and pre-determined schemes, and the ability to develop new excellence in research and education, also through new *curricula vitae et studiorum*.

There are many barriers and challenges. A large critical mass is required in terms of both human and material resources. This implies a new way of co-operation, being more open, reinforced, transparent and verifiable.

5   TOWARDS A EUROPEAN RESEARCH AREA

"Technology push" is however not the message to be retained. It should be remembered that the 6th research and innovation Framework Programme of the European Community is characterised by a changing role of EU Research Programmes.

It is based on three clear "political" pillars:

- The "Lisbon" objective, stated in spring 2000: "to help Europe become the most dynamic and most competitive knowledge-based economy within 10 years".

- The "Göteborg" objective, stated in spring 2001, puts sustainable development, i.e. environment, health, economy, employment, at the top of the agenda of all EU activities.

- Finally, the development of a true European Research Area (ERA) is the driving force of the EU research policy. ERA, whose aim is to create conditions for strengthening the coherence of research activities and policies conducted in Europe, offers real new horizons for researchers. Integrating, reinforcing and structuring research efforts at European level is the objective, stimulating industrial investments, thus closing the loop with the "Lisbon" objective. It should be remembered that Europe lags substantially behind its competitors in terms of investment in research. The objective is to reach a level of investment in research in Europe equal to 3% of GDP by 2010.

The new instruments in FP6 will allow better co-ordination and effectiveness of research activities at EU level. Several objectives should be encompassed such as ensuring system competitiveness and sustainability, quality of life, sustainable employment, high quality education, and of course ethical aspects, in an integrated approach at European (even international) level. Research and innovation activities should be jointly present in projects to ensure the quick exploitation of research results.

Education and skill development should receive particular attention in view of the challenge of the "knowledge-based society" and "knowledge-based industries".

Integrated Projects (IPs) will focus on clear and quantifiable objectives. Integration issues should be tackled at least on four aspects:

- Multidisciplinarity of research.
- Integration of research, innovation, awareness and skills development activities.

- Integration of partnership along the value chain for creation of knowledge communities.
- Complementary funding from different sources, private, regional, national, and European.

This involves **far-seeing research actions** and effective research projects to rapidly spread results into practice. It involves also a strong presence and interaction of innovative enterprises and research organisations, throughout Europe, in research consortia. Such a challenge implies the acceptance of large and complex networks in which companies should accept sharing of knowledge and **co-operation**, i.e. co-operation in research activities, while continuing to compete in the market place.

Networks of Excellence (NoE) are another response to this need for networking. They will be promoted to assemble EU research capacities and strengthen Science and Technological excellence for the competitiveness of EU industry. Partnership in NoEs will be characterised by participants wishing a progressive and lasting integration of their research capacities, since today they are too fragmented to ensure the efficiency of research required at world level.

Support to Small to Medium-sized Enterprise (SMEs) aspects is key. Europe wide networks are required that give the smaller companies access to new possibilities in product and process technologies, therefore stimulating implementation of paradigm shifts in traditional sectors. At least **15% of the total budget** allocated to the priorities of the framework Programme will be dedicated to SMEs. Large possibilities will exist to include high-tech SMEs in consortia, or to consider specific "modules" for SMEs in research projects. The Directorate on industrial research is also launching a specific call for IPs for SMEs focusing on the support to the transformation of the European traditional sectors.

The role of industrial and research associations should be highlighted to diffuse information, stimulate participation, and ensure dissemination and good use of results. (For more information see: www.europa.eu.int.comm.industrial_research)
The overall message is to pursue in the direction of better and more efficient research activities at EU level, to help the transformation of whole sectors of industry.

6 TOWARDS AN "ACTION PLAN"?

An action plan is certainly needed for integrated research activities, but the question remains whether this plan should be oriented towards steady state, decline or growth of the European role? Whatever the focus is, there is a need for a long-term visio. In that sense the future is in our hands, but the absence of an overall strategy would be a key missing element. European and national industrial federations have a key role to play to help in defining this vision.

The new paradigm is the **emergence of knowledge-intensive communities**, that is when people and organisations, supported by information and communication technologies, interact in concerted efforts to co-produce new knowledge. These communities are the new agents of change in a knowledge society. A key parameter is the use of ICT and of networks for the codification of new knowledge. The advantages are synergy in the production of this knowledge, quality control by the group and rapid dissemination. That leads to new ways of organisation of work. This has obviously profound implications in term of education, research, innovation, definition of jobs, governance and, ultimately, the organisation of society.

Joint EU efforts are necessary since the fragmentation of European RTD efforts can only generate inefficiencies; **co-operation between different schemes**, e.g. the FP and

EUREKA, is strongly in need of promotion; Short-medium and long term actions should be looked at, with clear and visible benefits to all stakeholders;

Multidisciplinarity and better-integrated paths for innovation should be promoted (perhaps networks of specialised experience should be avoided in favour of more broadly-based collaborations between actors). **"Platforms"** should be promoted for technological, organisational and social innovation. **New approaches** such as nanotechnologies, intelligent materials, pro-active manufacturing processes, stimulating working environments, ambient intelligence and new consumption patterns, should be fostered. Value through Services should be sound, connected and integrated. Life cycle vision should be explained publicly and largely used by all industrial actors.

Research will however not provide every required solution; progress in R&D should be clearly integrated with other innovation-related actions such as those dealing with regulation or entrepreneurship, as well as with better education. A possible Action Plan should integrate the major EU policy actors, industry, research and education and work not only on each area but also on the **policy interfaces**.

Clearly, setting-up such an action plan is not easy. How will it combine the paradoxes that policy should be simple yet embrace complexity? How to sustain diversity yet offer some uniformity? How to offer a basis for wider participation without jeopardising the speed of innovation? How to allow for creation of new businesses yet preserving the traditional and necessary industrial base? "Think globally and act locally"

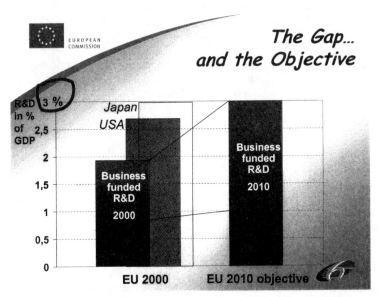

Figure 5   *EU objective for spend on R&D for 2010. Comparisons given with current levels of USA and Japan.*

6.1   Mobilisation

The Construction sector should mobilise itself, but not remain isolated. The role of national and European federations will be key in this process. They should reflect on strategic multi-disciplinary research activities as a key support to a clear industrial policy and the development of entrepreneurial spirit in an enlarged Europe.

For several years, networking activities have been carried out at EU level. Managed by ECCREDI, this action is today continued under the acronym E-CORE. This activity, funded by the Commission, has gradually opened itself to all national or private research projects, on voluntary basis. A simple visit to the database of this network (http://www.e-core.org) shows the important work realised.

However, is it sufficient? Clearly the industrial investment in research should be stimulated. This is the goal over the medium term of the European Union (Figure 5). The European research programmes will be there to help you! For further information: www.europa.eu.int/comm/research/industrial_technologies .

NANOTECHNOLOGY IN CIVIL ENGINEERING

K. P. Chong

National Science Foundation, Arlington, VA 22230, U.S.A.

1 INTRODUCTION

The transcendent technologies, which are the primary drivers of the twenty first century and the new economy, include nanotechnology, microelectronics, information technology and biotechnology as well as the enabling and supporting civil infrastructure systems and materials. Mechanics and materials are essential elements in all of the transcendent technologies. Research opportunities and challenges in mechanics and materials, including nanomechanics, carbon nano-tubes, bio-inspired materials, coatings, fire-resistant materials as well as improved engineering and design of materials are presented and discussed in this paper.

Nanotechnology is the creation of new materials, devices and systems at the molecular level - phenomena associated with atomic and molecular interactions strongly influence macroscopic material properties [according to I. Aksay, Princeton]; with significantly improved mechanical, optical, chemical, electrical... properties. Nobelist Richard Feynman back in 1959 had the foresight to indicate "there is plenty of room at the bottom". National Science Foundation [NSF] Director Rita Colwell in 2002 declared "nanoscale technology will have an impact equal to the Industrial Revolution".

- 10^{-12} m QUANTUM MECHANICS [TB, DFT, HF...]*
- 10^{-9} m MOLECULAR DYN. [LJ...]; NANOMECHANICS; MOLECULAR BIOLOGY; BIOPHYSICS
- 10^{-6} m ELASTICITY; PLASTICITY; DISLOCATION...
- 10^{-3} m MECHANICS OF MATERIALS
- 10^{-0} m STRUCTURAL ANALYSIS

ALL SCALES: *MULTI-SCALE ANALYSES & SIMULATIONS...*

---

*TB = TIGHT BINDING METHOD; DFT = DENSITY FTNAL THEORY;
HF = HATREE-FOCK APPROX.; LJ = LENNARD JONES POTENTIAL

Figure 1   *Range of scales and processes.*

In the education area to enable engineers to do competitive research in the nanotechnology areas, engineers should also be trained in quantum mechanics, molecular dynamics, etc. The following lists some of the key topics across different scales. The author has been advocating a summer institute to train faculty and graduate students to be knowledgeable in scales at and less than the micrometer (Figure 1).

The National Science Foundation has supported basic research in engineering and the sciences in the United States for a half century and it is expected to continue this mandate through the next century. As a consequence, the United States is likely to continue to dominate vital markets, because diligent funding of basic research does confer a preferential economic advantage.[1] Concurrently over this past half century, technologies have been the major drivers of the U. S. economy, and as well, NSF has been a major supporter of these technological developments. According to the former NSF Director for Engineering, Eugene Wong, there are three *transcendental* technologies:

- Microelectronics – Moore's Law: doubling the capabilities every two years for the last 30 years; unlimited scalability; nanotechnology is essential to continue the miniaturization process and efficiency.
- Information Technology [IT] – NSF and DARPA started the Internet revolution about three decades ago; the confluence of computing and communications.
- Biotechnology – unlocking the molecular secrets of life with advanced computational tools as well as advances in biological engineering, biology, chemistry, physics, and engineering including mechanics and materials.

Efficient civil infrastructure systems as well as high performance materials are essential for these technologies. By promoting research and development at critical points where these technological areas intersect, NSF can foster major developments in engineering. The solid mechanics and materials engineering (M&M) communities will be well served if some specific linkages or alignments are made toward these technologies. Some thoughtful examples for the M&M and civil engineering communities are (Table 1):

Table 1   *Example applications/technologies for nanotechnology in mechanics and materials and civil engineering.*

• Bio-mechanics/materials	• Simulations/modeling
• Thin-film mechanics/materials	• Micro-electro-mechanical systems (MEMS)
• Wave Propagation/NDT	• Smart materials/structures
• Nano-mechanics/materials Nano-electro-mechanical systems(NEMS)	• Designer materials Fire retardant materials and structures

MEMs can be used as platforms for NEMS. Another product of the nanotechnology is carbon nano-tubes [CNT] which are self-assembled in deposition from C-rich vapors, consisted of honeycomb lattices of carbon rolled into cylinders; nm in diameter, micron in length. CNT have amazing properties:[2,3,4]

- 1/6 the weight of steel; 5 times its Young's modulus; 100 times its tensile strength; 6 orders of magnitude higher in electrical conductivity than copper; CNT strains up to 15% without breaking.

- 10 times smaller than the smallest silicon tips in Scanning Tunnel Microscope - STM [CNT is the world's smallest manipulator].
- CNT may have more impact than transistors.
- Ideal material for flat-screen TV [~2003].
- Similar diameter as DNA.
- Bridge different scales in mesoscale - useful as bldg. blocks ~ e.g. nanocomposites; gases storage.
- Metallic or semiconducting.
- Single-electron transistor; logic gate.
- As RAM - on/off do not affect memory storage - no booting needed.

Considerable NSF resources and funding will be available to support basic research related to nano-science and engineering technologies. These opportunities will be available for the individual investigator, teams, small groups and larger interdisciplinary groups of investigators. Nevertheless, most of the funding at NSF will continue to support unsolicited proposals from individual investigator on innovative "blue sky" ideas in all areas including nano science and engineering technologies.

2 NANOTECHNOLOGY

2.1 Nanomechanics Workshop

Initiated by the author, with the organization and help of researchers from Brown [K. S. Kim, et al], Stanford, Princeton and other universities, a NSF Workshop on Nano- and Micro-Mechanics of Solids for Emerging Science and Technology was held at Stanford in October 1999. The following is extracted from the Workshop Executive Summary. Recent developments in science have advanced capabilities to fabricate and control material systems on the scale of nanometers, bringing problems of material behavior on the nanometer scale into the domain of engineering. Immediate applications of nanostructures and nano-devices include quantum electronic devices, bio-surgical instruments, micro-electrical sensors, functionally graded materials, and many others with great promise for commercialization. The branch of mechanics research in this emerging field can be termed nano- and micro-mechanics of materials, highly cross-disciplinary in character. A subset of these, which is both scientifically rich and technologically significant, has mechanics of solids as a distinct and unifying theme. The presentations at the workshop and the open discussion precipitated by them revealed the emergence of a range of interesting lines of investigation built around mechanics concepts that have potential relevance to microelectronics, information technology, bio-technology and other branches of nanotechnology. It was also revealed, however, that the study of complex behavior of materials on the nanometer scale is in its infancy. More basic research that is well coordinated and that capitalizes on progress being made in other disciplines is needed if this potential for impact is to be realized.

Recognizing that this area of nanotechnology is in its infancy, substantial basic research is needed to establish an engineering science base. Such a commitment to nano- and micro-mechanics will lead to a strong foundation of understanding and confidence underlying this technology based on capabilities in modeling and experiment embodying a high degree of rigor. The instruments and techniques available for experimental micro- and nano-mechanics are depicted in Fig. 1, courtesy of K. S. Kim of Brown University. One of the key instrument in nanotechnology is the atomic force microscope [AFM].

However, the limitation is the AFM scan speed ~100hz (takes ~30 min. for a small image of 20,000 pixels).

To improve the speed and performance of the atomic force microscope, more optimal control of the cantilever and more robust software to acquire and process the data as well as other improvements are needed. A portable AFM has also been developed by IBM recently.

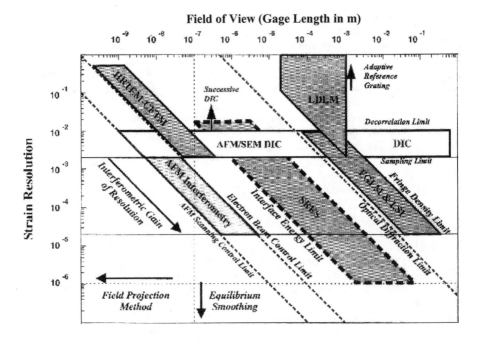

Figure 1 *Instruments and techniques for experimental micro and nano mechanics [courtesy of K. S. Kim of Brown University] where*

HRTEM	*High Resolution Transmission Electron Microscopy*
SRES	*Surface Roughness Evolution Spectroscopy*
CFTM	*Computational Fourier Transform Moire*
FGLM	*Fine Grating Laser Moiré*
AFM	*Atomic Force Microscopy*
LSI	*Laser Speckle Interferometry*
SEM	*Scanning Electron Microscopy*
DIC	*Digital Image Correlation*
LDLM	*Large Deformation Laser Moire*

The potential of various concepts in nanotechnology will be enhanced, in particular, by exploring the nano- and micro-mechanics of coupled phenomena and of multi-scale phenomena. Examples of coupled phenomena discussed in this workshop include modification of quantum states of materials caused by mechanical strains, ferroelectric transformations induced by electric field and mechanical stresses, chemical reaction processes biased by mechanical stresses, and changes of bio-molecular conformality of proteins caused by environmental mechanical strain rates. Multi-scale phenomena arise in situations where properties of materials to be exploited in applications at a certain size scale are controlled by physical processes occurring on a size scale that is orders of magnitude smaller. Important problems of this kind arise, for example, in thermo-mechanical behavior of thin-film nanostructures, evolution of surface and bulk nanostructures caused by various material defects, nanoindentation[5], nanotribological responses of solids, and failure processes of MEMS structures.

2.2 Nanoscale Science and Engineering Initiatives

Coordinated by M. Roco (IWGN, 2000)[6,7], NSF recently announced a second year program [NSF 02-148; see: www.nsf.gov] on collaborative research in the area of nanoscale science and engineering (NSE). This program is aimed at supporting high risk/high reward, long-term nanoscale science and engineering research leading to potential breakthroughs in areas such as materials and manufacturing, nanoelectronics, medicine and healthcare, environment and energy, chemical and pharmaceutical industries, biotechnology and agriculture, computation and information technology, improving human performance, and national security. It also addresses the development of a skilled workforce in this area as well as the ethical, legal and social implications of future nanotechnology. It is part of the interagency National Nanotechnology Initiative (NNI). Details of the NNI and the NSE initiative are available on the web at http://www.nsf.gov/nano or http://nano.gov.

The NSE competition will support Nanoscale Interdisciplinary Research Teams (NIRT), Nanoscale Exploratory Research (NER), Nanoscale Science and Engineering Centers (NSEC) and Nano Undergraduate Education (NUE). In addition, individual investigator research in nanoscale science and engineering will continue to be supported in the relevant NSF Programs and Divisions outside of this initiative. This NSE initiative focuses on seven high risk/high reward research areas, where special opportunities exist for fundamental studies in synthesis, processing, and utilization of nanoscale science and engineering. The seven areas are:

- Biosystems at the nanoscale.
- Nanoscale structures, novel phenomena, and quantum control.
- Device and system architecture.
- Nanoscale processes in the environment.
- Multi-scale, multi-phenomena theory, modeling and simulation at the nanoscale.
- Manufacturing processes at the nanoscale.
- Societal and educational implications of scientific and technological advances on the nanoscale.

The National Nanotechnology Initiative started in 2000 ensures that investments in this area are made in a coordinated and timely manner (including participating federal agencies – NSF, DOD, DOE, DOC [including NIST], NIH, DOS, DOT, NASA, EPA and others) and will accelerate the pace of revolutionary discoveries now occurring. Current request of Federal agencies on NNI is about $710 million. The NSF share of the budget is around $220 million [on NSE, part of NNI].

3 CHALLENGES

The challenge to the mechanics and materials and engineering research communities is: How can we contribute to these broad-base and diverse research agendas? Although the mainstay of research funding will support the traditional programs for the foreseeable future, considerable research funding will be directed toward addressing these research initiatives of national focus. At the request of the author [KPC], a NSF research workshop has been organized by F. Moon of Cornell University to look into the research needs and challenges facing the mechanics communities. A website and/or report with recommendations of research needs and challenges will soon be available.

Mechanics and materials engineering are really two sides of a coin, closely integrated and related. For the last decade this cooperative effort of the M&M Program has resulted in better understanding and design of materials and structures across all physical scales, even though the seamless and realistic modeling of different scales from the nano-level to the system integration-level (Fig. 2) is not yet attainable. *The major challenges are the several orders of magnitude in time and space scales from the nano level to micro and to meso levels.* The following list summarizes some of the major methods in bridging the scales, top down or bottom up, and their limitations.

- **FIRST PRINCIPLE CALCULATIONS - TO SOLVE SCHRODINGER'S EQ. *AB INITIO*, e.g. HATREE- FOCK APPROX., DENSITY FUNCTIONAL THEORY,...**
  - COMPUTATIONAL INTENSIVE, $O(N^4)$ – N is the number of orbitals.
  - UP TO ~ 3000 ATOMS

- **MOLECULAR DYNAMICS [MD] pico sec. range - DETERMINISTIC, e.g. W/ LENNARD JONES POTENTIAL**
  - MILLIONS TIMESTEPS OF INTEGRATION; TEDIOUS
  - UP TO ~ BILLION ATOMS FOR NANO-SECONDS

- **COMBINED MD & CONTINUUM MECHANICS [CM], e.g. MAAD[Northwestern U.]; LSU; BRIDGING SCALE; ...**
  - PROMISING, but still has limitations on bridging scales in space and time...

- **INTERATOMIC POTENTIAL/CM – HUANG [U. of Illinois – UC]: EFFICIENT; MD STILL NEEDED IN TEMPERATURE, STRUCTURAL CHANGES, DISLOCATIONS.**

In the past, engineers and material scientists have been involved extensively with the characterization of given materials. With the availability of advanced computing and new developments in material science, researchers can now characterize processes and design and manufacture materials with desirable performance and properties. One of the challenges is to model short-term nano/micro-scale material behavior, through meso-scale and macro-scale behavior into long-term structural systems performance, Table 1. Accelerated tests to simulate various environmental forces and impacts are needed[8]. Supercomputers and/or workstations used in parallel are useful tools to solve this scaling problem by taking into account the large number of variables and unknowns to project micro-behavior into infrastructure systems performance, and to model or extrapolate short term test results into long term life-cycle behavior[8,9]. Twenty-four awards were made totaling $7 million. A grantees' workshop was held recently in Berkeley and a book of proceedings has been published.[10]

Table 1 *Physical scales in materials and structural systems* [11,12]

MATERIALS		STRUCTURES		INFRASTRUCTURE
nano-level (10^{-9})	micro-level (10^{-6})	meso-level (10^{-3})	macro-level (10^{+0})	systems-level (10^{+3}) m
Molecular Scale	*Microns*		*Meters*	*Up to Km Scale*
*nano-mechanics *self-assembly *nanofabric a-tion	*micro-mechanics *micro-structures *smart materials	*meso-mechanics *interfacial-structures *composites	*beams *columns *plates	*bridge systems * lifelines *airplanes

Biomimetics or bio-inspired materials are to learn from nature's millions of years of evolution and optimization to see how to design and built stronger and high performance materials. An excellent example of nature's design is the nacre [or the shell of the red abalone] which has the nanostructured morphology and exceptional mechanical properties with just 2% of protein-binding agent. A finite element modeling of nacre was done by Katti, et al[13]. Researchers at Northwestern University and elsewhere are also investigating different aspects of nacre.

Tremendous progress is also being made in micro and nano sensors. Researchers at University of California – Berkeley have been working on "smart dust" sensors – with a target of a size of a cube 1mm on each side, capable of sensing [e.g. cracks in materials] and transmitting wirelessly the data collected.

Researchers at IBM – Zurich and Basel have developed the artificial nose to detect minute analytes and air quality in the air [Figure 3].

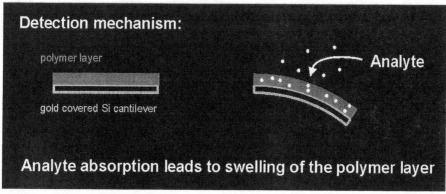

Figure 3 Cantilever detection mechanism [IBM – Zurich]

Researchers at NIST and other places have been building nano-clay filled polymers with natural clay form platelet [less than 1 nm in thickness] and 1 to 5% by volume. These nano-clay filled polymers can drammatically improve fire resistance as well as mechanical properties. Metal oxide nanoparticles have also been used in coatings for protection of UV light, self-disinfecting surfaces, solar cells, indoor air cleaners, etc.

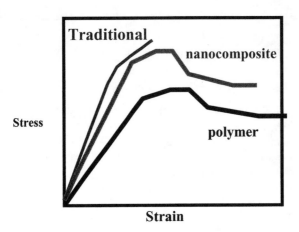

Figure 4 *The stress-strain characteristics of nano-composites (courtesy of l. schadler, rpi).*

4 SUMMARY, ACKNOWLEDGMENTS AND DISCLAIMER

An overview of research opportunities and challenges of nanotechnology in engineering mechanics and materials, including nanomechanics, carbon nano-tubes, bio-inspired materials, coatings, fire-resistant materials as well as improved engineering and design of materials are presented and discussed.

The author would like to thank his colleagues and many members of the research communities for their comments and inputs during the writing of this paper. Information on NSF initiatives, announcements and awards can be found in the NSF website: www.nsf.gov. Parts of this paper have been presented elsewhere[14,15]. Information pertaining to NIST can be found on its website: www.nist.gov. The opinions expressed in this article are the author's only, not necessarily those of the National Science Foundation.

References

1   Wong, E., An Economic Case for Basic Research. *Nature* 1996, **381,** 187-188.
2   Carbon Nanotubes- the first 10 years, *Nature,* 2001.
3   Mechanical Engineering, *ASME,* November 2000.
4   Technology Review, *MIT,* March 2002.
5   Asif, S. A. S., Wahl, K. J., Colton, R. J., and Warren, O. L., Quantitative Imaging of Nanoscale Mechanical Properties Using Hybrid Nanoindentation and Force Modulation. *J. Appl. Phys.* 2001, **90,** 5838-5838.
6   *Nanotechnology Research Directions,* Interagency Working Group on Nano Science, Engineering and Technology (IWGN). Kluwer Academic Publ. 37-44, 2000.
7   Nanoscale Science and Engineering. *NSF 02-148,* National Science Foundation, Arlington, VA, 2002.

8   Long Term Durability of Materials and Structures: Modeling and Accelerated Techniques, NSF 98-42, National Science Foundation, Arlington, VA, 1998.

9   Chong, K. P., 'Smart Structures Research in the U.S.' Keynote paper, in *Smart Structures, Proc. NATO Adv. Res. Workshop on Smart Structures*, Pultusk, Poland, Kluwer Academic Publ. 37-44 , 1999.

10 Monteiro, P. J. M., Chong, K. P., Larsen-Basse, J. and Komvopoulos, K., eds. *Long-term Durability of Structural Materials*, Elsevier, Oxford, UK, 2001.

11 Boresi, A. P. and Chong, K. P., *Elasticity in Engineering Mechanics*, John Wiley, New York, 2000.

12 Boresi, A. P., Chong, K. P. and Saigal, S., *Approximate Solution Methods in Engineering Mechanics*, John Wiley, New York, 2002.

13 Katti, K. S., Katti, D. R., Tang, J., and Sarikaya, M., *Proc. Mat. Res. Soc., Symp. AA, 2001*.

14 Chong, K. P., 'Research and Challenges in Nanomechanics' 90-minute Nanotechnology Webcast, ASME, Oct. 2002; archived in *www.asme.org/nanowebcast*

15 Chong, K. P., 'Nano-scale Mechanics Research and Challenges', *Plenary Paper, Study of Matter at Extreme Conditions Conference*, Miami Beach, FL, 2003

NANOTECHNOLOGY: BUSINESS AND INVESTMENT OPPORTUNITIES

Del Stark

Institute of Nanotechnology, Stirling University Innovation Park, Stirling, FK9 4NF, Scotland

ABSTRACT

Nanotechnology offers new opportunities for solving old problems across many industrial sectors. One key area of opportunity is in medical devices. The market for medical devices is very large and growing quickly, which, according to Eucomed (the trade association for the European medical devices industry) presently stands at a staggering $140 billion and is forecast to rise by 25% by 2005.

The realisation of inbuilt intelligence or smartness is being actively pursued by researchers in the quest to create structures and systems that respond in a meaningful way to the body's environment. Functional engineering at the nanoscale is a route to developing the smart materials and capabilities they seek that will make a new generation of medical devices and implants possible.

Another key area is construction. According to the September 2002 Engineering News-Record, domestic construction activity topped $819 billion in 2000 in the US and global construction activity was worth over $3.4 trillion. Currently, sol-gel chemistry and nanotechnology have been combined to develop smart self-cleaning windows. Coatings are being developed for stain resistant clothing and "smart paints". Car companies such as Audi are activity researching paint coatings that resist dirt and may one-day "self-heal" scratches.

Research in nanotechnology will provide nanocomposite materials, which can be used as improved building materials; with lighter weight materials, ductile cements and ultrahigh strength structures. The Scottish Centre for Nanotechnology in Construction Materials is working on various projects including self-settling underwater cements.

This presentation will discuss the opportunities offered by nanotechnology to variety of economically and socially critical industries while highlighting exciting technologies spinning out from new emerging nanotechnology companies. Activities of the Institute of Nanotechnology will be briefly highlighted as well as the Special Interest group for Construction and Nanotechnology.

The Institute of Nanotechnology in partnership with the Scottish Centre for Nanotechnology in Construction has established a Special Interest group to encourage new links between the construction industry and the nanotechnology community.

The utilisation of nanotechnologies in the construction sector are hitherto scarce and one of the first aims of the Special Interest Group will be to survey the current nanotechnology-related research and development in construction industry. Secondly, the Special Interest Group will assess the applicability of nanotechnologies to research & development in construction sector. Identifying potential areas of the construction industry where nanotechnologies have the potential to produce a significant impact on future of construction R&D.

The Special Interest Group is aiming to assist the construction industry in raising the awareness of the potentials of the emerging enabling technologies, thus improving the general perception of the construction as a 21st century modern industry.

INTEGRATION OF EUROPEAN NANOTECHNOLOGY RESEARCH IN CONSTRUCTION

Antonio Porro

Centre for Nanomaterials Applications in Construction, LABEIN - TECNALIA, Spain

1 INTRODUCTION

When the physicist and Nobel laureate Richard Feynman presented, in late 1959 at the California Institute of Technology the lecture "There is Plenty of Room at the Bottom", he outlined what today is known as nanotechnology. He stated that "..when we have some control of the arrangement of things on a small scale we will get an enormously greater range of possible properties that substances can have, and of different things that we can do." Years later the chemist and Nobel laureate Roald Hoffman, at Cornell University, stated that: "Nanotechnology is the way of ingeniously controlling the building of small and large structures, with intricate properties. It is the way of the future, a way of precise, controlled building, with incidentally, environmental benignness built in by design."

The physics and chemistry of nanomaterials are fundamentally different from those of their macro-scale counterparts. Dramatic changes and improvements are brought about not only by size reduction but also by new properties only apparent at the nanoscale.

An example of nanotechnology[1], offered by nature, is those of the abalone shell. These molluscs construct super-tough shells with beautiful, iridescent inner surfaces. They do this by organising calcium carbonate into tough nanostructured bricks. For mortar, abalones concoct stretchy goo of protein and carbohydrate. Cracks that may start on the outside rarely make it all the way through. The structure of the shell forces a crack to take a tortuous route around the tiny bricks, which dissipates the energy behind the damage. The stretchy mortar adds to the damage control. As a crack grows, the mortar forms resilient nano-strings that try to force any separating bricks back together. The result is a Lilliputian masonry that can withstand sharp beaks, teeth, even hammer blows.

This clever engineering of the abalone shell reflects one of nanotechnology's most enticing faces: by creating nanometer-scale structures, it's possible to control the fundamental properties — like colour, electrical conductivity, melting temperature, hardness, crack-resistance, and strength — of materials without changing the materials' chemical composition.

2 NANOTECHNOLOGY AND INNOVATION

Nanotechnology describes the creation and utilisation of functional materials, devices and systems with novel functions and properties that are based either on geometrical size or on material-specific peculiarities of nanostructures. This scale has become accessible both by application of new physical instruments and procedures. Also structures of animated and non-animated nature were used as models for self-organising matter. Only if the mastery of this atomic and molecular dimension succeeds, can the prerequisites for the optimisation of product properties be developed.

With the discovery of techniques to organise, characterise, and manipulate individual elements of matter the world-wide industrial conquest of nanoscale dimensions began. With the conquest of this nanometer dimension the speed of innovation achieved in the meantime led to the situation that physical fundamentals are still being investigated while first product groups are already entering the world markets. Their sales impacts are caused by the implementation of nanoscale architecture with new macroscopic functions.

The emerging fields of nanoscience and nanotechnology are leading to unprecedented understanding and control over the fundamental building blocks of all physical things. This is likely to change the way almost everything is designed and made.

Nanotechnology is discussed as one of the key factors in future global markets. New technologies contribute to safeguarding the economic terrain and to the solution of socially relevant problems. Early evaluation of innovative approaches and future market options is therefore vital in order to be able to play an active role in shaping competition in future markets.

The diffusion and impact of technological innovations is often as much a function of the development of complementary technologies and of a network of users as it is of the introduction of the discrete technology.

In the global context of nanotechnology, Europe is still in a competitive position but it is becoming vulnerable. Other non-EU investors are increasingly aware of the potential commercial returns in this area. It is essential for Europe to face such external challenges to maintain or improve its technological position.

3 THE CONSTRUCTION CASE

It is widely recognised that nanotechnology and its interaction with ICT are leading our Society towards a new industrial revolution. Nanotechnology is offering huge opportunities to develop new materials/components with new and multiple functionality.

At the same time there is a sector of great economic activity, the construction industry, dealing with coarse components consumed in large amounts seeking necessary innovations. The construction industry is also a great integrator of solutions offered by other industrial sectors i.e. a building integrates water supply, heating, ventilation and communication systems, etc.

The industrial investment in research, on the case of construction, remains lower than in other industrial sectors. Construction is unique, in that its products - in the form of individual building and civil engineering structures (mostly non-repeatable and custom built) - are typically constructed from a vast combination of both conventional and new 'high-tech' materials, using a relatively limited number of processes. Advances in productivity of the construction industry in the twentieth century were slow, and development of its technology (materials and processes) lagged behind that of other industrial sectors. With the European construction sector representing about 11% of the

European GDP, this slow advance affects the whole of European society, not only by holding back economic growth but also by an unacceptably slow adjustment to increasingly important requirements of environmental safety and sustainability. It is essential that an activity as significant as construction is not left behind in this twenty first-century industrial revolution. Put simply, it needs to catch up with other industries; in this context, it needs to make full use of nanotechnology as the powerful emerging enabling tool.

Nanotechnology is a route to achieving real competitive and sustainable growth and innovation within the construction industry. This scientific approach is essential if the potential for a new generation of materials that are both of higher performance and more economically viable is to be realised.

The Construction Industry is a conservative one. It is a classic (the classic?) example of a resource-based activity of limited sustainability. The challenge is to assist in its transformation to a knowledge-based economic activity where information and knowledge are the sources of value-added outcomes. Apart from the use of Information Technologies, cutting-edge technologies such as nanotechnology are needed to drive this revolution and reshape this industry. The technological revolution, based on nano-science, relies on a bottom-up approach, in which new high-performance materials can be produced from basic knowledge. Up to now, innovation has relied on variation of phenomenological parameters, treating systems as black boxes, and has been dependent on substantial narrowly focused investment. A different approach is now necessary: scientific and knowledge-based to stimulate real innovation and to produce high performance, sustainability and competitiveness.

The basic knowledge generated by the nanoscience can be transformed into specific applications to create a new class of multifunctional high performance construction materials. Multi-functionality means the emergence of properties additional to those that normally define a material. By nano-modification of traditional materials or incorporation of functional nanostructures in them, a material can extend its range of applications.

Interaction between materials and environment, between materials and user and between materials themselves (e.g. fibre-matrix interaction) occur at surfaces and interfaces. By modifying the surface of a material these interactions can be changed and new or dramatically improved functionalities achieved. Modification of surface layers of constituents of composites can benefit interaction and enhance bulk performance. Surface treatments and addition of thin films and coatings, often at nano-scale can produce improvements in materials' responses to changing environmental conditions and in improving people's environments.

Today there are interesting developments, based on nanotechnology, available in the market that can be used by the construction industry. There are glazing with special functionalities like:

- Solar-control glazing with a few nanometer thick metallic layers
- Photocatalitic self-cleaning glass
- Water repellent glazing
- Renewable energy systems, that can contribute to reduce dependence from fossil fuel energy use in housing, also can take benefits from nanotechnology:
- Hydrogen storage systems for fuel cells
- Quantum dots concentrators for photovoltaic cells

Other possible applications of nanotechnology are only envisaged. There are products like nanoparticles and nanostructures, such as carbon nanotubes or nanostripes, ceramic nanopowders, nanolayers, which could be incorporated in bulk construction materials or deposited on surfaces as functional reinforcements or films. They can confer new or improved properties, like polymer-based nanocomposites with higher fire resistance and durability, or having exceptional mechanical properties.

Nanotechnology offers also possibilities for development of:
- Less energy-demanding new cements based on the use of industrial waste residues by enhancing the pozzolanic properties of some of them.
- Synthesis of entirely novel binders (less energy demanding, environmentally friendly, cheaper) to partly replace Portland cement in concrete.

By this way there are reductions in both natural resources consumption and CO_2 emissions to the atmosphere. The nanoparticles use in cementitious composite lead to an improvement of bulk concrete properties and development of super strong and ductile cements.

4 LABEIN ACTIVITIES ON NANOTECHNOLOGY

In the year 2000 LABEIN, that is a private non-profit research foundation, decided to create a specific centre to develop the potential of nanotechnology into the construction sector. The Centre for Nanomaterials Applications in Construction (NANOC) was conceived as a platform for the development of the key technological capacities, allowing the use of nanotechnology as a competitive tool for the construction industry.

The NANOC objectives are:
1. The development of ultra high performance materials for the construction
2. To advance on the materials microstructure knowledge aiming to the progress on the nano-macro properties connection.
3. The multi-scale materials modelling.

These objectives are aligned with the European Union Policies contained on Lisbon and Göteborg agreements. The Lisbon Agreement to make the EU the world's leading knowledge based economy in 10 years term. It implies a necessary transformation of the European Union industry towards knowledge based one. The Göteborg Agreement to promote sustainable development. There are multiple implications behind the sustainable development concept, like: raw materials rational use, energy consumption reduction, CO_2 emissions reduction, materials and components recovery and recycle, etc.

The NANOC fields of activity are both multifunctional knowledge based materials and multi-materials, including active-adaptive materials, polymeric and cementitious matrix nanocomposites.

5 HOW TO ASSEMBLE THE NECESSARY SKILLS?

From the early beginning it was clear that the investment effort level required is enormous. The human and technological resources needed are too big for a single institution in such a broad field. The necessity of co-operation agreements between organisations is evident.

The first approach was, consequently, to develop bilateral contacts. Those contacts were established at a Regional level. The result is NANOMAT, a research platform leaded by LABEIN and composed by DIPC, INASMET, LABEIN, and POLYMAT (UPV-EHU) operating at Basque Country level, with the support of the Basque Government.

Later on, those contacts were extended to National level. Collaboration was established with some institute of the Spanish CSIC and also with the Universidad Pública de Navarra. There are now two research projects underway, partly sponsored by the Spanish Government.

Finally the third step was performed at a European level. Contacts were established with NANOCOM (University of Paisley), the Belgian Building Research Institute, the Centre Scientifique et Technique du Bâtiment and VTT.

Very soon it was realised that the FP 6 and the new instruments are offering an extraordinary opportunity to transform the incipient network of bilateral agreements into a Network of Excellence.

Now we have in front of us the possibility to create a single legal entity capable to co-ordinate a European network of working-centres. NANOCOM (University of Paisley), BBRI, CSTB and LABEIN are working, with other European partners, on the creation of this Network of Excellence. The questions to be solved are to maintain each member identity and to accept a progressive integration.

6 THE TWOFOLD CHALLENGE

On the way towards the creation of the Network of Excellence there is a great challenge. This challenge is twofold. From one side it is necessary to convince the public. From the other side it is the implementation of the necessary actions.

It is necessary to change the public perception. The association of both nanotechnology (the leading edge of technology) and construction (low technology perception) is still considered by many people as contradictory. The generation of awareness is also absolutely necessary. The construction activity comprises a high SME's percentage (97% in the EU). If the construction industry has to take the benefits of nanotechnology, the SME's must be aware of the situation as a necessary step. A lot of work is still to be done on this sense.

It is stated also that there is a challenge in implementing the necessary actions. The reason is clear. The construction industry includes a wide number of activities and processes. Consequently the applications of nanotechnology are also extremely wide. The implementing actions can include the following:

1. To develop the necessary basic knowledge. Although the science is offering the basic knowledge that is being used on the nanotechnology development, there are many questions to be answered by the scientific community.
2. To transform this knowledge into macro-scale applications. That means to introduce the concept of nanotechnology as a tool to develop new materials and consequently new components with new properties or functionality.
3. To develop the necessary processes. The macro-scale applications have to be obtained through new production processes.
4. To achieve an integration on the construction processes. Once developed, the new materials, components and their production processes have to be integrated on the construction industry value chain.
5. To promote specific start-up actions. There will be necessary to develop some specific start-up action to assist the undertaking of the nanotechnology by the industry, in particular by SME's.
6. Training and dissemination. Both aspects are crucial for the adequate deployment of nanotechnology. Once the knowledge, applications, processes are developed and integrated on the construction value chain, the industry will need technologist and technicians skilled to deal with them.

7 CONCLUSIONS

From all the above mentioned, some final remarks can be highlighted:

1. The construction industry is a European Union relevant economic sector that must use the available possibilities as to improve its business position.
2. Nanotechnology, and its exploitable applications, are a tool contributing to the construction industry sustainability and competitiveness.
3. High performance materials and components is one of the ways, but not the unique, to change the construction industry towards a more value added sector.
4. Integration of capabilities is absolutely needed to promote the new paradigm.
5. Recognition and the best use of available funding opportunities are essential to achieve the success of the full process.

References

1   *Amato "Nanotechnology – Shaping the World Atom by Atom" (NTSC report) 1999.*

APPLICATION OF NANOTECHNOLOGY IN CONSTRUCTION – CURRENT STATUS AND FUTURE POTENTIAL

Wenzhong Zhu, John C Gibbs and Peter JM Bartos

Advanced Concrete and Masonry Centre, University of Paisley, Paisley, PA1 2BE, UK

ABSTRACT

This paper examines current development, perceptions and future potential of nanotechnology in the construction industry by analysing results of a survey of construction professionals and leading researchers in the industry and by studying publications and reports on worldwide research development, commercialisation activities and future potential/directions of nanotechnology in relevant fields.

The study appears to indicate that interests in nanotechnology R&D in the broad field of construction and the built environment have been driven largely by governments, foresight institutes and academics/researchers working in the field. The application is still a very small, fragmented pursuit and largely unknown outside the scientific circle. However, there are already some nano-based materials and products that are used or ready to be adopted by the construction industry, and many others are coming to the market. The rate of development and application of nano-related materials and products is expected to accelerate with the very substantial increases in funding for nanotechnology around the world. The reluctant and very limited involvement of construction firms in nano-related research and development and the lack of focused technology transfer effort are believed to be among the main barriers to the rapid adoption of nanotechnology in the construction sector. There is, therefore, a need for the construction industry to change its perception/attitude toward new innovations and products enabled by nanotechnology so that involvement and investment in new technology is seen as beneficial to business success and essential for the future of construction.

1 INTRODUCTION

Nanotechnology has recently become one of the hottest areas in research and development worldwide, and attracted considerable attention in the media, the investment community and elsewhere. Definitions of 'nanotechnology' vary, but it generally refers to understanding and manipulation of materials/elements on the nanoscale, say, from 0.1 nm to 100 nm. Nanotechnology is not just about miniaturising things. The significance and importance of controlling matter at the nanoscale is that at this scale different laws of physics come into play (quantum physics); traditional materials such as metals and ceramics show radically enhanced properties and new functionalities, the behaviour of surfaces starts to dominate the behaviour of bulk materials, and whole new realms open up for us.[1] Gaining control of structures at the nanoscale can also lead to truly extraordinary materials such as carbon nanotubes, with a tensile strength often quoted as 100 times that of steel.

More than anything else, nanotechnology is an enabling technology, allowing us to do new things in almost every conceivable technological discipline. At the same time, its applications will lead to better, cleaner, cheaper, faster and smarter products and production processes. Just as other enabling technologies, such as electricity and microelectronics, have transformed lives, nanotechnology is likely to have a similar impact and the transformation will be much quicker due to the development of powerful computers, global communication and many other technological advances. With the backing of unprecedented funding from governments, businesses and investors, nanotechnology is rapidly emerging as the industrial revolution of the 21st century.[2]

The objectives of this paper are first to provide a brief overview of the major nanotechnology R&D and commercialisation activities in general, and then to examine in more detail current development, perceptions and future potential of nanotechnology in the construction industry. The study was carried out by analysing results of a survey of construction professionals and leading researchers in the industry and by studying publications and reports on worldwide research development, commercialisation activities and future potential/directions of nanotechnology in relevant fields.

2 NANOTECHNOLOGY IN GENERAL – A BRIEF OVERVIEW

2.1 Nanotechnology Initiatives and Funding

There has been an explosive growth of interest in nanotechnology for the last few years. The potential for enormous industrial developments led by nanotechnology research has been increasingly recognised by many national/regional governments, industry and investment communities. The National Nanotechnology Initiative in the US in 2000 was the catalyst to rapidly increasing funding and activities in this area. Global government spending on nanotechnology has more than doubled in the last 2-3 years, and now grown to well over $2 billion a year.[3] There is also substantial money flowing into nano-related research from multinational corporations and venture capital investments.[1,4] Many of the world's largest companies, such as IBM, Intel, Motorola, Hewlett Packard, Lucent, General Electric, Boeing, Hitachi, Mitsubishi, NEC, Pfizer, Corning, Dow Chemical, 3M, General Motors, Ford, etc. have all had significant nano-related research projects going on, or launched their own nanotech initiatives.

As a result of the increased funding and government initiatives, nano-related research and development in universities around the world has received a big boost over the past few years. By 2002, more than 30 universities had announced plans for nanotech research centres in the US. Similarly, in Europe over 80 nanotechnology networks had been set up with mainly national and EC funds.[1] More nano research centres and networks or such plans are at present being created around the world.

2.2 Research Activities, Publications and Patents

Scientific publications and patents provide a useful indication of the nature and extent of research activities and technological advances in particular areas. By conducting searches using terms 'nano*' on the Science Citation Index Expanded database (at http://wos.mimas.ac.uk), trends of nano-related publications were determined. Figure 1 presents the number of nano-related articles over the years. It is shown that the number of refereed publications has risen from a few hundred in 1990 to over 18000 a year in 2002, and is still increasing rapidly.

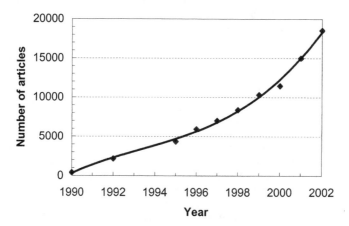

Figure 1   *Nano-related publications during 1990-2002*

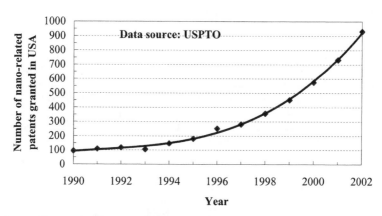

Figure 2   *Nano-related patents issued in US during 1990-2002*

Parallel to the study of SCI publications, searches for nano-related patents were also carried out with the keyword 'nano' in the title or the abstract on the USPTO database (United States Patent and Trademark Office). Figure 2 gives the number of nano-related patents granted in the US over the years. The results in Figure 2 showed a rapid increase of patenting activity in the past few years, but the increase was less dramatic compared to that for the nano-related journal publications in Figure 1. This difference may be due to the fact that much of the recent work on nanotechnology has focussed mainly on long-term or fundamental aspects rather than on application or product development.

The levels of scientific activities in nanotechnology among the major regions (Europe, Japan and US) were studied and compared by the World Technology Evaluation Centre, Loyala College in Maryland, USA in six broad areas, as shown below.[2]

- Synthesis and assembly: the United States appears to be ahead, followed by Europe and then Japan.
- Biological approaches and applications: the United States and Europe appeared to be on a par, followed by Japan.
- Nanoscale dispersions and coatings: the United States and Europe are again at a similar level, with Japan following.
- High surface area materials: the United States is clearly ahead of Europe, followed by Japan.
- Nanodevices: Japan leads quite strongly, with Europe and the United States following.
- Consolidated materials: Japan is a clear leader, with the United States and Europe following.

Europe appears to be ahead of the United States and Japan in terms of networking activities. A survey was carried out by the European Commission on European wide networks in the field of nanotechnology research. By November 2001, a total of 82 such networks were identified, which involved nearly 2000 institutions.[5] The work of the networks identified was found to be largely concentrated in seven application areas, i.e. Structural applications, Information applications, Nano-biotechnology, Chemical applications, Sensor applications, Long-term research and Instruments and equipment.

2.3 Commercialisation of Nanotechnology

From current activities in nanotech development and applications, it is evident that the commercialisation of nanotechnology is being pursued in a very broad spectrum of industry segments and geographies. A business of nanotechnology survey conducted by NanoBusiness Alliance in 2001 found that, contrary to common perceptions, a significant number of nanotechnology companies today is producing revenue derived from nanotech products.[6] Figure 3 shows the stage of product development of over 150 companies surveyed in 2001. The results indicate that over a fifth generates revenues, and an additional 16% also has products on the market.

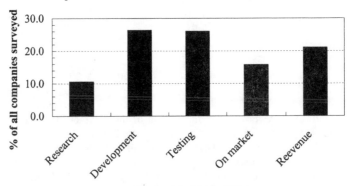

Product-to-Market Status

Figure 3   *Stage of product development of nanotech companies (data source: Ref.6)*

The same survey also revealed that about a third of the companies surveyed are mainly in the Materials and Manufacturing sector which covers producing ceramics, nanocomposites, nanoparticulates, nanotubes, etc. The second largest sector is in the Life Sciences (~22%), which comprise pharmaceuticals and healthcare/medicine, etc. Another survey of 227 companies worldwide conducted by an investment-consulting firm In Realis during 2000-2001 appeared to indicate a similar trend, as plotted in Figure 4.[7]

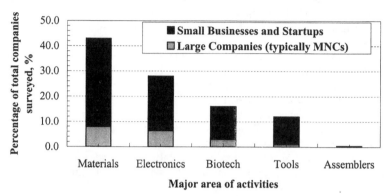

Figure 4 *Profile of current nanotech company focus (data source: In Realis[7]) Notes: Materials category includes efforts not specifically addressing one of the other four areas, focussed on developing and applying unique physical properties of passive nanostructured materials. Assemblers category cover the long-term objective of molecular nanotechnology, e.g. fabrication through self-replication.*

There are also several other surveys and reports which highlight the current commercialisation efforts and opportunities, such as those in the references.[2, 9-15] To gauge the exact number of companies active in nanotechnology and the market value of the nano-products appears to be difficult since huge variations in the nano company numbers (from 250 to over 1300) and in the existing market sizes (from $100 million to $30 billion) have been reported. The very fast speed of new nanotechnology products and companies being created and coming to the market is thought to be one of the main reasons for the difference in statistics. New products/nanotechnologies which seemed a far-off vision only a short time ago are with us almost before we can blink.[14] Different methods and extent of the surveying, analysing and categorisation used also contributed to the discrepancy reported. Some general common conclusions, however, could be drawn from all these surveys and reports. These are summarised and presented below:

- Contrary to popular belief, in the vast field of nanotechnology, many industries already produce/employ products which are either nano-sized or exploit nano effects, and are generating substantial revenues. Many of the application of nano-materials, e.g. advanced ceramics, composites, paints & pigments, thin films & coatings, fillers & additives, catalyst, adhesives, and biocompatible materials can be found today in industries including medicine, telecom, construction, electronics, aerospace, automotive and defence.[15] New viable products are being introduced almost daily, and the possibilities of nano materials/applications seem endless.

- The rate of commercialisation and growth of nanotechnologies, however, varies widely from industry to industry. Materials, Manufacturing and Tools/instrument sectors are expanding exponentially and generating most of the nano-related products/services revenues, while the Assemblers or molecular nanotechnology sector, in spite of significant research effort, has seen virtually no commercial activity. With some exceptions, most of the nanotechnologies are still at the early level of development, generally comparable to the computer industry of the 1960s, or the biotech industry of the 1970s.

- Understanding properties and processes is a significant focus of research (typically connected to an advancement in theory and in process control), highlighting that there is still much to learn about the nature of the new materials that can be engineered at the nanoscale. This may also result in the situation that most of the current nano-related businesses and start-ups are originated from or clustered around research and development efforts.

- The National Science Foundation (NSF) of the US predicts that nanotechnology products and services will generate revenues of $1 trillion per year by 2015.[11] Sanity-checking that projection with a detailed look at the industries most likely to be influenced by nanotechniques and natural adoption rates for new technologies suggests the figure is plausible, and quite possibly an underestimate.[7]

3 NANOTECHNOLOGY IN CONSTRUCTION

Construction was among the earliest of industrial sectors in identifying nanotechnology as a promising emerging technology, e.g. in the UK Delphi survey of the early 1990s,[16] and in foresight reports of Swedish and UK construction.[17-18] Two sectors of the construction industry, namely ready mixed concrete and concrete products, were also identified as among the top 40 industrial sectors likely to be influenced by nanotechnology within 10-15 years.[7] However, the application of nanotechnology in construction has lagged behind other industrial sectors. For example, unlike in the automobile, chemicals, electronics and biotech sectors where nanotechnology R&D attracts significant interest and investment from large industrial corporations and venture capitalists, in the construction industry much of the growth of interest and activities in nanotechnology has been driven largely by governments, foresight institutes and scientists working in the field.

This section examines the current development, awareness and perceptions of nanotechnology in construction by surveying those involved in the construction and related field. More specifically, summaries of the analysis of a survey of construction professionals and leading researchers, as well as results of a supplemental desk study of existing publications/reports are presented. Furthermore, examples of materials and products that are either on the market or ready to be adopted in the construction industry are provided. Finally, the long term potential and implications of nanotechnology development in construction are discussed.

3.1 Current Status of Nanotechnology in Construction – Survey Results

Recognising the huge potential and importance of nanotechnology to the construction industry, the EU approved funding for the Growth Project GMA1-2002-72160 "NANOCONEX" - Towards the setting up of a Network of Excellence in Nanotechnology in Construction, in late 2002. As part of this project, a survey was carried out to evaluate the current development, awareness and perceptions of nanotechnology in construction, and to identify potential partners for the EU network of excellence to be proposed.

With the Nanoconex project and proposing the EU NOE in mind, the survey, in the form of questionnaire and e-mail, targeted approximately 400 leading researchers, construction professionals, large companies involved in construction and nano-materials/tools manufacturers. About 90% of the targeted were from European countries, and the remaining 10% mainly from US, Canada, Australia, Japan and China. Out of all the recipients of the questionnaire, only 83 responded, a return rate of 21%. Due to few responses outside Europe, the survey results are thus incomplete, and likely to reflect mainly the current situation in Europe. Table 1 shows the distribution of respondents in different sectors related to construction and the built environment.

Table 1   Distribution of the survey respondents in various sectors related to construction

Core business sectors	Number of respondents	% of total respondents
University	28	34%
Research establishment	26	31%
Industry – materials and equipment supplier	15	18%
Industry – all sectors except materials and equipment supplier	9	11%
Others not yet related to construction	5	6%

As the results in Table 4 show, the majority of the respondents are from Universities and Research establishments, and only 11% are from a large number of construction industrial sectors other than materials/equipment supplier (including: contracting, design, survey, service, project management, etc.). Results for the awareness and main areas of involvement of the respondents, as shown in Figures 2 and 3 respectively, appeared to suggest that the low rate of industrial response was mainly due to the very limited knowledge/awareness of nanotechnology development and lack of interest in getting involved at this time among the construction firms.

The results in Table 2 appeared to show that Understanding phenomena at nanoscale and Nano particles, filler and admixtures were the two activities/applications in construction known to most of the respondents. The awareness of nano-related activities was generally much higher among the academics/researchers than the industrial personnel. Furthermore, the areas of activities which the respondents were aware of are those of the materials development and fundamental research nature. Development in devices/systems (e.g. control/monitoring, energy supply, etc.), which could have significant implications to construction industry in the medium/long term, were less known to the respondents.

The results in Table 3 appeared to be consistent with the findings of Table 2, i.e. the major activities of the respondents were in the areas of High performance materials, Understanding phenomena at nanoscale and Multifunctional materials/components. It is noted that except for limited involvement in the area of High performance structural materials, the industrial personnel have little involvement in nano-related activities.

Table 2   *Results of survey of awareness of nano-related applications in construction*

Awareness of nano-related research and applications	% of those responded	
	Academics/ researchers	Industrial personnel
Understanding phenomena (e.g. cement hydration) at nanoscale	82	58
Nano particles, fillers, fibres, and admixtures	80	37
Nanostructure modified materials (e.g. steel, cement, composites)	73	26
New functional and structural materials	61	26
Surface/interface assessment, engineering	55	21
Special coatings, paints and thin films	45	21
Integrated structural monitoring and diagnostic systems	39	11
Self-repairing and smart materials	31	11
New thermal and insulation materials	20	11
Intelligent construction tools, control devices/systems	22	11
Energy applications for buildings - new fuel cells and solar cells	24	0
Biomimic and hybrid materials	20	0

Table 3   *Extent of involvement in nano-related activities by the survey respondents*

Involvement in nano-related activities which are potentially applicable to construction	% of those responded	
	Academics/ researchers	Industrial personnel
High performance structural materials	80	37
Understanding phenomena at nanoscale	69	21
Multifunctional materials/components	40	11
Modelling/Simulation of nanostructures	38	5
Nanoscale engineering techniques/instruments	31	11
Smart materials and intelligent systems	29	16

Another important finding coming out of the survey was the perception/attitude of the construction industry in general on nanotechnology developments and potential applications in construction. All, but two, respondents agreed that the public is not properly informed, or is confused about nanotechnology in general, and the situation is among the worst in construction industry. Particularly, as pointed out by some respondents, the information on nanotechnology applications is not generally appreciated in the construction industry – being a traditional, low-tech industry which tends to be very conservative and not always receptive to new ideas/innovations. A typical attitude of a construction industrial professional is 'wait and see', and that we can now effectively do it without comprehending the complicated, cutting edge sciences. Nanotechnology is, if anything, perceived as expensive and too complex to explain to clients who want a structure built as soon as possible and as cheap as possible.

A positive note, perhaps, is the very rapid growth in awareness of and interest in nanotechnology development in many industrial sectors. In 2001, a report by Lux Capital found that less than 2% of Fortune 1000 executives were able to define nanotechnology, and less than 5% had ever even heard of the word. Just two years later, nanotech has become the buzzword in international politics, global financial markets, and on the pages, airwaves and screens of mainstream media.[19]

3.2 Current Status of Nanotechnology in Construction – Desk Study

The desk study of publications and reports on worldwide research development and commercialisation activities of nanotechnology relevant to the broad areas of construction and the built environment was carried out as a supplement to the e-mail survey. The results of this desk study are summarised and presented in this section.

As a results of the unprecedented increase of funding in the area of nanoscale science and technology, interests and activities in nanotechnology R&D has been increasing substantially across industrial sectors. The research and development relevant to construction is not an exception.

In Europe, an analysis of EoIs (expression of interest) submitted to the EC FP6 in 2002 found that there were 20 EoIs related to the nanotechnology application in construction. The analysis was based on results published on the E-Core (European Construction Research Network) database (http://www.e-core.org/frames/index_database_eoi_info.html) which had a total of 250 EoIs related to construction applications. The EoIs covered a wide range of interests/activities, namely understanding & modelling of phenomena at nanoscale, developing nanoscale particles, tubes/fibres & nanostructure modified materials/components, functional materials, thin films & coatings/paints, energy efficient devices, and smart materials & integrated systems incorporating nano sensors/actuators.

In the USA, several research programmes directly linked or relevant to construction and the built environment have been funded under the NNI and by the National Science Foundation (NSF) to study the fundamental aspects of materials, creation of new materials/functionality, deterioration science, sensing/diagnostic technologies, renewal engineering, multiscale modelling, simulation & design of materials, and construction automation etc.[20] Some research and development activities aiming at commercial application/products based on nanotechnology are also evident, and mainly funded/conducted by start-ups and large industrial cooperation.

In Australia, the new Australian National Nanotechnology Network announced last year its first collaborative effort – the Nano House Initiative. The nano house to be built will represent the current best practice in sustainable and environmentally friendly housing using the most recently developed nanotechnology based materials and components. The initiative is to demonstrate applications of nanotechnology and how these applications interact with each other and conventional materials, in order to stimulate the diffusion of nanotechnology into industries.[21]

In Canada, an internal report of the Institute for Research in Construction (IRC), National Research Council Canada, has also recognised the importance and possible applications of nanotechnology to the construction industry.[22] More recently, the IRC has initiated a multi-researcher project to develop new technologies and products for the construction industry based on nanotechnology, with an emphasis on cements, cement-based products, admixtures and concrete.[23]

The rapidly increasing interest in nanotechnology is also occurring in Japan, China and many other countries, and has been reflected in many publications and reports on worldwide research development and commercialisation activities. Several nano-enabled products, for example, high performance structural and/or functional materials have been developed specifically for applications in construction and the built environment, and some of them already have sales.

3.3 Nano-enabled Commercial Products and Applications in Construction

3.3.1 High Performance Structural Materials.   The most significant example is perhaps the nanostructure modified steel reinforcement – MMFX2 Steel, manufactured by MMFX Steel Corp. USA.[24] The steel, developed through the use of nanotechnology by Professor Gareth Thomas, University of California, Berkeley, has corrosion resistance similar to that of stainless steel, but at a much lower cost. Compared to conventional carbon steel, the MMFX steel has a completely different structure at nanoscale – a laminated lath structure resembling 'plywood' as shown in Figure 5. Due to the modified nanostructure, MMFX steel also has superior mechanical properties, e.g. higher strength, ductility and fatigue resistance, over other high-strength steels. These material properties can lead to longer service life in corrosive environments and lower construction costs.

MMFX STEEL – MICROSTRUCTURE

Transmission Electron Microscopy (TEM)

**Microstructure of Nano Sheets of Austenite
In Carbide Free Lath Martensite**

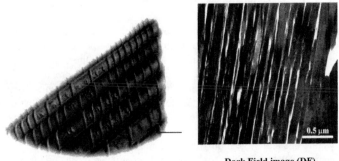

Dark Field image (DF)

Figure 5   Nanostructure modified steel reinforcement – MMFX Steel Corp. USA[24]

Since corrosion of steel is one of the most serious and costly problems facing the construction industry today, the implications of highly corrosion resistant steel, such as MMFX 2, are far reaching. So far, the MMFX steel has gained essential qualification/certification for use in general construction throughout the US. It has attracted considerable interest from Federal Highway Administration, US Navy and Departments of Transport of various states. Projects using the MMFX steel are currently underway or in the federally funded submission stage in 22 states of the US. MMFX is also concentrating efforts on the approximately 12 million ton (US, Mexico and Canada) general construction reinforcing steel markets.

Several other nano-enabled high performance structural materials, such as high strength ductile ceramics, glass and polymer nanocomposites, which have been developed for automotive and military applications, also have the potential for construction applications. For example, hard and tough ceramics could be used as cutting tools, high strength springs and wear parts; and ductile glass and polymer nanocomposites could be used in construction components for earthquake or ballistic/impact resistant structures.

3.3.2 Functional Coatings and Thin Films. Incorporating certain nanoparticles into transparent coatings/thin films can provide enhanced performance and additional functionalities. Currently, several products have been developed and marketed for construction and the built environment, including: hard, protective or anti-corrosion coatings for components; self-cleaning, thermal control/energy saving, anti-reflection coatings for glass/windows; easy-to-clean, anti-bacteria coatings for work surfaces (i.e. for kitchens, bathrooms, and door handles, etc.); and more durable paints and anti-graffiti coating for buildings/structures.

For example, Pilkington developed its 'Activ' self-cleaning glass which uses a special 50 nm thick coating.[25] The special coating works in two stages. First, using a 'photocatalytic' process, nano TiO_2 particles in the coating reacts with ultra-violet rays from natural daylight to break down and disintegrate organic dirt. Secondly, the surface coating is hydrophilic, which lets rainwater spread evenly over the surface and 'sheet' down the glass to wash the loosened dirt away.

Furthermore, nanostructured coatings can be used to selectively reflect and transmit light in different wavebands. These coatings can be produced by either using nanoparticles or forming multiplayer thin films metals and dielectrics.[26] These coatings can be used on windows as radiant heat reflectors, with many energy saving applications. There are also coatings incorporating nanoparticulate silver powder to provide antibacterial action, and coatings using nanoparticulate gold to destroy toilet odours.[27] Special coatings developed can also make the applied surface both hydrophobic and aleophobic at the same time. These could be used for anti-graffiti surfaces, carpets and protective clothing, etc.

3.3.3 Nano Fillers, Additives and Admixtures This area is perhaps the most commercially developed/researched and the use of these nanoscale materials has been the basis for many current and potential applications in various industrial sectors, such as electronics, healthcare, chemical, defence, energy, automobile and construction, etc. There have been a large number of companies producing many different types of nano particulate/fibre materials, such as carbon nanotubes, nanoclay, metallic and non-metallic oxides, etc.

With particular relevance to construction and the built environment, several nanoparticulate materials (e.g. TiO_2, SiO_2, $CaCO_3$) have been widely used as fillers/additives in coatings/paints, adhesives and sealants. For example, carbon black nanoparticles have been added to rubber to increase wear resistance for many years.

Nanoclay and nanotubes/fibres are increasingly used as reinforcement in high performance composites. Dispersion/slurry of amorphous nanosilica powders has also been produced as concrete admixtures, which can be used to improve segregation resistance for fresh self-compacting concrete, or to enhance strength/durability performance of hardened concrete.[28-29] Nanoparticulate materials have also been used in environmental friendly anti-corrosion coatings for steel reinforcement and waterproofing admixtures for concrete. Furthermore, nanoparticles, carbon nanotubes and nanofibres offer the potential for developing new functional materials and devices/systems with much enhanced performance for use in construction and the built environment.

3.3.4   *Environment and Performance Monitoring Sensors.*   Microscale sensors and microscale sensor-based devices/systems have already been used in construction and the built environment to monitor and/or control the environment conditions (e.g. temperature, moisture, gas/smoke, noise, etc.) and the materials/structure performance (e.g. stress, strain, vibration, pressure, flow, corrosion, deterioration, etc.).   Nanotechnology approaches are not only enabling sensors/devices to be made much smaller, more reliable and energy efficient, but also opening up new possibilities beyond the reach of mere micron-scale manufacture (e.g. biomimetic sensors and carbon nanotubes systems).[22]

Being very small, inexpensive and low energy consuming, lots of sensors/devices can be widely used and positioned to give a high degree of coverage. For example, strain measurement devices could be scattered throughout a wall or bridge deck, providing real time results that show the strain in the whole structure, rather than at selected individual points.[22] In addition, nanoscale sensors should also allow measurements to be made with minimal interference with the structure being measured.

Nanotechnology enabled sensors/devices also offer great potential for developing smart materials and structures which have 'self-sensing' and 'self-actuating' capability. The device used for air bags in cars is such an example. Siemens and Yorkshire Water are developing autonomous, disposable chips with built-in chemical sensors to monitor water quality and send pollution alerts by radio.[30] Potential applications in construction may include building materials (e.g. concrete) with self-healing ability at cracking/damage, tools with embedded intelligence and buildings/infrastructure able to adapt themselves to changing service and environment or weather conditions.

3.3.5   *Other New Products and Applications.*   Many other nanotechnology enabled developments and materials/products are fast emerging or already exist in bulky and expensive forms in different industrial sectors. With the continuing reduction of size and affordability and improvement of performance, these products and developments are also likely to impact on construction process and the built environment in the short to medium term.   These may include ever improving computers, electronic and communication devices, tracking/tagging devices, energy efficient lightings and fuel cells, and novel (e.g. biomimetic, self-assembly) materials, just to name a few.

Furthermore, progress in development and application of nanotechnology-based instrument and modelling/simulation tools will results in our improved understanding of properties of traditional materials (e.g. cement, concrete, bitumen) and their interaction with the environment at the nanoscale. This will lead to improvement of performance of such materials and development of new approaches to extend the life of existing structures or to prevent/arrest deterioration in the new build. Advances in development of design tools (with the use of modelling and simulation) will also allow structures to be built with smaller design margins.

3.4 Future Trends and Potential of Nanotechnology in Construction

The survey of construction professionals and the examination of current worldwide nano-based activities/development relevant to construction applications appear to indicate that the following areas will have a significant impact on construction and the built environment within the next 10-15 years.

- Understanding phenomena at nanoscale
- Nano-structure modified materials
- High performance structural materials
- Special coatings, paints and thin films
- Multi-functional materials and components
- New production techniques, tools and controls
- Intelligent structures and use of micro/nano sensors
- Integrated monitoring and diagnostic systems
- Energy saving lighting, fuel cells, communication and computing devices.

The advances within this time frame are likely to be incremental on existing materials and technologies. With improved understanding of structure-properties relationships and ability to control materials structure on the nanoscale, major breakthroughs in materials science are expected. Such breakthroughs will enable the 'materials by design' approach to replace the traditional 'trial and error' one to tailor a material for a specific requirement.

One of the ultimate goals (and the most debated area) of nanotechnology is molecular manufacturing (or molecular nanotechnology) based on self-assembly or self-replicating nano machines (thus the vision of buildings that build themselves). Though there is increasing evidence indicating that molecular manufacturing is possible, full molecular nanotechnology capability is unlikely to be developed for at least one or two decades.[31] However, extensive effort toward that goal, and particularly biomimetic study to better understand and replicate/mimic nature's version of molecular nanotechnology (i.e. the way living organisms produce food, mineral and composite materials) will have huge potential for new materials and processes applicable to construction in the next 10-15 years.

For example, new coating development based on replicating lotus leaves could lead to self-cleaning surfaces for buildings. Uncovering the secrets of strong and tough seashells could offer not only new approaches to composites structure design, but also potential new manufacturing processes with much improved energy efficiency and sustainability. Also, by mimicking geckos' hairy feet, researchers have produced prototype super-sticky adhesive tape which is extremely strong, waterproof and reusable.[32]

4   CONCLUSIONS

Based on results of the survey of construction professionals and the desk study of current worldwide nano-based activities/development relevant to construction applications, the following conclusions could be drawn.

- Nanotechnology R&D in the broad area of construction and the built environment lags behind other industrial sectors, and has been driven largely by governments, foresight institutes and academics/researchers working in the field.
- Nanotechnology application in construction is still a very small, fragmented pursuit and unknown outside the scientific circle. However, some limited commercial

activities have started to emerge in the materials and equipment supplier sectors. As a result, some nano-based materials and products are now used or ready to be adopted by the industry, and many others are coming to the market.

- One of the main barriers for the rapid adoption of nanotechnology in the construction sector appears to be the lack of awareness and negative perceptions of nanotechnology among construction professionals. To overcome some barriers, integrated actions are needed for targeted R&D, for technology watch and technology transfer in construction industry.
- Huge potential has been predicted for nanotechnology applications in construction, and even minor improvements in materials and processes could bring large accumulated benefits considering the scale of activities involved in the construction industry.

References

1   CMP Cientifica, NANOTECH: The Tiny Revolution, November 2001.
2   Siegel, RW, Hu, E and Roco, MC, Nanostructure Science and Technology: A Worldwide Study, National Science and Technology Council, Committee on Technology, IWGN, September 1999.
3   UK Advisory Group on Nanotechnology applications, New Dimensions for Manufacturing: A UK Strategy for Nanotechnology, June 2002.
4   Garcia, B.E., Nanotechnology takes off, *The Miami Herald*, Oct. 21, 2002.
5   European Commission, Survey of networks in nanotechnology, Release 2, Nov. 2001.
6   Tinker, N, 2001 business of nanotech survey, Nanobusiness Alliance, Oct. 2001.
7   In Realis, A critical investor's guide to nanotechnology, Feb. 2002.
8   Saxl, O, The Institute of Nanotechnology, http://www.nano.org.uk/applications.pdf.
9   CMP Cientifica, The Nanotechnology Opportunity Report, May 2002, http://www.cmp-cientifica.com/.
10 Makoto, W, Special report: A Future Society Built by Nanotechnology, *Journal of Japanese Trade and Industry*, Sept./Oct. 2001.
11 National Science Foundation, Societal Implications of Nanoscience and Nanotechnology, NSF, Arlington, Virginia, USA, 2001.
12 Targart, G, Nanotechnology: the technology for the 21st century, APEC Industrial Science and Technology Working Group, August 2002.
13 Fowler, C, Smaller, Cleaner, Cheaper, Faster, Smarter: Nanotechnology Applications and Opportunities for Australian Industry, Commonwealth Department of Industry, Tourism & Resources, June 2002.
14 Institute of nanotechnology, Eureka and Department of Trade and Industry, Nanotechnology: The huge opportunity that comes from thinking small, Published by Findlay Publication Ltd. UK, 2001.
15 FirstStage Capital, Nanotechnology today: Reality or Hype? Nov. 2001. www.firststagecapital.com.
16 Gann, D, A Review of Nanotechnology and its Potential Applications for Construction, SPRU, October 2002.
17 Flanagan, R, et al., htthp://www.bem.chalmers.se/sh/forskning/vision2020article.pdf.
18 DTI, Constructing the future, Foresight report, London, 2001.
19 Lux Capital, The nanotech report 2003: investment overview and market research for nanotechnology (volume II), 2003.

20 Chong, KP, Presentation at the 1st International Symposium on Nanotechnology in Construction, Paisley, Scotland, 23-25 June 2003.

21 Masens, C, Briefing document: Business Plan – UTS/CSIRO/SGI Nanohouse Initiative, June 11, 2003, nano@uts.edu.au.

22 Makar, J, Implication of nanotechnology for the construction industry, Internal report, Institute for Research in Construction, National Research Council Canada, January 2002.

23 Institute for Research in Construction, Building envelope and structure, *Construction Innovation*, Vol.7, No.4, Dec. 2002.

24 MMFX Steel Corporation of America, http://www.mmfxsteel.com/.

25 Pilkinton, http://www.activglass.com/.

26 Placido, F, Presentation at the 1st International Symposium on Nanotechnology in Construction, Paisley, Scotland, 23-25 June 2003.

27 Nano Composite, http://eng.nanocomposite.net/.

28 Bigley, C and Greenwood, P, Concrete, Feb. 2003, pp43-45.

29 Akzo Nobel, http://www.colloidalsilica.com.

30 CRISP/OST Foresight Briefing Paper, Foresight/CRISP workshop on nanotechnology: What is nanotechnology? What are its implications for construction? June 2001.

31 Phoenix, C, FAS Public Interest Report, Spring 2003, pp.14-17.

32 Powell, K, Gecko glue round the corner, *Nature Science update*, 28 August 2002.

NANOTECHNOLOGY FOR CONSTRUCTION: BEYOND THE IMAGERY

Bob Cather

Arup Research & Development and Arup Materials Consulting, London

EXTENDED ABSTRACT

How can we convey, something of Nanotechnology and its relevance to society and to industries and to people? There's so much to think about. I can only start to prompt, from a personal perspective and that is essentially all that I will try to do. Over the coming days, months and years, more discussion and exchange should help all of us to think widely and freely.

It is easy to get dazzled by the glitz and the headlines of Nanotechnology. By images of:

- Nanobots and nanomachines that self-assemble.

- Atomic manipulation.

- By popular presentation (in novels, films, TV and the press).

Nanotechnology, like many other movements such as Smart Materials or Sustainable Development, means different things to different people.

There are of course specific definitions for Nanotechnology but delving into that is not particularly fruitful in trying to see the bigger picture and understand something of the potential.

There may also be a tendency for everything to acquire the NT label; to catch the latest bus of R&D fashion. We may be, and I think I was, somewhat surprised to see Self-Compacting Concrete being described as nanotechnology – surely a dose of post-rationalisation.

In my day-job interacting with design teams, clients and contractors I am principally interested in the nature and performance of materials and how we design and construct with them. With a fuller materials science understanding we can improve the performance of current materials and know how to assess, adopt and design with new materials or meet new desires.

Over history we have tended to use a materials selection process that has parallels with natural selection (aka 'trial and error). We use things and those that work we use again. Those that don't work we discard – at least until we've forgotten why we discarded them!

In more recent times our capabilities in materials selection and performance are considerably increased and enhanced by our understanding of microstructure and of chemistry – Or Materials Science.

With varying rates of progress for different materials, our understanding of microstructure and hence materials performance has been greatly aided by the advances in microscopy and imaging that have evolved, particularly in the past 20 - 30 years or so.

We have used these techniques not just for the new 'hi-tec' materials but also for the every day, for steel, timber, masonry and concrete.

Now we have even better capabilities, to see individual atoms and atomic level structures. These are some of the keys to the doors of the nanotechnology world.

More excitingly, there are the beginnings of the ability to manipulate atoms and structures to create new structures and materials. I'm not suggesting that this is or should be commonplace but the idea of the potential is very stimulating.

We might start to see this as the materials equivalent of moving from natural selection to GM. We should always take care and show concern in the way we deploy any new technologies. Hopefully we can take up the possibilities without the negative connotations that GM foodstuffs manage to generate. Although we are already fighting the battle of rational argument versus newspaper sales.

Many may be at this conference because of the newness of the technology and Nanotechnology becoming the newest new-kid on the block. However it is not new and unexploited. Research into nanotechnologies has been underway for 20 years or more. It is here now and products are in the market. Progress is substantial and so is the potential.

One major question for this conference is "Is construction getting enough"? This parallels the question and angst often expressed about construction's reluctance to adopt 'New Materials'. There is a separate presentation here 'New materials – what do you mean by *new*?'.

There are different threads to Nanotechnology, conventionally described as 'top-down' and 'bottom-up'. I don't disagree with the notation but have found more benefit in creating a 3-way broad strand of activity. These are:

- 'Passive conventional NT' – solids or surfaces dependant on behaviour at NT (active)
- Micro/nano electronic devices – continued miniaturisation (chips and sensors)
- Manipulation and building from individual atom or molecules (bottom up)

At present we are clearly seeing more products in the passive conventional materials and in continued miniaturisation. With applications in many industries, such as efficient catalysts for chemical processes, pollution reducing or sieving systems to control the effects of chemical plant and as anti-wrinkle and sun-screen compounds for the cosmetics industry. Perhaps interestingly, at a Nanotechnology lecture at the IEE it was claimed that, based on a count of research paper citations and filed patents, the most active nanotechnology company was L'Oreal. The impact of very fine particulates on the optical properties of the skin apparently having considerable advantage.

For other industries, including construction, the interaction of fine fibre and fillers, with binder systems and with the environment, be it UV light, corroding gases, excessive humidity, offer good opportunities for coatings systems – both protective and decorative. Or, perhaps we'll take the bigger step of just modifying the surface characteristics to avoid the need for coating.

More directly into construction we can see direct impact such as in the self-cleaning glass described in the press. At present the real applications that have popular recognition are rather limited, but more are surely to arrive soon. In the development and adoption of new materials and products there may be many drivers. Some will come from the idea itself, we have a new idea or technology, how can we best use it. Others will come from a desire to solve a desire or problem "I wish we could ……..." Sometimes innovation, development and progress come from 'the big idea'. 'To send a man to the moon and return him safely to Earth' was clearly a 'big idea' that had considerable technology payback both directly and indirectly. The range of activities under the banner of Nanotechnology may well, paradoxically, be seen as the next big idea. If we as an industry are to get most out of the Nanotechnology explosion, we must find ways of increasing the cross dialogue between those investigating and developing the technologies and those wanting to solve problems or meet desires in a new more effective way.

I am not a specialist or particularly hands on experienced in nanotechnology, but I am tremendously excited by the potential. There is a considerable amount of interest and energy but we must find ways of making these energetically charged clouds coalesce into the constructive lightning strikes of products and processes that will benefit us all.

There may be a danger of an overly 'techy' image and language created around nanotechnology generating negativity or blockages that may be off-putting to those outside the club. This would be very unhelpful and quite unnecessary.

Designers and constructors don't necessarily need to understand the fuller detail of the science, just as the great majority of people using PCs have very little idea of what is actually going on inside the box.

So we don't need to understand the nano-science of the interaction of particle size and visible light, of surface promoted catalytic breakdown of dirt or, of hydrophilic surfaces. We can just benefit from having self-cleaning windows, with benefits for building appearance, for reduce environmental impact of less detergent use and for H&S issues of maintenance access.

What is important is to create a head-up dialogue so that designers and clients can formulate and pass on the thought "I wish we didn't have to keep on cleaning windows" and the research base can say "funny you should say that...". Of course within manufacturing industries this sort of process occurs regularly, but in the typically more dispersed design and construction set-up we'll have to work harder at facilitating the dialogue.

So where in construction should Nanotechnology be adopted? I can only begin to respond to that. We've talked electronic systems for data handling, visualisation and control. Also sensing devices. Will they become so cheap and commonplace that every component in a building is tagged with information and sensors that not only identify its original condition and properties, but also adds real in-service data to update?

We've mentioned coatings and surface modification.

We will need to consider the drivers for change, for example sustainable development, may benefit from lighter, stronger, lower embodied energy materials. It may benefit from new generation fuel cells, from enhanced battery technology.

We may get closer to the long held desire for thermally efficient transparent materials that are practical to use. Although even here, there is now an interesting new product on the market.

These suggestions for possible applications are tending to be technology driven- 'got a great new material' - which have benefited from research funding. But we need to balance this with positive action encouraging the identification and development of applications. There is much to discuss in how best to do this.

Looking back from say, 20 years in the future, what technologies and materials performance will we see that we cannot imagine now? This leads us to two applications of the 'new' science. New science plus conventional thinking – that we can imagine now:

- self healing cracks
- paint-on sensors
- strong metal alloys
- corrosion antibodies
- corrosion reversal nanobots
- porosity modifying
- auto-modulus correctors

and, New science plus new thinking or imagining that we cannot see now:

- new ways of thinking
- new ways of imagining
- ideas we cannot yet be aware

There is a perception that the UK is great at the basic science but that we fall short in properly exploiting that knowledge. In the context of today we can also say that we have considerable world-class design capability with a reputation for innovation. Surely bringing these together can produce really useful and successful applications for nanotechnologies.

A further suggestion to conclude. There is a danger of limitation and exclusion if we just see nanotechnology just as a technology with defined attributes and classifications, however correctly phrased. There is merit in taking a wider looser view, a way of thinking or even a way of seeing. In this vision we explore the nature of materials through their structure, their constituents, the physical processes taking place at surfaces and the energies being exerted, of defects and failure initiation at whatever scale is relevant.

In this context with the importance and manipulation of particulate size and shape and solid fluid interactions, even nanotech self compacting concrete doesn't seem quite so far fetched!

Part 2: Techniques and Instrumentation

FOCUSED ION BEAMS (FIB) – TOOLS FOR SERIAL SECTIONING OF NANOINDENTATION SITES IN CEMENTITIOUS MATERIALS

P. Trtik

Institute for Building Materials (IBWK), ETH – Hönggerberg, CH-8093, Zürich, Switzerland
*formerly Scottish Centre for Nanotechnology in Construction Materials (NANOCOM), University of Paisley, Paisley, Scotland, UK

1 SUMMARY

The paper reports on a particular application of focused ion beam (FIB) techniques applied for the enhancement of nanomechanical test methods. Nanomechanical testing techniques, such as nanoindentation, are increasingly important tools for the characterisation of mechanical properties of materials. By utilising the nanoscale resolution of the ion column, the FIB techniques show a large potential for the better understanding of processes which occur during micro/nanoindentation test methods. FIB techniques can be used for investigation of micro/nanoindents. An indent could hitherto be observed only from "above", as an impression into the surface of the tested material. In its cross-sectioning mode, the FIB workstation enables the material to be milled away in such a manner that the previously unavailable - but very important - observation of the deformation of material underneath the indent can be carried out. Such cross-sections reveal entirely new information about the 'third dimension' of the micro/nanoindents and about the deformation induced on the material in vicinity of the indent. The paper presents an outcome of investigation that shows that by means of serial sectioning of the indentation sites a quasi-3D picture of subsurface deformations induced by instrumented indentation could be revealed in cement-based materials.

2 INTRODUCTION

In order to fully understand the mechanical behaviour of cement-based materials, it is necessary to characterise their response to nanomechanical test procedures, such as instrumented indentation technique. The mechanical properties of basic constituents of the cement pastes can be connected with the compositional properties of the indentation sites (see Hughes & Trtik[1]). However, these correlations are based 'only' on the top down 2D observations of the indentation sites in a scanning electron microscope. As the cementitious materials are intrinsically 3D disordered materials, a need for 3D characterization of indentation sites arises. Due to the sub-10nm resolution of the ion beam, the FIB workstation can be readily utilised for such tasks for various materials characterised (e.g. Wu et al.[2]).

The principal area of utilisation of focused ion beams (FIB) is well established in microelectronics, namely in the semiconductor and integrated circuits industries, where ion beams are commonly used for the modification of prototype integrated circuits. The operating principles of FIB workstation is described elsewhere[3]. However, recently the investigation using FIBs spread into many other fields of materials research. The progress in utilisation of focused ion beam technology for micro and nanoengineering applications has been recently summarised in a literature review by Langford *et al.*[4]

Figures 1a shows an example of a very extensive deformation induced by indentation testing, for which the Figure 1b clearly reveals the true subsurface extent of that in the z-direction.

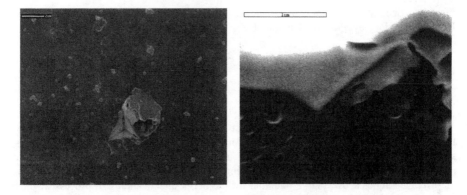

Figure 1   *(a) top-down view of an 1200nm-deep indent in OPC cement paste (scale bar corresponds to 2μm), (b) FESEM image of the corresponding cross-section prepared by FIB single sectioning of the indentation site (tilted view 45 degrees, scale bar equals 1μm)*

The single sectioning of the indentation site helps to understand the extent of the subsurface deformation (see Trtik et al.[5]). However, as the state of deformation underneath an indent in cement-based materials is not expected to be symmetrical, the single sectioning of such indentation sites is not representative. In order to receive full picture, such an area must be "mapped" by serial sectioning of the site. The technique itself has been pioneered by Inkson et al.[6], who utilised it for "slicing" nanoindents in Cu-Al multilayers. In the comparison with this case, the serial sectioning of indentation sites in the cement-based materials is more complicated, due to the insulating nature of these materials. Thus, the imaging capabilities of the ion beams are significantly lowered - especially on the vertical walls of the milled sites. The next paragraph presents an attempt to overcome this problem.

3   SERIAL SECTIONING OF INDENTATION SITE

In its initial phases the serial sectioning technique is very similar to the single sectioning technique. It consists of three basic parts: (i) site set-up, (ii) single sectioning part and (iii) serial sectioning loop.

In the 'site set-up' the large positioning indents are found on the sample (e.g. using optical microscope) and then the sample is placed into the chamber of FIB workstation in such a manner, so that the view of the grid of nanoindents corresponds to the view which has been adopted in instrumented indentation. This arrangement prevents confusion in the indent numbering to occur. By using single scan image updates the area of interest is found in FIB. In the next step, a protective layer (approximately 15 x 15 x 1μm) of platinum is deposited onto the region of interest (i.e. one indentation). In order to assure good deposition conditions, the platinum deposition rate should be slower than 200 nm per min. Figure 2 shows the image of the test site on sample of cement paste with two large positioning indents and grid of small indents.

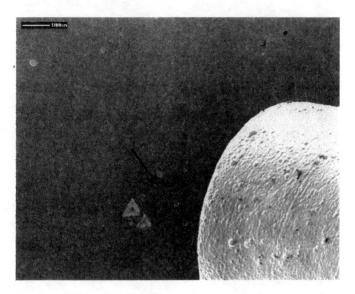

Figure 2 *Image of the region of interest on the cement sample showing large two positioning indents, grid of 25 nanoindents (see the end of the arrow) and tube for delivering the precursor gas for platinum deposition (scale bar corresponds to 100μm)*

In the second part of the procedure – the single sectioning – a large rectangular block (app. 15 x 30 x 6 μm) is milled away at 2700pA in the position adjacent to the 'platinumed' area. The relatively high milling current is chosen in order to shorten the otherwise extensive milling time. The lower milling current would – on the other hand – produce a sharper section profile. Figure 3 shows the grid of 25 indentations with one indent prepared to be sectioned. In the next step, two parallel "marker trenches" are milled in the platinum protection layer (approximately 0.5 x 15 x 0.5μm in size). These marker trenches later assist with positioning of the individual images in the 'quasi-3D map' of the indentation site. At this point of the procedure the site is prepared for serial sectioning.

The serial sectioning loop starts with a clean-up mill of the face of the region of interest by 150pA beam current. At this stage, the sample stage is tilted about several degrees (app. 5 degrees) and a platinum layer is deposited onto the vertical wall of the region of interest for very short deposition time (approximately 2-3 seconds).

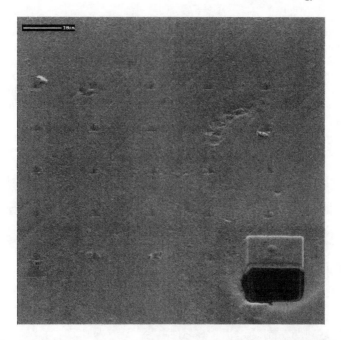

Figure 3 *Image of the region of interest on the cement paste sample grid of nanoindents and one indentation site prepared for the serial sectioning loop (scale bar corresponds to 10μm). Note the low contrast on the vertical wall of the milled block.*

This step seems to improve the contrast of the image on the vertical wall of the region of interest, while not masking the microcracks induced by indentation. The time of the platinum deposition has to be small enough not to mask the details of the subsurface damage, while being long enough in order to sufficiently improve the contrast of the image on the vertical wall of the mill. Then, the sample stage is tilted to 45 degrees and the image of the vertical wall of the region of interest is taken using low (12pA) beam current. In the last step several hundreds of nanometres thick "slice" of the region of interest is milled away and the entire loop of the serial sectioning starts from the beginning and is applied until the region of interest is "mapped". The result of the above mentioned serial sectioning is shown in Figures 4 to 10 for the case of an indentation, which was carried out using a corner of cube diamond probe. The depth of the indentation at the end of the loading part of the indentation test has been set to be 1200nm. While Figures 4 to 6 show the progress of the slicing through the indent itself thus revealing the shape of the indent, Figures 7 to 10 show the region "behind" one of the indentation corners. These images reveal the third dimension of a radial microcrack which apparently initiates from the corner of the indentation. In this particular case, it is necessary to mention that the radial cracks stemming from the other two corners of the nanoindent were not revealed. In general, it should be highlighted that the microcracks that are positioned in parallel direction with the individual slices are likely not to be revealed, as such a feature is likely to be milled away during one serial section.

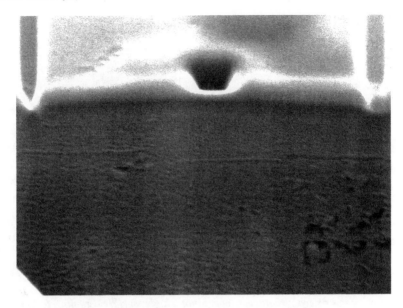

Figure 4 *Image of cross-sectioned indentation site by serial sectioning technique (width of the image is approximately 7.3μm)*

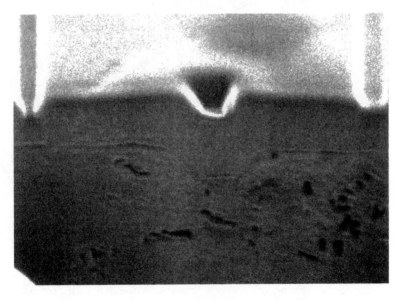

Figure 5 *Image of cross-sectioned indentation site by serial sectioning technique (width of the image is approximately 7.3μm)*

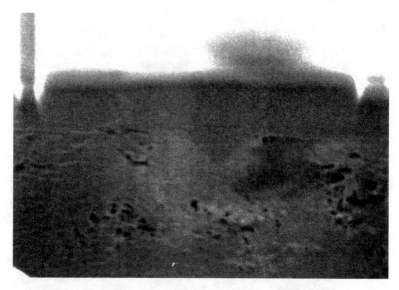

Figure 6 *Image of cross-sectioned indentation site by serial sectioning technique (width of the image is approximately 7.3μm)*

Figure 7 *Image of cross-sectioned indentation site by serial sectioning technique (width of the image is approximately 7.3μm). The arrow shows the third dimension of a radial microcrack stemming from the corner of the indentation.*

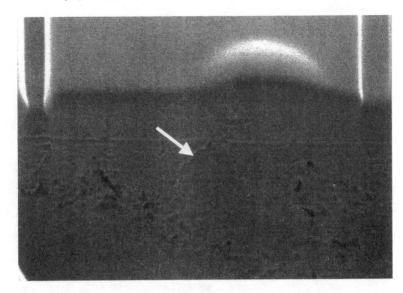

Figure 8 *Image of cross-sectioned indentation site by serial sectioning technique (width of the image is approximately 7.3μm). The arrow shows the third dimension of a radial microcrack stemming from the corner of the indentation.*

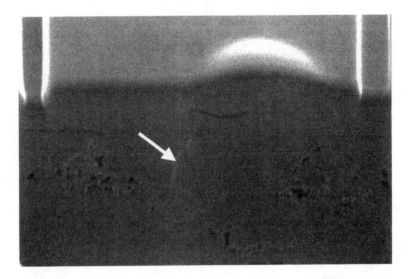

Figure 9 *Image of cross-sectioned indentation site by serial sectioning technique (width of the image is approximately 7.3μm). The arrow shows the third dimension of a radial microcrack stemming from the corner of the indentation.*

Figure 10 *Image of cross-sectioned indentation site by serial sectioning technique (width of the image is approximately 7.3μm). The arrow shows the third dimension of a radial microcrack stemming from the corner of the indentation.*

4 CONCLUSION

The serial sectioning shown above has been carried out using FEI FIB 200 workstation. It is necessary to highlight that serial sectioning technique using the ion beam only is quite time consuming. This is due to the fact that the beam current is repeatedly changed between high and low values (i.e. between milling and imaging mode of work). Thus, refocusing of the beam is necessary after each such change. Moreover, the need of repeated repositioning of the specimen stage into different tilt positions also slows down the procedure.

It has been shown that FIB is able to reveal the sub-surface deformations induced into cementitious materials by instrumented indentation technique and that it is possible to follow these deformations by means of the serial sectioning technique of an indentation site. It has also been shown that FIB technique used on its own is applicable, but - due to the extensive time required - not the ideal tool for investigation of deformations induced by the instrumented indentation technique in the cementitious materials. However, with the current progress in instrumentation, a system combining the focused ion beam with electron microscopy (i.e. dual beam system) is to be much more efficient, as in these systems the ion beam is used only for the milling part of the serial sectioning, while the electron beam is used for imagining the site.

Acknowledgement

It is the author's pleasure to acknowledge Mrs Margaret Corrigan, University of Paisley, for help with taking images by scanning electron microscope and Mrs Sue Bond for assistance with FIB work at MIAC, University of Edinburgh.

Author would like to express sincere thanks to Mrs Margaret Nocher, ACM Centre, University of Paisley, for the proof-reading the paper.

The last but not the least, Scottish Higher Education Funding Council (SHEFC) is acknowledged for providing resources for project NANOCOM, thus funding the research reported in this paper.

References

1   Hughes, J.J. & Trtik, P., Micro-mechanical properties of cement paste measured by depth sensing nano-indentation: a preliminary correlation of physical properties with phase type, *accepted for publication in Materials Characterisation*

2   Wu, H.Z., Roberts, S.G., G. Mobus, and B.J. Inkson, Subsurface damage analysis by TEM and 3D FIB crack mapping in alumina and alumina/5vol.%SiC nanocomposites, *Acta Materialia*, (2003) **51**(1) 149-163

3   Prewett, P.D., Focused ion beams-microfabrication methods and applications, *Vacuum* (1993) **44** 345-351

4   Langford, R.M., A.K. Petford-Long, M. Rommeswinkle, and S. Egelkamp, Application of a focused ion beam system to micro and nanoengineering, *Materials Science and Technology*, (2002) **18** 743-748

5   Trtik, P., Reeves, C. M., Bartos, P. J. M., Use of focused ion beam (FIB) for advanced interpretation of microindentation test results applied to cementitious composites, *Materials & Structures*, (2000) **33** 189-193

6   Inkson, B.J., Steer, T., Möbus, G., Wagner, T., Subsurface nanoindentation deformation of Cu-Al multilayers mapped in 3D by focused ion beam microscopy, *Journal of Microscopy*, (2001) **201**(2) 256-269

MICRO – AN INTERMEDIATE STEP TO NANO LEVEL ANALYSIS IN CONCRETE LIKE COMPOSITES

J. Kasperkiewicz

Institute of Fundamental Technological Research, Polish Academy of Sciences, Warsaw, Poland. e-mail: jkasper@ippt.gov.pl

1 INTRODUCTION – GENERAL REMARKS

Observations on microhardness of cementitious materials are reported since almost half a century[6]. In the nineteen-seventies a technique that was developed around 1920, and later on became standardised for testing of metallic materials, was applied effectively by Grudemo[3], who in his structural observations and tests on cement paste components was using Vickers tip and conical tip indenters. Due to technical difficulties with processing and interpretation of the results the microhardness approach did not become as popular as it deserves. In consequence, among about 30 papers found in this topic by the present author more than 60% have been published only in the last decade. The recent revival of the interest in the cement paste indentation experiments seems to be related to the availability of modern equipment for micro and nano testing and observations.

However, the typical peculiarity of the concrete materials remains unchanged, and the rich new possibilities (at the nano-level) of getting more high quality data are introducing additional complications. The challenge concerns the interpretation of results. Problems of this kind have already been noticed in the past. For example Grudemo was discussing effects of the volume of material affected by fracturing during indentation, but the issue has never found a more rigorous, quantitative description.

Concrete, like other similar brittle matrix composites, with their specific, inhomogeneous structure, substantially differ from the materials like steel. Not only in concrete materials present are cracks, air voids and other cavities, regions of increased bleeding, flaws and various special discontinuity effects, but in the cement paste the typical atomic bonds are non-metallic. These do not allow plastic deformations, and stress concentrations cannot be quietly released, as all ionic and covalent solids are naturally brittle due to their bonding. Various microstructural elements in concrete behave differently under mechanical actions and serious inhomogeneity can be observed during indentation experiments in any scale – from macro- to sub-micro.

Additionally, certain discontinuities are typical only for concrete-like materials, as in case of the microcracks, which cannot be classified easily whether they are natural (i.e. appearing during fabrication) or they result from various exploitation actions, which can be of mechanical, environmental or chemical nature. The mechanical role of all such discontinuities in the micro scale may be quite varied, and their characteristics in the nano scale seem to have never been studied, yet.

The discussion above touches quite a general problem: in all indentation tests aimed at assessment of material properties it is very important that there is a correct quantification of the structural geometry and structural topology, based on images of the material at different levels of magnification. The structural features of different importance to the quality of concrete are:

- air voids, especially their distribution and morphology,
- morphological characteristics of the ITZ, (Interfacial Transition Zone),
- particular hydration process regions and their distribution,
- dispersion, mean distance and orientation of aggregate particles, of non hydrated cement grains, etc., (in general: their distribution),
- cracks, microcracks, their size, orientation and topology.

It can be added that the mechanical influence of all the above features on the results of indentation tests may be related also to the technique of specimen preparation. A simple example may be regions of scratches and locally tilted surfaces remaining after a poor quality grinding, (e.g. when abrasives used in the polishing of the tested surface are limited to the mesh number not more than 600). Another example might be macro-inhomogeneity due to poor compaction, which can be observed in samples of impregnated, reground surfaces in ultraviolet light. Such observation can be done only on large areas analysed at very low magnification scale (visible features of the order of millimetres and centimetres), and the same effect could not be practically perceived at high magnifications.

Still another problem in rational characterization of the microstructure of cementitious matrices is that all these materials are continually changing in time, not only due to their internal physical-chemical processes, but also according to new developments in concrete technology. A few examples are appearance of air entrainment admixtures, of lightweight aggregates, plasticizers and superplasticizers, fibres and microfibres, as well as introduction of new technologies like HPC, self compacting concrete, SIFCON, SIMCON, etc. The observations on concrete materials, collected in the past, may be of rather limited importance in case of modern components, and many experiments must now be repeated applying a higher quality equipment and new concepts.

Also the observations from the past have certain technical limitations. For example in the broadly cited microindentation test results according to Lyubimova and Pinus[7], a typical region of interface phenomena (ITZ) was estimated to be of order of 20 μm, at the force of indentation about 500 mN. This was an average value obtained in case of cement pastes of microhardness about HV = 200÷800 MPa. The result can be discussed in view of the typical indentation depth and diameter dimensions for different loadings - Table. 1, (also cf. Figure 1). It can be seen that the size of the imprint - 35 μm - was almost twice the above-mentioned size of the ITZ area width that was to be measured. The reliable estimate of a similar ITZ area would require a nanoindentation technique, like one described by Trtik and Bartos[11].

Table 1   *Typical numerical characteristics of Vickers indentation test in concrete*

indentation force [mN]	indentation depth [μm]	indentation diameter [μm]	size of the corresponding features
45000	48	335	air void size
5000	16	112	crack width, fine sand grain size
500	5	35	non hydrated cement grain size

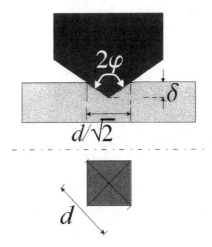

Figure 1   *Vickers indenter geometry: depth – diameter relationship, ($\varphi = 68^{o}$, $d \cong 7.00006 \cdot \delta$)*

The rationale for the application of a nanoindentation approach is obvious in such a simple case as the above described analysis of the ITZ region. Instead, to obtain a reliable estimate of the hardness of the cement paste, with all its different phases, the size of the region of interaction of the tested material with the indenter - the contact area – should be one or even two orders of magnitude larger than in the previous case, (Table 1). Otherwise, depending on the hardness of the tested material and on the magnitude of the loading force, working regions of the systems of Vickers, Berkovich, Knoop, etc., may be too small compared to the main structural components of cement matrix, and this would not allow correct, average estimates to be obtained. In consequence, in particular indentation studies the loading force must be above a selected lower limit value.

Certain conclusions on considering a lower limit of the loading force for a given microindentation system can be worked out from the analysis of the proportions in Figure 1. Many loading force – indenter displacement diagrams ($F - \delta$) have a form like that in Figure 2. The figure on the left is an idealisation, but the same form of the diagram can be encountered in many reported tests on micro- and nanohardness of cementitious materials, (also cf. the example diagrams in Figure 5, below). The diagram on the right

Nanotechnology in Construction

hand side has been calculated taking into account the geometrical relations in Figure 1. It can be seen that the resulting microhardness is practically indeterminate at the initial branch of the $F - \delta$ curve – Figure 2b, and becomes more stable only later on.

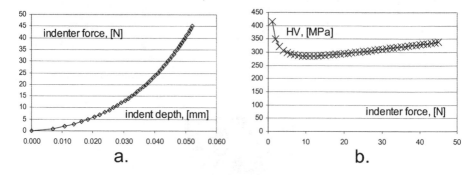

Figure 2   *Idealised Vickers indentation $F - \delta$ diagram (a), and corresponding microhardness values calculated at different loading (b).*

The suggestion that there should be a lower limit of correct indenter loading seems to be also in accord with the observation that the microhardness tester tip has always a certain, finite curvature. From the atomic force images it can be believed that the radius of curvature of the tip may be about one micron. Assuming that the value of 1 μm is a correct estimation of the real tip curvature, a conclusion can be drawn, that for an indenter manufactured with such precision, a really 'Vickers test configuration' can be considered only when the indentation depth (δ) is more than 2 or 3 μm. This means that the diameter of the imprint should be on the level of at least some 20 μm, (see Figure 1). Similar conclusion can be drawn looking at the pictures in the paper dedicated to possibility of reengineering of the tip of indenter[10]. It is supposed from the above discussion that many microhardness results for cement materials might be reported with uncertain accuracy.

Experiments dedicated to measuring microhardness of concrete at the 'nano' level, (nanohardness), without appropriate analysis of the structure of the material may easily become a kind of a blind search. The available microhardness diagrams reveal variations of the measured value within certain limits, e.g. diagrams by Tamimi[9], or – as can be observed in case of the modulus of elasticity – Figure 4 in the paper by Trtik and Bartos[12] – the property of material is represented by a whole cloud of experimental values differing as much as 1 to 6 over a distance of 100 μm. Such effects seem to be related to ill-defined nature of the structural components (phases) in the indentation area.

It is not possible to create an ideal (mathematical) plane section of the concrete sample, and what is submitted to nano- or micro-indentation as a plane surface is actually a more or less rough surface resulting from synergetic effects of the material structure, of the material's current physical state, it's cutting, grinding and polishing damages, and of the observation technique applied, which occasionally may introduce artefacts into images. It is usually assumed that appropriate grinding and polishing of the surface of concrete guarantee the quality of a surface appropriate for thorough observation. This is done with silicon carbide (SiC) powders, sometimes using silicon carbide papers, and during the final treatment, diamond sprays. The surface to be tested of concrete is polished applying a sequence of decreasing grain size media, like SiC powders of 100, 220, 320, 400, 600,

1200 mesh numbers and then - depending on needs - also using diamond polishing pastes of 6, 3, 1 and 0.25 µm.

From the mesh number of the abrasive, that is derived from the openings per linear inch in the control sieving screen, it can be calculated that powders # 100 and # 600 correspond to the abrasive grains sizes of 130 µm and 9.3 µm, respectively. The finest mesh size #1200 corresponds to about 3 µm. The roughness of the polished surface is related to the fineness of the polishing medium, so - at a typical preparation - maximum differences in surface (relief) are expected to be of the order of 1 µm, but even bigger artefacts may sporadically be encountered.

The quality of the preparation can be checked to obtain a quantitative estimate of the surface roughness, for example as the average deviation from the mean. This can be done by different methods like profilometry, stereophotogrammetry, atomic force microscope (AFM) or confocal scanning optical microscopes, to certain extent also by a SEM (Scanning Electron Microscope) or using an acoustic microscope. Different methods have various possibilities and limitations, but the whole important problem is discussed very rarely.

The visual analysis system for evaluation of the imprint should, therefore, guarantee a proper recognition of the working region of the surface. The working region size is related to

(A) surface nature (type of the material: pure cement paste, mortar or concrete, with or without additives, admixtures, etc.),
(B) surface preparation processing (by grinding, polishing, cleaning, application of surface protection layers),
(C) actual material state (whether it is or was loaded, unloaded, dry, wet, warm, frozen, etc.).

All of the above factors may be of importance for the results of the indentation test. So far there are no standardisation suggestions of the procedure. The accuracy of the measurement is defined only by the possibility of measuring correctly the indentation loading and depth. And these two are related to all the three above factors (A, B, C).

2 EARLY CONCRETE MATERIAL INDENTATION TESTS

In spite of various objections as discussed above results of various microhardness tests reported in the past seem to be reasonable enough. A review of more than 30 references shows that all the microhardness results reveal a similar tendency[5]. In individual papers the relations are even more distinct. Microhardness magnitudes obtained by Russian and Swedish investigators[2, 3, 4, 6, 7] seem to be well correlated with concrete - strength or with its w/c ratio (water-cement ratio), except that the quality of the cement and concrete used at that time was probably hard to compare with the contemporary products. An example concerning laboratory tests on cement pastes, and showing an excellent correlation can be seen in Figure 3.

It is interesting to note, that in the opinion of Grudemo who experimented with several forms of the indenters, (Vickers, Knoop, straight edge, conical tip)[2, 3, 4], the hardness values obtained using different tools on the same material are coming out nearly the same. The interesting suggestion deserves now to be checked with the help of modern nanoindentation experiments.

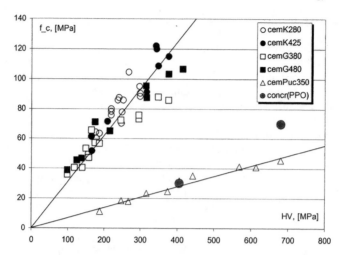

Figure 3 *Relation between Vickers hardness (HV) and compressive strength of hardened cement paste according to Lyubimova et al.[6]. Two points described as concr(PPO) - are added from the recent trials of the loading system described further below.*

3 RECENT INDENTATION TESTS RESULTS

The mechanical testing system applied in the experiments (Lloyds) has been purchased to enable evaluation of the concrete strength from microindentation experiments. The system being of DSI type (Depth Sensing Indentation) has a limited accuracy, and encompasses the range next to the 'nanotechnology', as recently defined[11]. The meaning of the DSI qualification is that the system is able to function in such way that during the test on indentation, the whole $F - \delta$ diagram can be recorded with an accuracy better than 0.1 μm for the deflection and better than 10 mN for the loading.

To evaluate whether the precision of the system is sufficient for the microhardness tests on concrete materials a multiple series of imprints have been produced using a Vickers indenter on several samples of cement pastes and concrete. The surface of all the specimens have been typically prepared as for the standard air voids analysis of air-entrained concrete. A number of the imprints that hit the edge of an aggregate or area of a larger air void were discarded and on the remaining indentations the diagonals were measured by microscopic observations, with help of an automatic image analysis system, (Image ProPlus).

The applied procedure was then at first to make a series of indents and to observe later and to classify the geometry of the observed imprints. The stereoscopic microscope applied during observation was set to a medium magnification of about 50×. This provides accuracy inferior to what is offered in modern commercial systems dedicated to hardness testing of metals (usually accuracy better than 1 μm).

It was found from the experiments that the size of the diagonal calculated from the displacement of the indenter, δ, is a fair enough estimate of the value observed visually (Figure 4). To calculate the diagonal applied should be formula $d \cong 7.00006 \cdot \delta$. It was also observed that a better accuracy would not be of much help, as the outline of a typical indentation is rather unclear, cf. Figure 6. It has been finally assumed that the sizes of diagonals in the Vickers test can securely be evaluated from the indication of the loading machine – the displacement of the indenter δ.

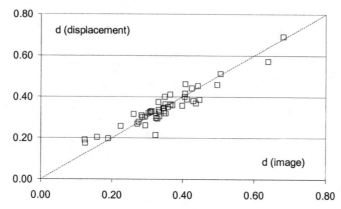

Figure 4 *Vickers indentation diagonal - d(image) measured directly in the image and its corresponding value calculated from δ, (Figure 1), as d(displacement), in [mm]; coefficient of correlation: r=0.924*

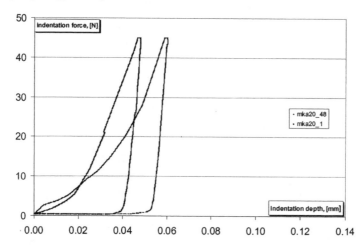

Figure 5 *Two extreme curves encircling all the other Vickers indentation $F - \delta$ diagrams from the same concrete specimen series.*

Preliminary experiments have been conducted on samples of four types of concrete and - more systematically, on samples of several cement pastes [1]. The microhardness values, (HV), have been calculated from $F - \delta$ records, typically as average of up to 20 indents, (selected from among 50), classified as "approximately pure cement paste". The remaining indents classified as "positively aggregate grains" and "air voids regions" have been discarded. Some of these results and results reported by other investigators are shown in Figure 7. In the diagram a characteristic inverse relationship between the hardness *HV* and *w/c* can be seen.

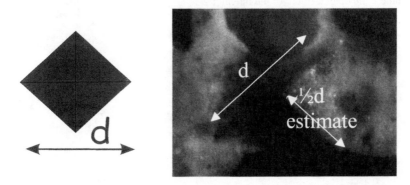

Figure 6 *An example of difficulties with evaluation of the diameter of the Vickers imprint in the cement paste. On the left - an idealized imprint. On the right hand side - a real, poor visibility imprint, of approx. d=350 μm, obtained at the loading force 45 N.*

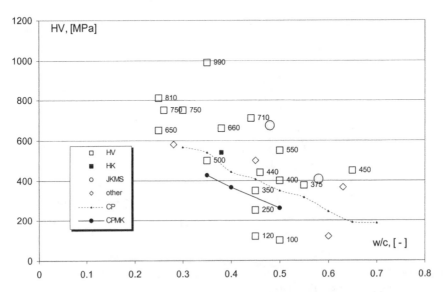

Figure 7 *A selection of different kind of microhardness indentation tests results, (Vickers, Knoop, Berkovich), for various cementitious matrices, reported in different sources, in function of water-cement ratio of concrete.*

4 DISCUSSION

Even assuming an "ideal" surface preparation technique there will be certain number of distinguishable objects in the field of vision of different importance for the testing results. So their presence should be taken into account. Certain objects are of size 50 µm or more, and to obtain a reliable estimate of the material properties, the imprint area should cover a sufficient number of such features.

The situation is illustrated in Figure 8, which is an idealised microphotograph of the properly prepared sample of concrete (the preparation of the sample was for backscattered electron technique observations [8]). There is a fine aggregate grain with visible secondary mineral inclusions on the right side to the dashed line. In the cement mortar to the left of this line fine grains of sand and a few air voids can be seen. Even the Vickers imprints, (white rhomboids) seem to be too small to characterize properly the inhomogeneity of the structure although, naturally, they would be too large for appropriate analysis of the interface phenomena. The interface characteristics can be studied with the nanohardness approach, (white triangles).

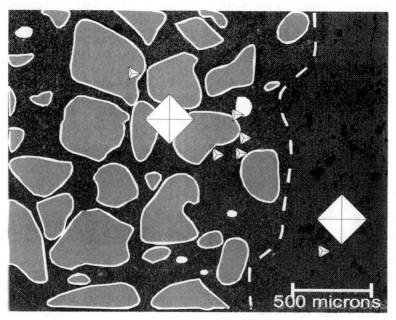

Figure 8 *Microhardness (Vickers) v/s nanohardness (Berkovich) observations on a sample of concrete – a symbolic illustration based on actual BE image [8].*

The appropriate size of the microindentation imprints in the scale of Figure 8 should be even larger - about 300-500 mµ (Vickers imprints). Much smaller Berkovich form imprints could be used for investigation of ITZ and bond phenomena, but further complications may appear at higher magnifications of the surface of the specimen.

Certainly, there is a need to combine both types of observations (the micro- and the nano-) in one complementary testing program.

5 CONCLUSIONS

1. Various historical reports from microhardness testing all indicate similar inverse relationships between the microhardness and the water-cement ratio of concrete. The exploratory recent experiments using a DSI loading system corroborate such correlation.
2. At the first approximation the relation seems to be little dependent on the form of the indenter, (Vickers, Knoop, Berkovich or else).
3. Applying the testing system as described above, in Vickers indentation tests the diagonal of an imprint can be calculated with a sufficient accuracy from the magnitude of the loading tip deflection.
4. For complete information on the physical properties of concrete a series of indentation experiments is needed in different scales, covering different structural features. Exactly nano-level hardness tests can be a valuable source of information concerning interface region phenomena, for which the microhardness tests are not enough precise.
5. Experiments of indentation at various geometrical scale levels should be accompanied by appropriate morphological observations. Care should be taken to avoid introducing unnatural states during the preparation of the tested surfaces, (the artefacts). On the other hand extremely high magnification (e.g. during SEM observations) may provide too rich information, which can hardly be exploited fully.
6. It is expected that the microhardness tests combined with structural observations are quantitatively and qualitatively related to various properties of concrete, among others - also to its durability.

Acknowledgements

This paper was prepared with a support from the NATO SfP Project 97.1888 – Concrete Diagnosis.

References

1   M.A. Glinicki, J. Kasperkiewicz, M. Sobczak, M. Zielinski, Testing of the structure of concrete by Vickers indenter (in Polish), submitted to the *Proceedings of 49-th KILW & PZITB Engineering Conference, Krynica 2003.*

2   A. Grudemo, Modified-heat cement – some structural data, *CBI Research*, fo 3.85, Stockholm 1986, 17 pp.

3   A. Grudemo, Strength – structure relationships of cement paste materials. Part 2. Methods and basic material data for studying strength and failure properties, *CBI Research*, fo 8.79, Stockholm 1979, 130 pp.

4   A. Grudemo, Strength vs structure in cement pastes, *Paper presented at 6th Int. Symp. on Chemistry of Cement, Moscow,* Sept. 23-27, 1974. *CBI Reports*, 13:75, Stockholm 1975, 15 pp.

5   J. Kasperkiewicz, Report: *Summary of the initial experiments at IFTR PAS on estimating concrete properties by Vickers needle indentation.* IFTR PAS, Warszawa, May 2002, 16 pp.

6   T.J. Lyubimova, P.A. Agapova, Application of microhardness approach in investigation of the structure of cement paste in concrete, (in Russian), *Beton i Zhelezobeton*, 1959, 7, 229-303.

7   T.J. Lyubimova, E.R. Pinus, Crystallization processes in the contact zone between aggregate and matrix in cement concrete, (in Russian), *Kolloidnyi Zhurnal*, 1962, **24**, 5, 578-587.

8   P.E Stutzman., J.R. Clifton, Specimen preparation for scanning electron microscopy, in: *Proceedings of the Twenty-First International Conference on Cement Microscopy*, L. Jany and A. Nisperos, Eds., April 25-29, 1999, Las Vegas, Nevada, USA, pp. 10-22; available also at: http://ciks.cbt.nist.gov/~garbocz/icma1999/icma99.htm

9   A.K. Tamimi, The effects of new mixing technique on the cement paste-aggregate interface. *Cement and Concrete Research*, 1994, **24**, 7, 1299-1304.

10 P. Trtik, C.M. Reeves, P.J.M. Bartos, Use of focused ion beam (FIB) techniques for production of diamond probe for nanotechnology-based single filament push-out tests, *Journal of Materials Science Letters*, **19**, 2000, 903-905.

11 P. Trtik, P.J.M. Bartos, Nanotechnology and concrete: what can we utilise from upcoming technologies?. In *'Cement and Concrete – Trends and Challenges' - Proc. of the Anna Maria Workshop Nov. 7-9, 2001*, Ed-s A.J. Boyd, S. Mindess and J.Skalny, Am. Ceram. Soc., Westerville 2002, 109-120.

12 P. Trtik, P.J.M. Bartos, Micromechanical properties of cementitious composites, *RILEM Materials and Structures*, June 1999, **32**, 388-393.

APPLICATIONS OF DUALBEAM IN THE ANALYSIS OF CONSTRUCTION MATERIALS

Steve Reyntjens

FEI Company, Applications Laboratory, P.O. Box 80066, 5600 KA Eindhoven, the Netherlands

1 INTRODUCTION

Combining the power of scanning electron microscopy (SEM) and focused ion beam (FIB) technologies on a single platform, DualBeam gives material scientists access to an extended range of analytical information. While the electron beam images a selected area, the focused ion beam acts as a micro-surgical tool, allowing the user to locally modify the specimen for a better understanding of its internal structure. This overcomes the limits of traditional top-surface imaging and provides a three-dimensional subsurface view of the material's microstructure.

In practice, DualBeam is used extensively in failure analysis and advanced materials research: samples can be opened in-situ for genuine three-dimensional inspection. When used for defect review, surface and sub-surface defects are conveniently analysed by sectioning with the FIB and subsequent high-resolution imaging using the SEM. Alternatively, thin specimens for transmission electron microscopy (TEM) analysis are readily prepared. The DualBeam provides three-dimensional analysis and characterization of a wide range of relevant materials, including important construction materials such as metals, alloys, ceramics, plastics, glass, concrete and wood (obviously, this list is far from exhaustive).

This paper offers an introduction to SEM and FIB technology, and more specifically their combination in the form of the DualBeam. We present different analysis techniques, such as cross sectioning, chemical analysis using energy-dispersive X-ray spectroscopy (EDX or EDS), orientation imaging microscopy (OIM) using electron backscattered diffraction (EBSD) techniques, preparation of thin transmission electron microscopy (TEM) samples and scanning transmission electron microscopy (STEM) analysis of these samples in the DualBeam. These techniques are illustrated by examples that are relevant to the nanometer-scale analysis of construction materials.

2 EQUIPMENT AND TECHNIQUES

2.1 Scanning Electron Microscope (SEM)

Electron microscopy is mainly divided into two families, notably scanning and transmission electron microscopy (SEM and TEM, respectively).   Scanning electron microscopy being an integral part of the DualBeam, it is briefly situated within the broad field of microscopy.

The scanning electron microscope (SEM) generates a beam of electrons in a vacuum. The beam is collimated by electromagnetic condenser lenses and focused by an objective lens. The beam is scanned across the surface of the sample by electromagnetic deflection coils, hence the name *scanning* electron microscope. The primary imaging method is by collecting secondary electrons that are released by the sample on a detector. By correlating the scan position with the resulting detector signal, an image is formed (figure 1 (a)). Where the resolution of optical microscopes is limited to 200 nm by the diffraction of light, the limitations of electron microscopes arise primarily from spherical and chromatic aberrations.[1]

The DualBeam system incorporates a state-of-the-art SEM with a field emission gun (FEG). This allows imaging and analysis of materials at low beam energies and down to nanometer resolution. Low voltage imaging and analysis is especially useful for the inspection of charging materials with little or no sample preparation at all. Besides imaging, also more advanced techniques such as EDX and EBSD are available for chemical analysis and crystal orientation analysis.

2.2 Focused Ion Beam (FIB)

A focused ion beam (FIB) uses accelerated Ga^+ ions in a finely focused beam to enable imaging as well as maskless micro- and nanomachining.[2,3]

The structure of the ion column is similar to that of an SEM, the major difference being the use of gallium ions (Ga^+) instead of electrons. The ion beam is generated from a liquid-metal ion source (LMIS) by the application of a strong electric field. The ion beam is collimated by an electromagnetic condenser lens and focused by an objective lens.

On most FIB's, a system is available for delivering a variety of gases close to the sample surface. This is achieved by using fine nozzles inside the vacuum chamber at the end of small gas containers. The gases are used for faster and more selective etching, as well as for the deposition of materials.

FIB is an imaging tool in its own right. While electron beams generally produce higher resolution images, the ion-beam imaging contrast is often stronger. Although most FIB images are acquired using secondary electrons, secondary ion imaging can provide additional image information. For example, more topographical information can often be gained from a secondary ion image. The FIB imaging principle is illustrated in figure 1 (b); note the strong similarity to SEM imaging.

It should be mentioned that FIB imaging inevitably induces some form of damage to the sample. Most of the Ga^+ ions enter inside the sample material; thus ion implantation occurs. The depth of this implanted region is related to the ion energy and the angle of incidence. A typical value is 27 nm (silicon, perpendicular incidence, 30 kV).[4] Besides implantation, some milling occurs when the ion beam is scanned across the sample surface. Of course using a low ion current drastically reduces this milling effect.

Besides imaging, the high current density beam allows precise material removal. This is achieved using a high ion current. The result is a physical sputtering of sample material, as illustrated schematically in figure 1 (c). By scanning the beam over the substrate in a user-defined pattern, an arbitrary shape can be etched. The resolution of the milling process is a few tens of nanometers. In order to speed up the milling process, or to increase the selectivity towards different materials, an etching gas can be introduced into

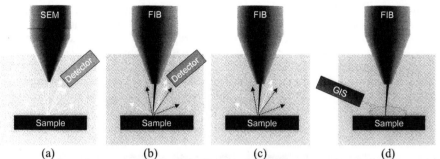

(a)   (b)   (c)   (d)

Figure 1   *Basic principles of (a) SEM imaging; (b) FIB imaging;(c) FIB milling;(d) FIB deposition with gas injector and precursor gas.*

the work chamber during milling. This enhances the etching process by chemically facilitating the removal of reaction products.

FIB also enables the localized maskless deposition of both metals and insulators by chemical vapor deposition (CVD).[5] The metals that can be deposited using commercially available precursor gases are platinum (Pt) and tungsten (W). As an insulating material, SiO_2 is available. The deposition process is illustrated in figure 3 (d): the precursor gases are sprayed onto the surface by a fine needle (nozzle), where they adsorb. The incending ion beam locally decomposes the adsorbed precursor gases. The volatile reaction products desorb from the surface and are removed through the vacuum system, while the desired reaction product remains fixed on the surface as a thin film. The deposited material is not fully pure however, since organic contaminants and Ga^+ ions are inevitably included.

To conclude this paragraph, a note about typical machining dimensions is in order. Although fundamentally not limited, the size of FIB-machined sites is typically in the 10 to 50 micrometer range, with processing times from a few minutes to half an hour. Usually, these volumes are sufficient for microanalysis of materials. Machining larger areas is possible, but processing times may become unacceptably long.

2.3 The DualBeam advantage

Figure 2 shows a typical high-resolution SEM, a FIB and a DualBeam. It is clear that the basic structure of a DualBeam is similar to an SEM and an FIB. However, it is much more than just the combination of a SEM and a FIB column on a bigger process chamber.

The fundamental DualBeam advantage is that the entire instrument is built around a so-called "eucentric" height, i.e. the normal working position where the electron and the ion beams coincide, as well as the gas injectors, the electron, ion and EDX detectors. In practice, this results in a number of distinct operational benefits. With the sample surface at this coincident point, the advantages of the DualBeam are apparent.

As illustrated in figure 3, the SEM can be used to monitor the FIB milling process, e.g. during cross sectioning (see further for a detailed description). The DualBeam is therefore extremely well suited for precisely positioning cross sections through highly localized areas of interest. This real-time monitoring of the FIB process has the additional advantage of high-accuracy end point detection.

Further advantages include the possibility to prepare (in-plane or cross sectioned) surfaces for in-situ chemical or crystal orientation analysis (EDX or EBSD, respectively). This prevents the sample surface from oxidizing or contaminating during the transfer from

FIB-preparation to SEM-analysis. Alternating between surface preparation and analysis allows full three-dimensional analysis of material properties and structure on a set of slices.

As mentioned earlier, we also stress the fact that electron and ion beam images often reveal complementary sample information. Although the electron beam is considered the primary imaging tool in a DualBeam, the images from the ion and electron beams are often complementary. An example is shown in figure 4, where the ion image shows clear and explicit grain orientation contrast on a cross-sectioned copper wire, while the (secondary) electron image better reveals the surface topology.

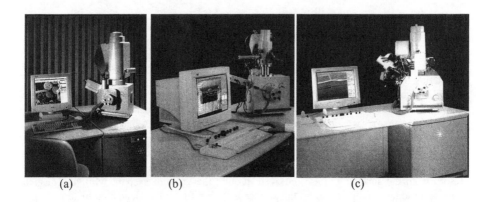

(a)   (b)   (c)

Figure 2 *Examples of (a) a scanning electron microscope (FEI Sirion), (b) a focused ion beam (FEI Strata FIB 201) and (c) a DualBeam (FEI Strata 235 DB).*

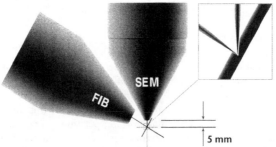

Figure 3 *A typical DualBeam system configuration. The vertical SEM column and the tilted FIB column have a single "coincident" point on the sample surface. SEM imaging during FIB milling enables real-time monitoring of the FIB-process (inset).*

(a)   (b)

Figure 4   *(a) Ion beam and (b) electron beam image of a coated Cu wire cross section.*

2.4 Equipment

The FEI Strata 235 DualBeam, which is used for the analyses presented in this paper, has a range of ions currents available: from 1 pA for nanoscale milling and/or imaging, up to 20 nA for fast bulk milling. The nominal acceleration voltage is 30 kV, but voltages down to 5 kV may be used for specific applications. At optimum beam conditions, a spot size as small as 7 nm can be obtained.

Different metal and insulator deposition chemistries are available, as well as enhanced etch gases for insulators, metals and polymers. For FIB work on samples with extremely low conductivity such as glass, an electron floodgun can be used. It neutralizes the build-up of positive charges on the sample surface by flooding it with low-energy electrons.

The acceleration potential of the FEG SEM column ranges from below 500 V to 30 kV with electron currents varying from 1 pA to around 9 nA. Features as small as 2 nm can be resolved at a working distance of 5 mm using a 5 kV acceleration voltage.

3 APPLICATIONS

In this section, we present typical DualBeam analysis cases of construction materials: site-specific cross sectioning (single or sequential), EBSD and EDX preparation and analysis, preparation and analysis of TEM lamellas.

3.1 Site-specific cross sectioning. EDX analysis and map

Precise cross sectioning is used in many applications, including defect analysis, process and materials characterization, and process monitoring and control. The main advantages of DualBeam are the site-specific character of the section and the locality of the preparation process: the location can be chosen very precisely, and in principle the rest of the bulk sample remains unaffected.

After locating the area of interest, the cross section procedure usually consists of three steps:

- The optional deposition of a protective Pt strap for protection and planarization of the area of interest. This may be omitted in the case of a planar and uniform sample surface;
- A bulk milling giving access to the material below the surface of the sample, this is usually accomplished using a relatively high ion current for speed;
- A polish mill using a smaller ion current to obtain a smooth and flat cross sectioned face.

This procedure being completed, the SEM is used to produce high-resolution images of the sectioned face (alternatively, a low ion current beam is used to exploit different contrast mechanisms).

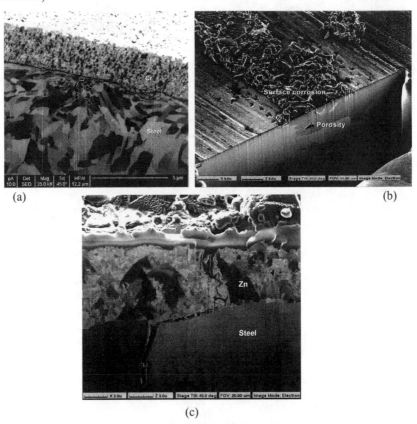

(a)   (b)

(c)

Figure 5   *DualBeam cross sections of different types of steel: (a) a Cr coated steel wire, (b) a weld on a stainless steel and (c) a galvanized steel.*

Figure 5 shows some examples of such cross sections through different steel samples. All three are secondary electron (SE) images, but with the ion beam as the primary beam. In figure 5 (a), a cross section through a chromium coated steel wire is shown. Note the explicit grain contrast, both in the steel and the chromium layer (on top). The Fe-Cr interface may be studied from this image, as well as the grain size distribution and other relevant material parameters. Figure 5 (b) shows a weld in stainless steel. The internal porosity as well as some corrosion on the surface are clearly visible. Although essential to the internal structure, note that (SEM) surface imaging alone would never reveal this porosity. In figure 5 (c), a cross section through a galvanized steel is shown. The bright, smooth layer on top of the zinc is a Pt layer that is deposited using localized FIB deposition, prior to the sectioning process. This is done to preserve the top surface of the sample completely intact, and to smooth out the surface topography (which otherwise may cause artefacts in the section, known as "curtaining").

Not only is the SEM useful for high resolution imaging of cross sections, it also enables the use of EDX. This technique analyses the x-rays that escape from the sample upon electron irradiation. Since the x-rays from different chemical elements exhibit characteristic energy peaks, their detection yields compositional (elemental) information about the materials in the cross section. Of course, also top surfaces can be examined using the same technique.

The technique can be applied for distinguishing layers in stacks consisting of different materials or to analyse the presence and distribution of particles in a matrix.

As an example, figure 6 (a) shows a cross section through a stainless steel, coated with a titanium nitride layer. The SE image allows distinguishing different material layers and some inclusions in the bulk material. By recording an energy spectrum at each image point, element maps are obtained, displaying the intensity for the selected element at that specific image point. Relevant element maps are shown in figures 7 (a) – (f).

These individual maps are then overlaid on each other, yielding an elemental composition of the cross section of figure 6 (a). For easy comparison, the result is shown in figure 6 (b). The iron bulk clearly contains isolated Cr/V inclusions. The titanium-rich TiN layer on top of the bulk is clearly correlated to the dark band in the SE image, while the intermediate thin grey layer in the SE image appears to be a molybdenum top surface. The Pt layer is obviously the ion-beam deposited protective strap.

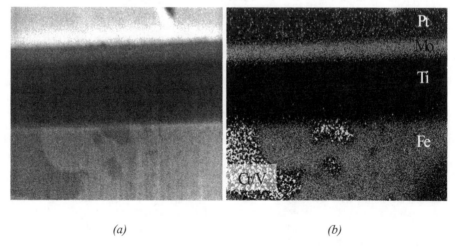

(a) *(b)*

Figure 6   *DualBeam (a) SE cross section image through a stainless steel with titanium nitride coating and (b) elemental composition of the same cross section.(a) (b*

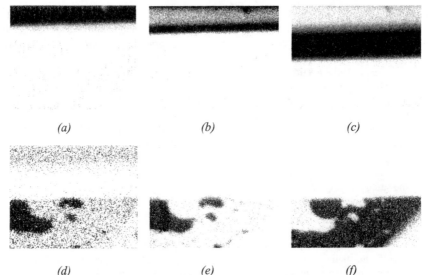

(a)   *(b)*   *(c)*

(d)   *(e)*   *(f)*

Figure 7   *Elemental maps of the cross section of figure 6 (black = high element intensity): (a) Pt; (b) Mo; (c) Ti; (d) V; (e) Cr; (f) Fe.*

3.2 Sequence of cross sections or "Slice and View". Three-dimensional OIM analysis by sequential EBSD

As shown in the previous sections, the ability to use the electron beam to record images of milled cross sections is a key benefit of the DualBeam system. A more advanced technique is the milling of successive slices through a region, using the SEM after each milled slice to record an image. This "Slice and View" technique can be automated, producing a complete series of SEM images through an area of interest. Automation allows the easy and unattended acquisition of datasets consisting typically of 30 up to more than 100 images (typical run time 2 hours). Because an image is collected automatically after each slice, it is impossible to "miss" any unique features. The full set of images can be viewed in an animated movie, showing progression through the sample material. This allows the user to better understand the three-dimensional nature of the given feature.

Not only is it possible to make electron beam images of the individual slices, but also more advanced analysis techniques such as EDX and EBSD can be applied to each or some of the slices. EBSD allows crystal orientation to be determined at each point on a polished sample surface, using the diffraction patterns produced by forward scattered electrons (the diffraction is caused by the interaction with the crystal lattice). As an example, fifteen slices through an aluminium sample are milled and after each slice an EBSD analysis is performed. Using OIM software, a colour is assigned to each individual data point in a slice. The colour corresponds to the specific lattice orientation at that point in the sample; thus points of a similar colour form complete grains. Figure 8 (a) shows a set of such "orientation maps".

Once an array of images has been created, the three-dimensional volume can be explored in other ways than just by looking at the individual slices or at the animated sequence of slices. Using dedicated software, sections can be made through the sample volume in other directions than the originally milled sections, as shown in figure 8 (b).

The images show orthogonal sections through the analysed volume. Alternatively a 3D solid model can be calculated.

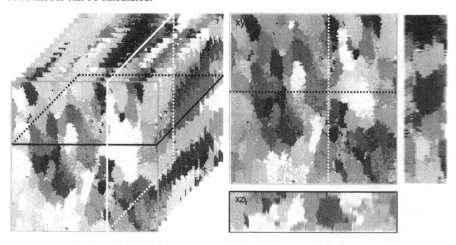

Figure 8   *Slice and View on Al sample with sequential EBSD analysis: (a) set of EBSD maps of 15 slices. The slices are approximately 200 nm thick and have a surface area of 20 x 25 μm^2; (b) orthogonal slices through the same dataset. yz- and xz-slices are based on calculations on the complete set of maps.*

3.3 TEM sample preparation

TEM's (transmission electron microscopes) accelerate electrons through an extremely thin sample (typical sample thickness ranges from below 50 nm to 350 nm). This enables material observation at atomic scale. Using traditional techniques, extensive and complex sample preparation is necessary to produce these extremely thin samples. However, TEM samples with a surface area of a few tens of μm^2 are conveniently prepared in the DualBeam.

Two methods are currently used. The "pre-thinning" or "trench" technique starts from a preprocessed block of material containing the feature of interest, typically no more than 50 μm thick. For the "lift out" method, the complete bulk sample is placed in the DualBeam (as far as it fits the chamber), without any prior preparation. Intermediate steps and the final result of a typical TEM "lift out" preparation process are shown in figure 9:

- The first step is to deposit a metal strap for protection and planarization of the top surface of the bulk sample. This is analogous to the milling of a standard cross section (see section 3.1). The optional milling of crosses to the left and the right of the area of interest is used in the case of an automated process (see further) – figure 9 (a);
- On both sides of the strap a relatively large amount of bulk material is milled away to access to the material underneath the protective strap – figure 9 (a-b);
- The sample is then tilted towards the ion beam, and the approximately 1 μm thick membrane is cut free at the bottom and sides, leaving it attached to the bulk material only with two small bridge on one or both sides of the membrane – figure 9 (c);

• The final polishing uses a relatively low ion current to thin the membrane down to its final thickness, typically ranging between 50 and 250 nm – figure 9 (d-e), whereafter the membrane is released completely from the bulk – figure 9 (f).

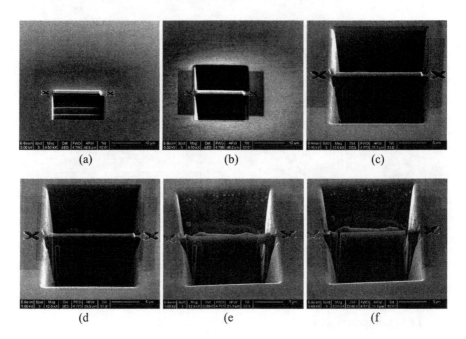

(a)   (b)   (c)

(d   (e   (f

Figure 9   *The TEM sample preparation process illustrated: (a) deposition of protective strap, and milling of alignment crosses for automation; (b) bulk milling in front and behind the region of interest; (c) release milling of the bottom and sides of the membrane; (d)and (e) final thinning; (f) complete release from the bulk sample.*

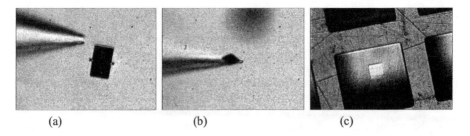

(a)   (b)   (c)

Figure 10   *The ex-situ liftout process illustrated (optical microscope images): (a) TEM preparation site and electrostatic glass probe; (b) the TEM lamella on the tip of the glass probe; (c) the TEM lamella on a carbon TEM grid.*

After this process, the sample is placed under an (external) optical microscope. There the TEM membrane is "lifted" out of the bulk using a glass probe to which the lamella electrostatically adheres (figures 10 (a) and (b)). With this probe the lamella is transported to a carbon grid, on which it is laid down (figure 10 (c)). The sample is now ready for inspection in the TEM.

Alternatively, the sample can be lifted out before the final polishing, using a micromanipulator inside the DualBeam chamber. This is called "in-situ" lift out. The lamella is then placed on a dedicated type of TEM grid, where it is thinned down to the desired thickness, using normal low-current FIB milling.

The preparation of TEM samples is an important application of FIB and DualBeam.[6] An obvious benefit is the complete absence of pre-FIB mechanical sample preparation (or, very limited preparation in the case of the pre-thinning technique): the entire bulk sample can be placed directly in the FIB/DB, and after a couple of hours maximum, a TEM sample is prepared and ready for inspection in the TEM. This drastic speed advantage is complemented by the strong site-specificity of the preparation. The position of the lamella in the bulk sample can be determined very precisely (50 nm typical placement accuracy). Besides this, the remainder of the bulk sample is preserved quasi-intact. This also enables the (automated or manual) milling on multiple sites on one and the same bulk sample.

A wide range of materials can be prepared using FIB techniques, including very fragile ones like polymers and materials under mechanical stress, as well as material combinations with large hardness differences. The latter are impossible to prepare using mechanical techniques or argon ion mill because of smearing and differential etching. In the DualBeam, the electron beam is used to monitor sample preparation, allowing the section to be positioned even more precisely through the point of interest.[7,8] This can be realised without any damage to the sample surface. In addition, the electron beam can be used to monitor the thickness of the sample and to verify that the protective layer on the sample remains intact.

Drawbacks of FIB preparation are the (limited) damage that is caused to the surface of the TEM lamellas by Ga^+ ion implantation. Related to this also some amorphization is observed in crystalline materials. For silicon a typical damaged layer thickness is 27 nm (this can be reduced to below 10 nm by low-energy cleaning). It has not caused any restrictions in e.g. the recording of lattice interference fringes in the TEM, as is clearly illustrated by the example below.

As an illustration, figure 11 (a) shows a top surface of a concrete sample. This brittle material is difficult to handle using traditional techniques. The preparation of a TEM lamella from this is relatively easily done using FIB. The result is shown in figure 11 (b). A TEM phase contrast image of the same lamella is shown in figure 12. The upper half of the image reveals the lattice structure of a dolomite crystal. It is separated from the surrounding matrix by a grain boundary. Further analysis enables e.g. the study of the grain boundary and possible intermediate layer properties.

Preparation of a TEM sample from this type of material would obviously be very difficult if not impossible without the use of FIB/DualBeam techniques.

Thanks to the excellent stability and repeatability of DualBeam performance, a number of tasks such as TEM sample preparation can be automated.[9,10]

The required accuracy is achieved by the use of image matching (also known as image or pattern recognition). This is extremely important on charging materials, which often cause image drift. Dedicated "fiducials" are milled into the sample to this end (the "crosses" in figure 9). The automated routines use these marks to compensate for the slightest of drift, allowing samples with thicknesses below 150 nm to be prepared without operator intervention. This significantly increases the repeatability of the sample results, and enhances the tool's productivity, since large batches of multiple samples can be ran overnight, unattended.

It must be mentioned that very thin samples (< 100 nm) require manual final thinning in most cases, because of the extremely high-accuracy required to correctly end-point the thinning process. However, even in this case, the majority of the preparation process can run unattended using an automated routine and thus an important amount of tool time is gained.

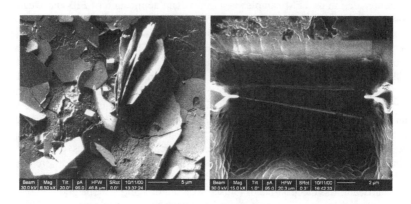

Figure 11   *Bulk concrete (a) top surface and (b) FIB-prepared TEM sample.*

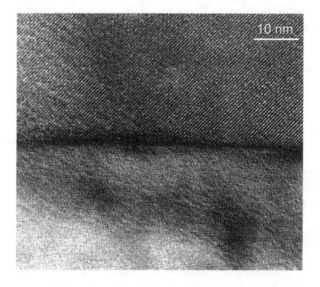

Figure 12   *TEM phase contrast image from a concrete sliver showing lattice structure and grain boundary.*

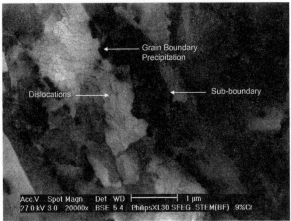

Figure 13 *27 kV STEM image on stainless steel foil (with 9% Cr) showing dislocations, grain boundaries and precipitation.*

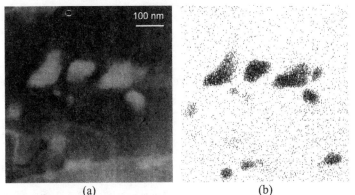

      (a)   (b)

Figure 14 *High-magnification (a) STEM image and (b) EDX map (for Cr) revealing the Cr precipitates on the grain boundaries (images on the same scale, see scale bar).*

3.4 STEM imaging and STEM-EDX analysis

Scanning transmission electron microscopy (STEM) requires the similar thin samples, but can be performed in-situ in the DualBeam, using a dedicated detector. The increased imaging and analytical information available in thin sections is thus used to produce high-contrast images. 100-300 nm thin samples are irradiated with a scanning and focused electron beam, and the transmitted electrons are collected. Resolutions on the order of 1 nm for imaging, and below 25 nm for EDX can be obtained.

    Figure 13 shows an example of a thin section of a stainless steel, revealing such material details as grain boundaries and sub-boundaries, precipitation along boundaries and even dislocations. In figure 14, a higher-magnification detail of the same sample is displayed, along with a chromium-EDX map of the same area. This result clearly demonstrates the chromium precipitation along grain boundaries. Note the high resolution of the EDX map; particles about 25 nm in diameter are easily revealed.

4 CONCLUSION

In this paper, it is demonstrated that SEM, FIB and especially DualBeam open up new possibilities for researchers and material scientists, taking their investigations beyond the sample surface and examine structural composition, layer morphology, and material defects or failures in all three dimensions.

We have attempted to give an overview of the most important analytical techniques that these tools embrace; their relevance to construction materials is demonstrated by multiple examples. Although this overview is far from exhaustive, we believe it offers a glance into the numerous application possibilities. Important opportunities lay ahead, especially in the field of nanomaterials research and in the development, production and failure analysis of new and existing construction materials.

Acknowledgements

We would like to thank René de Kloe (EBSD application specialist with EDAX) for the excellent cooperation on the EBSD work. Also FEI colleagues are acknowledged for contributing images, samples and information, especially Francis Morrissey, Remco Geurts and Dong Tang.

References

1   M. W. Davidson, M. Abramowitz, Optical microscopy, published electronically on the Florida Stata University "Molecular Expressions" website, 1999 (available at http://micro.magnet.fsu.edu/primer/opticalmicroscopy.html)
2   Young R.J., Vacuum, 1993, **vol. 3/4**, 353
3   J. F. Walker, D. F. Moore, J. T. Whitney, Microelectronic Engineering, 1996, **vol. 30**, 517
4   J. F. Ziegler, J. P. Biersack, U. Littmark, The stopping range of ions in solids, Pergamon N.Y., 1985
5   M. J. Vasile, L. R. Harriott, J. Vac. Sci. Technol. B, 1989, **vol. 7**, 1954
6   L.A. Giannuzzi, F. A. Stevie, Micron, 1999, **vol. 30**, 197
7   Ma Z., Davies B., Brandt J., Baker B., Headley K., and Miner B., Microscopy and Microanalysis proceedings, 1999, **vol. 5**, supplement 2, 904
8   Xu Y., Schwappach C., and Cervantes R., Microscopy and Microanalysis proc., 2000, **vol. 6**, supplement 2, 516
9   Young R.J., Carleson P.D., Hunt T., and Walker J.F., Proceedings of 24th ISTFA Conference, 1998, 329
10 Young R.J., Microscopy and Microanalysis proc., 2000, **vol. 6**, supplement 2, 512

SYNCHROTRON-RADIATION X-RAY TOMOGRAPHY: A METHOD FOR THE 3D VERIFICATION OF CEMENT MICROSTRUCTURE AND ITS EVOLUTION DURING HYDRATION

Lukas Helfen[1,2], Frank Dehn[3], Petr Mikulík[1], Tilo Baumbach[1]

[1]Fraunhofer Institute of Nondestructive Testing (IZFP/EADQ), Krügerstr. 22, D-01326 Dresden, Germany
[2]European Synchrotron Radiation Facility (ESRF), BP 220, F-38043 Grenoble Cedex, France
[3]Institut für Massivbau und Baustofftechnologie, Universität Leipzig, D-04109 Leipzig, Germany

ABSTRACT

New possibilities opened up by synchrotron radiation could present x-ray tomography as an experimental imaging method to study the microstructural evolution during cement hydration.

The present work demonstrates the potential of the method for the investigation of cement structure after solidification. Quantitative determination of the spatial absorption-coefficient distribution due to the use of monochromatic radiation and the achieved high spatial resolution allows us to characterise the hydration process of cement volume on a micrometer scale in three dimensions. In particular, effects as the formation of microcracks and pores related to autogeneous shrinkage involved with the hydration process could be observed for the first time with down to one micrometre resolution.

By image analysis of the acquired three-dimensional data sets the temporal evolution of microporosity for different water-cement ratios of the initial cement paste could be determined.

1 INTRODUCTION

Up to now the inherent structural complexity of concrete leaves a variety of scientific issues not fully understood. Prominent examples are the influence of the mixture parameters on the hardening process and the relationship between the resulting structure and the macroscopic properties of the liquid and solid concrete.

The basic cement powder is a particulate system of a wide size range and several mineral phases. Its chemical and physical properties control the hydration behaviour of the cement paste. In particular for high-performance concretes (HPC), mineral admixtures and chemical additives are employed to influence the hydration process.

Moreover, structural complexity extends over scales ranging from nanometres (the gel structure of the hydration phases) over micrometres (the scale of the cement particles) up to millimetres (the scale of aggregates in the concrete). Finally, the macroscopic properties employed by engineering are decisive on the length scale of the final building components, *i.e.* on length scales of metres. A complete description of the relationship between structure and macroscopic properties would therefore require the consideration of up to nine orders of magnitude in length scales. Therefore, computer modelling and simulation techniques[1,2] based on structural models have been used to explore this relationship.

Different experimental techniques have been applied to the structural investigation of concrete on the nanometre and micrometre scale. *Scanning electron microscopy* (SEM) was employed[3,4] to scan the surface of a sample with resolutions down to some *nanometres*. SEM with a special sample environment[5] (*Environmental SEM*) permits the investigation of hydration products at the sample surface. For volume investigations, the sample preparation is tedious and destructive because the samples have to be cut. These methods are therefore not suited for repeated volume investigations. *Small-angle x-ray scattering* (SAXS), *small-angle neutron scattering* (SANS) and *nuclear magnetic resonance* (NMR) can yield statistical information about the specific surface area of the internal boundary[6] between solid phase and the pore system. Generally speaking, SANS[7,8] is sensitive to pore sizes ranging from the nanometre scale up to the micrometre scale whereas SAXS is able detect pore sizes up to some hundreds of nanometres, depending on the x-ray energy employed. Furthermore, the structural properties of water were determined by means of *quasi-elastic neutron scattering*[9,10,11] where the state of water molecules can be classified as structural, constrained and free. During hydration water transforms from the free state to the structural (*i.e.* by chemical reactions) or constrained (*i.e.* water in the cement gel layers or adsorptively bound to the gel surface) state, which can be quantified by the application of this technique.

We will apply x-ray imaging, *i.e. computed tomography* (CT), to resolve the 3D structure of cement specimen during hardening. X-ray tomography does not demand special sample preparation. It investigates the sample non-destructively such that experiments can be repeated at the same sample volume. The resulting experimental data are 3D images of the sample structure. Tomography with *synchrotron radiation* (SR) has already been applied to building materials,[4] with a spatial resolution in the order of 7 μm to mortar[12] and bricks[13] where conclusions about the transport properties could be drawn and compared to existing models.

The recent improvements of synchrotron imaging set-ups predestine x-ray tomography as a tool for *micro*structural investigations of cement.[14,15] Monochromatic radiation allows a quantitative reconstruction of the absorption coefficient for distinction between pores, not completely hydrated cement and hydration products. We achieve a high spatial resolution of down to 1 μm, allowing an observation of the formation of microporosity. The pore structure appears in combination with macroscopic shrinkage of the cement paste, without any external load.

Among a number of factors[16] which are decisive for shrinkage (such as humidity of the environment of the hardening concrete) the so-called *autogeneous* or *chemical shrinkage* is inherently connected to the hydration process: the final reaction products of the hydration process have a higher mass density and therefore consume less volume than the initial products. In cement, macroscopic shrinkage is constrained by unhydrated remnants of cement particles which is thought to lead to stresses at the interfacial transition zone (ITZ) between unhydrated cement grain cores and hydrated cement matrix. In concrete, coarse aggregates are chemically inert, leading to amplified stress at the ITZ between the shrinking cement matrix and the aggregates. Therefore, crack formation may result especially in concrete.

The aim of the present paper is to demonstrate by examples of the hydration of Portland cement the potential of the proposed 3D investigations which could also be applied to the study of cementitious mixtures employed for the fabrication of HPC. In the latter case autogeneous shrinkage becomes an important issue[16], depending on the fabrication parameters as well as on the use of admixtures and additives.

In Section 2 we will first describe the studied samples and the experimental procedure. Then in Section 3 follow the results obtained by SR tomography and their discussion considering the temporal evolution of pore formation connected to autogeneous shrinkage.

2 SAMPLE AND EXPERIMENTAL SET-UP

The study was focussed on the hydration of ordinary Portland cement (OPC). According to the international nomenclature the samples investigated here are classified under the designation CEM I 42.5R-HS (EN 197-1). This cement provides a nominal cylinder compressive strength of 42.5 N/mm^2 after 28 days of hardening. The abbreviation R signifies "rapid hardening" entailing a high early strength. HS denotes a high resistance against acids containing sulphates by its low C_3A mineral contents ($\leq 5\%$). This cement was used for its excellent workability and the gained good experience in several HPC projects realised in Germany[16].

In general, upon mixing Portland cement with water, there begins a variety of chemical reactions determining the properties both of the non-hardened cement paste and of the hardened cement paste, mortar or concrete. The precise schedule of mechanisms depends on many parameters and could not be fully explored up to now. The hydration process is rather fast at its beginning and slows down asymptotically. Even after months or years this process may still be in progress. Therefore, for hydration times above approximately 4 hours, it is possible to determine by tomographic experiments the evolution of the cement structure in "real time", with respect to the moderate time scale of the hardening process.

Under sealed conditions (no water exchange with the environment) an average water mass part of 0.42 has to be added to the cement powder (w/c=0.42) in order to achieve complete hydration (*i.e.* for hydration times tending to infinity). Therefore, in normal concrete (NC) the hydration process slows down mainly due to growth of a hydration layer on the cement particles through which the water has to diffuse in order to enter chemical reactions with non-hydrated compounds. The reactivity is limited by water diffusion. In contrast, the limited water content in HPC leads to a more abrupt decrease of reactivity.

Depending on the w/c ratio used for HPC fabrication, the hydration process therefore slows down significantly already after about 20 to 30 hours. For w/c=0.3, for example, the relative humidity in the pore system of the hardened cement paste has decreased to 85 percent after a few weeks.[16] Besides the already mentioned abrupt slow-down of the hydration process, the loss of humidity causes capillary tensions and, as a consequence, the additional formation of porosity and, possibly, microcracks. In cement and concrete, porosity is present at different length scales. The investigations will show that high-resolution CT performed with SR is suited to observe capillary pores and shrinkage pores with a size larger than approximately 1 μm and the compaction pores[17].

In this tomographic approach we made preliminary attempts to distinguish between isolated pores and the pores forming a pore network on the micrometre scale: we characterise for different hardening times the open porosity which is, for example, important with regard to the transport properties of the hardened cement or concrete.

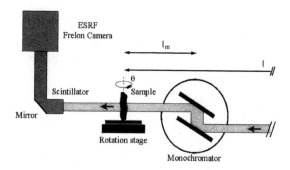

Figure 1 *Sketch of the experimental set-up used for real-time microtomography at the beamline ID19 of the ESRF. The scheme shows the set-up between monochromator and detector, where l_m is the distance between sample and monochromator. The synchrotron source is a multiple of this distance away from the monochromator (e.g. $l_m \approx 5$ m, $l \approx 145$ m for ID19). The ESRF Frelon CCD camera serves as a basis for the electronic detector system.*

A sketch of the experimental set-up implemented at ID19 of the ESRF is shown in Fig. 1. The low beam divergence of synchrotron radiation permits the use of monochromator optics. Monochromatic radiation enables the quantitative reconstruction of the spatial distribution of the linear attenuation coefficient $\mu(x,y,z)$ for the selected x-ray energy, avoiding artefacts due to beam hardening. In order to increase beam intensity for high imaging resolutions with pixel sizes down to 0.3 µm, we can employ instead of the sketched double-crystal monochromator a high-bandwith multilayer monochromator. A present limitation due to the detector width and reconstruction time is that with a pixel size of *e.g.* 1 µm we are in practice restricted to investigate cylindrical samples with diameters of around 1 mm. With this sample geometry we obtained best contrast at an x-ray energy of 18 keV, provided by a multilayer monochromator. As a consequence reconstruction artefacts are reduced and identifying spatial regions of slightly different absorption coefficient, *i.e.* hydration products and base materials, becomes possible.

The rotation stage sketched in Fig. 1 is used to turn the sample for the acquisition of 900 projection radiographs during the tomographic scan. From these, the 3D image of the sample attenuation coefficient was reconstructed off-line by means of a filtered-backprojection algorithm.[18]

A series of samples with different water-cement ratios w/c was prepared from the Portland cement. After filling the viscous cement paste into conical sample vessels made of plastics, the vessels were closed with wax or adhesive tape to prevent excessive water evaporation. If solidification is not sufficiently advanced the paste moves under the rotational and translational movements performed during the tomographic scan. Therefore the samples were left for between 4 and 8 hours in order to allow sample solidification. In this way we could obtain useful data for $w/c \approx 0.3$ after 4 hours whereas for $w/c \geq 0.5$ we had to wait up to 8 hours.

3 EXPERIMENTAL RESULTS

We report the observation of cement hydration and, as a consequence thereof, the formation of pores during hardening. We examine the temporal evolution of two selected samples with w/c=0.41 and a limiting case of w/c=0.97 for studying the formation of hydration products. The former sample corresponds well to a cement with no excess water such that after infinite hydration time virtually no capillary pores filled with water should remain. Its water content is in the range which is usually employed for HPC fabrication.

In the following, representative *slices* and *histograms* of the reconstructed 3D data sets will be presented. The histograms contain statistical, quantitative information about volume fractions and concentrations in the sample which are related to spatial features visible in the slices.

3.1   Investigation of unhydrated cement

For identification of unhydrated cement remnants in hydrated samples let us first investigate a dry cement powder (without any added water). For this we introduced the pure cement powder into a sample vessel and subjected it to an ultrasonic treatment in order to pack the particles as densely as possible. We performed a tomographic scan on the powder sample under the same experimental conditions as with the hydrated cements.

In Fig. 2, a reconstructed slice taken from the 3D data set through this unhydrated cement is given. The light grey regions with sizes up to approximately 100 μm are the cement grains. A white subtexture is visible which indicates the presence of at least two mineral phases of different linear attenuation coefficient μ. The space between the cement particles exhibits varying grey levels. It may be attributed to a varying packing density of cement particles which have sizes below the imaging resolution (fine dust).

The histogram of the entire 3D image gives statistical information about the volume fraction of the dust particles. The plot of Fig. 2 shows the histogram of the 3D data obtained for unhydrated cement (solid line). The second peak („cement particles") comprises voxels completely covered by the large cement particles (with dimensions near and above our resolution). The first peak (towards low μ values) corresponds to the space between the large cement particles. It is mainly due to air but comprises also the cement dust (with particle sizes below the image resolution) which causes the slight shift with respect to zero.

The observed peak shift of about 0.8 indicates that approximately 11 percent of the volume between the large cement particles must be cement. Expressed in relation to the total cement volume the fine dust must occupy a volume fraction of approximately 24 percent which is in rough agreement with literature values.[16]

For comparison we add in Fig. 2 plots of the histograms of the 3D data obtained for hydrated cement (w/c=0.97 and 0.41) after around 8.5 hours of hardening. As before the second peak represents the large cement particles. The first peak corresponds in these cases to the early cement paste in which early hydration products have already evolved. In the reconstructed slices it is difficult to distinguish between water (linear attenuation coefficient μ_{water}=0.74 cm^{-1} at an x-ray energy of 18 keV) and air (μ_{air}≈0) since the standard deviation of the background noise (as found in large air pores) amounts to approximately $\Delta\mu$=1.6 cm^{-1}. Obviously, for the hydrated cements, the peak shift is higher than for the dry cement powder since water (possibly with cement minerals in solution) and early hydration products (such as ettringite) are filling the space between the cement particles.

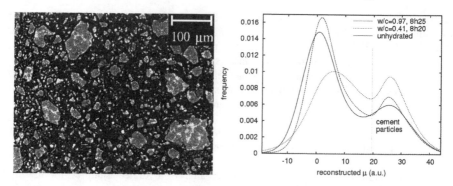

Figure 2 *Left: slice taken from the reconstructed 3D data of an unhydrated cement powder. Image size is 512×400 pixels of 1 µm size. Right: histogram of the 3D image for the unhydrated cement powder (solid line). The histograms for hydrated cements after about 8 hours are given for comparison. Phase-contrast features of the acquired images contribute mainly to the tail towards low µ of the first peak.*

3.2　Observation of the hydration process

Let us now investigate the temporal evolution of the two samples by comparing different hardening stages, first by examination of reconstructed slices, then by plotting the histograms of 3D data sets. By observation of the temporal evolution of the samples it is possible to observe a „transformation" of the sample structure due to hydration.

For one of the samples, identical slices taken from the 3D data sets at five different stages of cement hardening are given in Fig. 3. In the comparison between the images we notice – by the growing regions – pixels which lighten significantly with increasing hardening times (marked by letters „H"). This change in grey level from a dark grey to a lighter grey is equivalent to an increase in linear attenuation and corresponds to the transformation of the early cement paste to a more dense cement paste which is richer in the late hydration products. Simultaneously, we observe the increasing appearance of pores (black, see arrow) with proceeding hydration.

On a quantitative basis, the „transformation" of the early cement paste towards a more dense cement paste – which is accompanied by shrinkage and the involved appearence of pores – is given in Fig. 4 by the associated histograms of the 3D images. In the plots for $w/c=0.97$ we see that during the temporal evolution the second peak remains almost unchanged while the first peak decreases in height significantly. With increasing hardening time a third mode (called „hydration products") is starting to appear between the two peaks near $\mu \approx 15$. It represents the hardened cement paste which is richer in late hydration products. Parallel to the appearance of the central mode the tail of the first peak towards low μ (called „pores") becomes broader which conforms to the observed increase of porosity in the slices. During cement hydration the population of the early cement paste (first peak) decreases in favour of the populations of the hydration products (central mode) and the pores (left tail) .

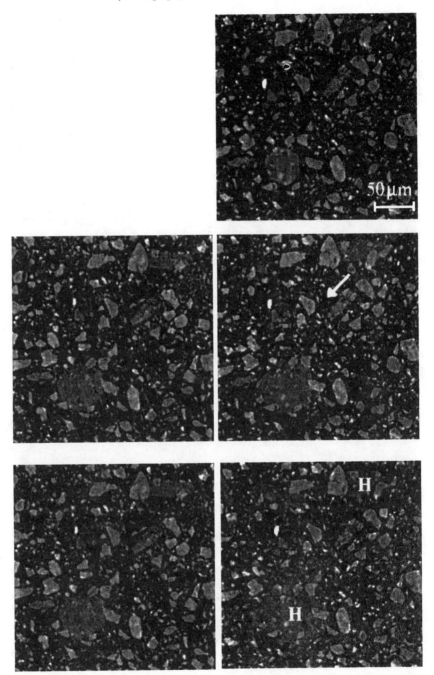

Figure 3 *Cut out slices of the 3D data sets through the sample with w/c=0.97 at different solidification times: 8h25 (top right), 10h50 (centre left), 12h30 (centre right),14h35 (bottom left), 16h30 (bottom right). Image sizes are 256×256 pixels of 1 μm size.*

Let us now turn to the histograms for *w/c*=0.41 of Fig. 4. In the discussion of Fig. 2 we concluded from the shift of the first peak that initially the cement paste in the sample *w/c*=0.41 is more dense than the paste of sample *w/c*=0.97. Also here the first peak decreases in height up to about 13 h after hardening which is accompanied by a shifting towards higher μ values. But after around 13 h the peak height seems to increase again continuing its shifting towards higher μ values. The peak shifting can again be explained by the increasing appearance of late hydration products at the formerly observed μ value of the central mode (*i. e.* $\mu \approx 15$) at the expense of the early cement paste. The effects responsible for this shifting behaviour are similar but the initial peaks („early cement paste" and „cement particles") are closer and their difference in height is not as pronounced.

This transformation observed through individual slices and the histograms of the reconstructed images allow us to identify 1) cement particles, 2) regions rich in late hydration products, 3) the early cement paste and 4) pores appearing due to the transformation of the early paste towards a more dense hydrated cement paste. Tomography with monochromatic radiation and with a resolution at the scale of one micrometre is therefore well suited to study non-destructively the hydration of cement in the sample volume.

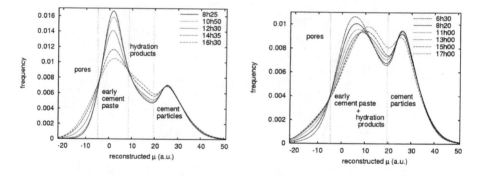

Figure 4 *Histograms of the 3D images at different hardening times for the sample with w/c=0.97 (left) and the sample with w/c=0.41 (right).*

3.3   Temporal evolution of microporosity

Additional information about the evolution of the pore structure with time was obtained by various 3D image processing techniques. One question of interest was to see whether there exists a large pore network or cracks at the micron scale. With absorption contrast by its own it would be not so easy to distinguish between early cement paste and pores due to their low contrast. Phase contrast[19] due to partially coherent illumination at synchrotron imaging set-ups, however, emphasises the pores.

For quantification of the pore sizes, pores were isolated within the images and classified. by digital image processing techniques. Firstly the pore structure was segmented by the hysteresis method which is a kind of a region-growing algorithm.[21] As with every segmentation procedure the parameters of the segmentation algorithms have to be adjusted carefully such that the segmented pore structure reflects the unsegmented image best.

From the resulting binary image of the complete pore structure we extracted and labelled connected pores. Then their sizes in voxels were determined.

For our application the labelling procedure serves to distinguish between a) the largest pore network and b) the remaining smaller pores. A spatially extended pore network contributes to the transport properties, *e.g.* of fluids, through the hardening cement paste. For instance, a large pore network facilitates the entry of attacking substances and, hence, determine the durability of the hardened cement paste.

In Fig. 5 the temporal evolution of the size of the segmented pore space is plotted on the left-hand side. The total porosity with a spatial extension above the micron scale increases with proceeding hydration time. For w/c=0.41 the porosity attains a maximum after approximately 16 hours where it seems to level out. For w/c=0.97 the increase in porosity seems to slow down also but attains around 50 percent higher values.

If only the size of the largest pore (forming a network) is considered we obtain an idea about the formation of a pore network within the cement paste. The plot on the right-hand side of Fig. 5 shows the fraction of porosity contributed by the largest pore. For w/c=0.41 we see between 6 hours and 17 hours of hydration almost a doubling of the fraction of pore space forming a pore network which levels out at around 85 percent of the total porosity above 12 hours. In the case of w/c=0.97 we see an increase in this fraction of a factor of around 10 up to values of more than 90 percent.

The porosity in sample w/c=0.97 attains higher values and finally forms a highly interconnected pore system which can be attributed to the higher volume fraction of water present in this sample. After complete hydration the excess water would remain in water-filled capillary pores which have already formed after 16 hours of hardening.

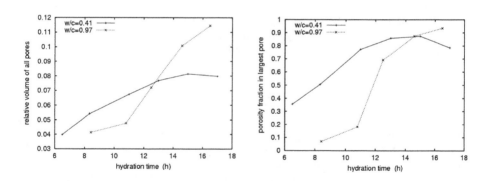

Figure 5 *Temporal evolution of porosity above the µm scale during hydration: relative volume of total porosity (left) and the fraction of porosity contained in the largest pore network (right).*

3.4 Discussion and interpretation

It is unlikely that calcium silicate hydrate (CSH), the compound which provides the intermediate and final strength of the concrete, is responsible for the stiffening necessary for performing a tomographic scan. More likely, early hydration products, such as ettringite or calcium hydroxide, are thought to provide sufficient early strength to allow performing a tomographic scan. For w/c=0.97 it takes a longer time for solidification since initially the cement particles are at a larger distance with respect to each other. The water film between adjacent particles is simply larger[21] such that longer ettringite crystallites have to be formed to bridge the films in the case of w/c =0.97.

CSH formation starts after about 4 to 6 hours of hardening. This suggest that the hydration product observed from 6 hours of hardening onwards is primarily CSH. It is commonly assumed[9] that the onset of hydration of CSH is governed by a dissolution-limited process in which Ca^{2+} and $H_2SiO_4^{2-}$ go into solution with water. In a second process dominating the late stages, a shell of hydration products has built up around the CSH particles and water has to diffuse through the resulting barrier.

When this layered barrier has been built up, the hydration process slows down and changes to a diffusion-limited behaviour in which the water becomes depleted in dissolved substances as mentioned above. Along with this, pore formation should slow down equally which has been observed.

The degree of hydration lies approximately in the range between 0.2 and 0.45 for hydration times between 6 and 18 hours.[21] The structural properties of water in OPC with w/c=0.4 have been determined by means of quasi-elastic neutron scattering.[9-11] After 16 hours of hydration at a constant temperature 20°C the fraction of chemically bound and constrained water was determined to approximately 0.5. After about 20 hours of hydration, the formation of constrained water was found to slow down, whereas the formation of chemically bound water was still proceeding at a reduced rate and closely following the exothermic heat output.

This slow-down is confirmed by our tomographic approach, in this case with respect to pore formation. For w/c=0.97 we see significantly higher porosity after 16 hours of hydration. For sealed curing conditions every unit mass of cement needs on average 0.42 unit mass of water for complete hydration. Approximating our sealing performed by adhesive tape as a complete sealing, it can be concluded that a higher fraction of capillary (water-filled) porosity will remain for the sample w/c=0.97.

4   CONCLUSIONS

First results demonstrate that CT performed with monochromatic radiation allow the identification of pores, the early cement paste, the hydrated cement paste and cement particles. The formation of hydration products from the early cement paste was studied by following the temporal evolution especially of water-rich cement pastes.

Therefore high-resolution microtomography is well adapted to the non-destructive investigation of the hardening and hydration process in the sample volume. It is suited to yield complementary information to the picture obtained by the other experimental methods listed in the Introduction.

The decrease of the volume fraction of the early cement paste in favour of the pores and the hydration-product-rich hardened cement paste could be investigated. The evolution of microporosity could be quantified: with proceeding hydration a large pore network on the micrometre scale develops which is much more pronounced in the case of water-rich cement pastes.

Acknowledgements

This work was supported by ESRF. The authors would like to thank the team of beamline ID19, especially P. Cloetens, J. Baruchel, E. Boller and F. Peyrin, for all the work they invested in the constant improvement of the micro-CT set-up. Support by the AIF (grant KF 0010502KWM9) is gratefully acknowledged.

References

1   E. Garboczi and D. Bentz, *Construction and Building Materials,* 1996, **10**, 293.

2   D. Bentz, *J. Amer. Ceram. Soc.*, 1997, **80**, 3.

3   U. Jakobsen and N. Thaulow, 'Combining optical fluorescence microscopy and scanning electron microscopy for the examination of deteriorated concrete' in *Proceedings of the 7th Euroseminar on Microscopy Applied to Building Materials*, eds., H. Pietersen, J. Larbi and H. Janssen, Delft University of Technology, 1999, pp. 191–202.

4   D. Bentz, P. Stutzman, C. Haecker and S. Remond, 'SEM/X-ray imaging of cement-based materials' in *Proceedings of the 7th Euroseminar on Microscopy Applied to Building Materials*, eds., H. Pietersen, J. Larbi and H. Janssen, Delft University of Technology, 1999, pp. 457–466.

5   J. Bisschop and J. van Mier, 'Quantification of shrinkage microcracking in young mortar with fluorescence light microscopy and ESEM' in *Proceedings of the 7th Euroseminar on Microscopy Applied to Building Materials*, eds., H. Pietersen, J. Larbi and H. Janssen, Delft University of Technology, 1999, pp. 223–232.

6   J. Thomas, H. Jennings and A. Allen, *Concrete Science and Engineering,* 1999, **1**, 45.

7   A. J. Allen, C. G. Windsor, V. Rainey, D. Pearson, D. D. Double and N. M. Alford, *J. Phys. D: Appl. Phys.*, 1982, **15**, 1817.

8   H. Baumbach, F. Häußler, M. Hempel, F. Eichhorn and A. Hempel, *Physica Scripta*, 1995, **T57**, 184.

9   S. FitzGerald, D. Neumann, J. Rush, D. Bentz and R. Livingston, *Chem. Mater.*, 1998, **10**, 397.

10   J. Thomas, S. FitzGerald, D. Neumann and R. Livingston, *J. Am. Ceram. Soc.,* 2001, **84**, 1811.

11   S. FitzGerald, J. Thomas, D. Neumann and R. Livingston, *Cement and Concrete Research*, 2002, **32**, 409

12   D. Bentz, N. Martys, P. Stutzman, M. Levenson, E. Garboczi, J. Dunsmuir and L. Schwartz, 'X-ray microtomography of an ASTM C109 mortar exposed to sulfate attack', in *Microstructure of Cement-Based Systems, Bonding and Interfaces in Cementitous Materials*, eds., S. Diamond, S. Mindess, P. Glasser, L. Roberts, J. Skalny and L. Wakeley, Boston, MA, Volume 370, 1998, pp. 457-466.

13   D. Bentz, D. Quenard, H. Kunzel, J. Baruchel, F. Peyrin, N. Martys and E. Garboczi, *Materials and Structures*, 2000, **33**, 147.

14   T. Baumbach, L. Helfen, F. Dehn and G. König, *Tomography on high-strength concrete*, Fraunhofer Institute for Nondestructive Testing (IZFP), Annual Report 2001, 2002, pp. 60–61.

15   D. Bentz, S. Mizell, S. Satterfield, J. Devaney, W. George, P. Ketcham, J. Graham, J. Porterfield, D. Quenard, F. Vallee, H. Sallee, E. Boller and J. Baruchel, *J. Res. Natl. Inst. Stand. Technol.*, 2002, **107**, 137.

16   G. König, N. V. Tue and M. Zink, *Hochleistungsbeton*, Ernst & Sohn, Berlin, 2001.

17   K. Wesche, *Baustoffe für tragende Bauteile*, Volume 2, Bauverlag GmbH, Wiesbaden, Berlin, 1993.

18   G. T. Herman, *Image Reconstruction from Projections*, Academic Press, New York, 1980.

19 P. Cloetens, M. Pateyron-Salomé, J. Y. Buffière, G. Peix, J. Baruchel, F. Peyrin and M. Schlenker, *J. Appl. Phys.*, 1997, **81**, 5878.
20 K. Castleman, *Digital image processing*, Prentice-Hall International Inc., Englewood Cliffs, New Jersey, 1996.
21 S. Röhling, H. Eifert and R. Kaden, *Betonbau*, Verlag Bauwesen, Berlin, 2000.

OBSERVATION OF THE NANOSTRUCTURE OF CEMENT HYDRATION BY SOFT X-RAY TRANSMISSION MICROSCOPY

Maria C.G. Juenger,[1] Paulo J.M. Monteiro,[2] Vincent H.R. Lamour,[3] Ellis M.Gartner,[4] Greg P. Denbeaux,[5] David T. Attwood[5]

[1]Department of Civil Engineering, The University of Texas, Austin, Texas 78712 USA; [2]Department of Civil and Environmental Engineering, University of California, Berkeley, California 94720 USA; [3]Département de Génie Civil, Ecole Normale Supérieure de Cachan, France; [4]Lafarge, Laboratoire Central de Recherche, 38291 St. Quentin Fallavier, France; [5]Center for X-Ray Optics, Ernest Orlando Lawrence Berkeley National Laboratory, Berkeley, California 94270 USA

1 EXTENDED ABSTRACT

Over 9 billion tons of concrete are produced yearly around the world – no other material, except water, is consumed in such tremendous quantities. Besides their technological significance, the relatively fast reactions (seconds to days) between calcium silicates and water are of scientific interest because they can provide valuable insight to similar geological reactions that occur over a longer period of time (centuries to millennia). Calcium silicate hydrates (C-S-H) resulting from the dissolution of calcium silicate cements are particularly interesting because they are responsible for the strength and durability of concrete and, more recently, for the feasibility of using concrete in nuclear waste containments. In spite of this, relatively little is understood about the hydration reaction responsible for the transition of a calcium silicate cement and water fluid mixture to a strong and durable solid. Much of the mystery surrounding C-S-H can be attributed to its variability, but also to the inadequacy of observation and analytical techniques. Direct observation is inherently flawed because microscopy techniques that allow high enough resolution generally require destructive drying of the sample. Environmental Scanning Electron Microscopy (ESEM) allows moist samples to be monitored, but it is difficult to get good resolution for very wet cement samples at early ages and observation is limited to surfaces, which are not representative. Atomic Force Microscopy (AFM) may be used to observe wet samples, but can only examine extensive solid surfaces. Models for C-S-H microstructure are primarily based on surface area measurements, which have their own artifacts associated with drying and geometric assumptions. A new technique, soft x-ray transmission microscopy, has been pioneered at the Center for X-ray Optics (CXRO) at the Advanced Light Source (ALS) at the E.O. Lawrence Berkeley National Laboratory (LBNL). The microscope, XM-1, was designed for the observation of biological cells in their natural, wet state at atmospheric pressure. It is now also being used for several materials science applications including cementitious and magnetic materials.

2 EXPERMIMENTAL METHOD

The soft x-ray microscope utilizes radiation generated by a synchrotron source. The radiation is filtered by a plane mirror and only that in the soft x-ray regime with wavelengths between 0.7 and 4 nm (300 to 1800 eV) is sent to a condenser zone plate, which then focuses the radiation onto the sample. X-rays transmitted through the sample are refocused with a micro zone plate onto an x-ray CCD camera (Figure 1). More detailed information on the microscope has been reported elsewhere.[1,2] The use of soft x-rays, with wavelengths more than 100 times shorter than visible light, leads to a maximum resolution of 25 nm. In these experiments, a wavelength of 2.400 nm (516.6 eV) was used. The magnification of the images is between 1600 and 2400x and remains fixed during a given session. The images have a circular field of view of approximately 10 μm in diameter. If a larger field of view is desired, several images are taken across a grid and a composite image is created by montage alignment using a custom autocorrelation process.[3] Many of the images presented here have been processed in this manner.

Samples are viewed through thin silicon nitride membranes which are sandwiched between stainless steel plates fitted with rubber o-rings to minimize evaporation and exposure to CO_2. The primary limitation of this instrument is that in order for x-rays to be transmitted, the maximum thickness of samples is 10 μm. The initial water-to-cement ratio by mass (w/c) used for this study varied between 5 and 50; these solutions were then centrifuged so that the only the smallest (<10 μm) cement grains remained in the supernatant, resulting in an actual w/c of 100-500. (Typical concrete w/c values are 0.2-0.6.) In order to prevent excessive dissolution of the cement grains, the solutions used in this study were saturated with respect to calcium hydroxide ($Ca(OH)_2$) and gypsum ($CaSO_4 \cdot 2H_2O$).

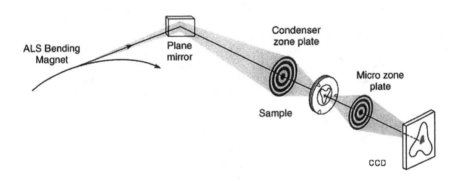

Figure 1 *Soft x-ray transmission microscope, XM-1*

3 EXAMPLE OF APLICATION

We are beginning to have an understanding of the details of the hydration process. At the first minutes (Fig. 2a), a quick burst of reaction products are formed in the reactive sites of the tricalcium silicate. Previous NMR work indicated that monomers are formed in the first stages of the reaction. After that there is a "dormant period" takes place as the pore fluid gets saturated. It is interesting to note that the original spikes of the monomer do not grow much over time, and that the new products are formed in the pores between the cement grains. There are three possible explanations for that: a) the new hydration products are dimmers (as indicated by NMR) that do not fit well with the original monomers, b) the three-dimensional network of the dimmers requires that they precipitate in the large spaces, c) the electrical charge of the cement grains are not conducive for the precipitation of the new hydration products.

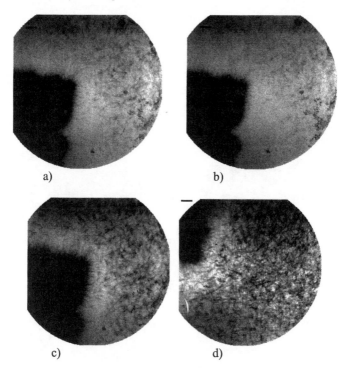

a)

b)

c)

d)

Figure 2 *Evolution of tricalcium silicate over time. The soft x-ray microscope (XM-1) located at beamline 6.1.2 at the Advanced Light Source is a remarkable tool that allows high resolution images of wet samples. The resolution is high enough for us to observe the formation of C-S-H and other hydration products around tricalcium silicate and in solution. Furthermore, since the sample is kept wet at all times, we can observe the progression of hydration over time. Scalebar is one micron.*

Acknowledgements

Funding for this project was provided by Lafarge, Laboratoire Central de Recherche, France. Research at XM-1 is supported by the United States Department of Energy, Office of Basic Energy Sciences under contract DE-AC 03-76F00098. The authors would like to gratefully acknowledge the work of Werner Meyer-Ilse, Kimberly Kurtis, and Angelic Pearson.

References

1. Meyer-Ilse, W. et al. *Synchrotron Radiat. News* **8**, 23-33 (1995).
2. Denbeaux, G. et al. Soft X-ray microscopy to 25 nm with applications to biology and magnetic materials. *Nucl. Instrum. Meth. A* **467**, 841-844 (2001).
3. Loo, B.W., Meyer-Ilse, W. & Rothman, S.S. Automatic image acquisition, calibration and montage assembly for biological X-ray microscopy. *J. Microsc - Oxford* **197**, 185-201 (2000).

STUDY OF POZZOLAN-CEMENT INTERACTION BY ATOMIC FORCE MICROSCOPY (AFM)

U Rattanasak, M Rotov, K Kendall

Department of Chemical Engineering, University of Birmingham, Edgbaston, Birmingham B15 2TT, UK

1 ABSTRACT

The interaction between a silica sphere and cement-calcium silicate phases (i.e. dicalcium silicate, C_2S, and tricalcium silicate C_3S) was observed under water. Atomic force microscopy (AFM) was used to measure the force versus distance relationship between the silica sphere and a flat cement clinker surface by monitoring the deflection of the cantilever spring onto which the particle was adhered as a function of displacement from the flat surface. Adhesion forces were also measured by AFM. The results showed that the adhesion forces in water between a silica sphere and different phases of clinker change with treatment time. With increasing treatment time, the adhesion force increased.

2 INTRODUCTION

The hydration reaction of cement to form calcium silicate hydrate (C-S-H = $(CaO)_x$-SiO_2-$(H_2O)_y$) is responsible for the main properties of final hydrated cement paste. Formation of C-S-H takes place by heterogeneous nucleation on the tricalcium silicate surface followed by growth on this same surface.[1] Such formation is occurring at the nano-scale. Some researchers[2-4] applied atomic force microscopy (AFM) to investigate the nanometre topography of pozzolan, cement components and hardened cement paste. In addition, published papers[5-9] have investigated the hydrate formation using AFM. However, the reactions involved in pozzolan- cement are very complex and still remain ill-understood.

Therefore, this paper presents the interaction observed by reaction between a sphere, using a pure silica particle, and the calcium silicate phases (i.e. dicalcium silicate, C_2S, and tricalcium silicate C_3S) in polished cement clinker under water.

It is believed that an AFM study of force acting between a silica sphere and different components of polished cement clinker should provide both quantitative and qualitative information of the appropriate probe/substrate system. Additionally, it is expected that the interaction between the silica and the cement phase at the nanometre level should be different. Therefore, this study aims to investigate the interaction between a silica sphere and the several phases identified in the polished cement clinker surface in order to obtain a better understanding of the hydrate formation.

3 ATOMIC FORCE MICROSCOPY

Atomic Force Microscope (AFM) is a mechano-optical instrument, which detects atomic-level forces on scale of nN through measurements of the movement of a very sensitive cantilever tip along surfaces. Typically AFM is operated in the Contact Mode, where the AFM tip remains in contact with surface (in this way some friction may be observed). Alternatively, the Tapping Mode may be used, which the tip is lifted from the surface at each point, moved over to the next point and lowered again. In this experiment Contact Mode AFM was used.

3 MATERIALS AND METHODS

3.1 Cantilever modified with silica sphere particle

Attachment of a silica sphere (6 μm diameter) was done by use of the atomic force microscope under the control of an optical microscope. One cantilever was used to deposit a tiny amount of glue onto the tip of the main cantilever. Then, another hydrophilic cantilever brought the silica sphere to attach onto the main cantilever. Figure 1 shows the final cantilever attached silica sphere.

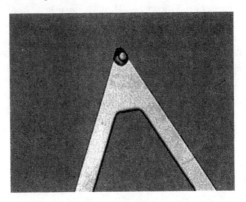

Figure 1   *Cantilever modified with silica particle (6μm dia.)*

3.2 Sample preparation

Cement clinker was obtained from Castle Cement Ltd. and divided into small pieces. Each slice was impregnated with low viscosity epoxy resin under a low vacuum for 1 hour. The vacuum was applied for 20 min, after which it was released for 20 min and then reapplied for further 20 min to remove any remaining air and to ensure that the majority of the void space had been eliminated. The resin was then left to harden for 24 hrs.

The sample was dry-polished with silicon carbide grit from 30 to 4 μm. Material dusts were then cleaned out in methanol using an ultrasonic method. Polishing was completed by using a 1-μm diamond paste with the addition of a lubricant.

3.3 Force measurement

AFM was used to measure interaction between the silica sphere and flat surface of C_3S and C_2S in polished cement clinker. Force was measured as a function of separation distance. Additionally, the effect of treatment time, i.e. the time of contact between the cement of water, was investigated.

4 RESULTS AND DISCUSSION

4.1 Imaging of polished cement clinker

The structure of the polished cement clinker was first examined by contact mode AFM imaging as show in Figure 2. The roughness of investigated clinker obtaining from AFM is 34.508 nm as also shown in the 3-D diagram (Figure 2C). The dimension of this area is 4x4µm. AFM reveals that cement clinker has a variable surface intersected by pores and significant the roughness.

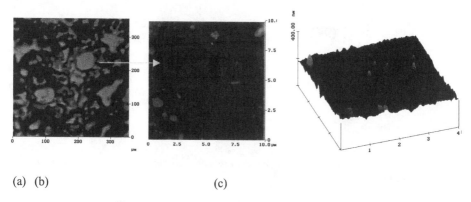

(a) (b)   (c)

Figure 2 *(a) An AFM image of polished cement clinker, (b) enlarged area, and (c) 3D showing roughness of highlighted area*

4.2 Interaction force between surfaces

This experiment used AFM to measure the force versus distance relationship between a silica sphere and a flat cement clinker surface by monitoring the deflection of the cantilever spring, onto which the particle is adhered, as function of displacement from the flat surface. The C_3S and C_2S components of cement clinker were investigated.

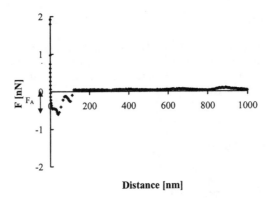

Figure 3 *Force versus distance curve for interaction between the silica sphere and C_3S surface for a treatment time of 10 min. F_A donates the adhesion force.*

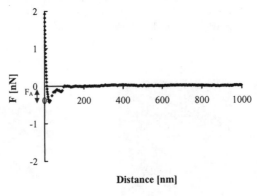

Figure 4 *Interaction between the silica sphere and C_2S surface for a treatment time of 10min.*

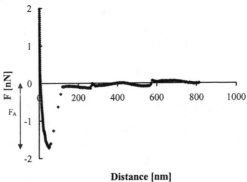

Figure 5 *Interaction between the silica sphere and C_3S surface for a treatment time of 24hrs.*

Figure 3 presents the interaction force as a function of distance between the silica sphere and C_3S in pure water. When the silica sphere approaches the C_3S surface at 10-minute treatment time, the force curve changes at surface distances of 50 nm and 100 nm before the probe adheres to the C_3S surface. The adhesion force is approximately 0.6 nN.

Force measurement was repeated on a C_2S surface at the same conditions as C_3S (Figure 4). It can be seen that adhesion force between silica sphere and C_2S surface is nearly the same as that of C_3S has.

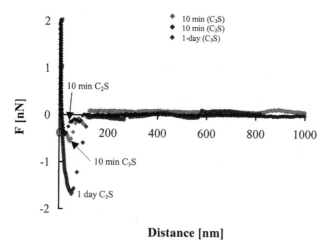

Distance [nm]

Figure 6 *Summarised force curve*

1.1 Adhesion with increasing treatment time

Force measurement was repeatedly investigated in water over a longer treatment time of 24 hours. Figure 5 demonstrates the interaction between the silica sphere and a C_3S surface. With increasing treatment time, the adhesion force increases substantially. When all the force-distance curves are plotted together as presented in Figure 6, it can be seen that the adhesion force of C_3S is nearly equal to that of C_2S at 10-minute treatment time and this force increases with treatment time. C_3S is the most reactive compound, however, rate of reaction bears no relation to final strength development. The calcium silicates provide most of the strength developed by Portland cement. C_3S provides most of the early strength (in first 3 to 4 weeks) and both C_3S and C_2S contribute equally to ultimate strength.[10]

5 CONCLUSION

AFM presents a new method for studying interaction between a particle and cement clinker components. This investigation also brings insight into how the adhesion occurs. It reveals that the adhesion forces between a silica sphere and different phases of clinker are equal at the same treatment time. Furthermore, with increasing treatment time, increased adhesion forces were obtained.

Acknowledgements

U Rattanasak is supported by a grant from Royal Thai Government.

References

1   Nonat A and Lecoq X, 1997, "The structure stoechiometry and properties of C-S-H prepared by C_3S hydration under controlled conditions", *Nuclear Magnetic Resonance Spectroscopy of Cement-based Material*, Springer-Verlag.
2   Demanet CM, 1995, "Atomic force microscopy determination of the topography of fly ash particles" *Applied Surface Science*, 89, 97-101.
3   Papadakis VG, Pederson EJ and Lindgreen H, 1999, "An AFM-SEM investigation of the effect of silica fume and fly ash on cement paste microstructure", *Journal of Materials Science*, 34, 683-690.
4   Yang T, Keller B and Magyari E, 2002, "AFM investigation of cement paste in humid air at different relative humidities", *Journal of Physics D: Applied Physics*, 35, L25-L28.
5   Gauffinet S, Finot E, Lesniewska E and Nonat A, 1998, "Direct observation of the growth of calcium silicate hydrate on alite and silica surfaces by atomic force microscopy", *Earth & Planetary Sciences*, 327, 231-236.
6   Lesko S, lesniewska E, Nonat A, Mutin JC and Goudonnet JP, 2001, "Investigation by atomic force microscope of force at the origin of cement cohesion" *Ultramicroscopy*, 86, 11-21.
7   Finot E, Lesniewska E, Goudonnet JP and Mutin JC, 2000, "Correlation between surface forces reactivity in setting of plaster by atomic force microscopy" *Applied Surface Science*, 161, 316-322.
8   Uchikawa H, Hanehura S and Sawaki D, 1997, "The role of steric repulsive force in the dispersion of cement particle in fresh paste prepared with organic admixture" *Cement and Concrete Research*, 27(1) 37-50.
9   Mitchell LD, Prica M and Birchall JD, 1996, "Aspects of Portland cement hydration studied using atomic force microscopy", *Journal of Materials Science*, 31, 4207-4212.
10 Mindess S and Young J F, 1981, *Concrete*, New Jersey: Prentice-Hall Inc., pp.

ESTIMATION OF THE DEGREE OF HYDRATION AND PHASE CONSTITUTIONS BY THE SEM-BSE IMAGE ANALYSIS IN RELATION TO THE DEVELOPMENT OF STRENGTH IN CEMENT PASTES AND MORTARS

S. Igarashi[1], M. Kawamura[1] and A. Watanabe[1]

[1]Department of Civil Engineering, Kanazawa University, Kanazawa 920-8667, Japan

1 INTRODUCTION

The total porosity, pore size distributions and specific surface areas have been used to characterize the pore structures in porous materials. In addition to these parameters, pore connectivity and tortuosity may be also important in the durability of concrete. Pore diameters in concretes range from nm to a few mm. The mercury intrusion porosimetry (MIP) method has been commonly used to evaluate pore size distributions in cement-based materials. However, it should be noted that the pore structure revealed by the MIP method is not always representative of the real structure in concrete because of the inappropriate assumption to be made for the complex network of pores.[1]

Another method to evaluate the porosity and characteristics of coarse capillary pore structures is the quantitative SEM-BSE image analysis.[2,3] Pores as one of constituent phases in an original BSE image are quantitatively evaluated by the procedures of image analysis. Based on the simple stereological principles, area fractions obtained for a 2D cross section are equal to volume fractions for the 3D real structure when materials have a completely random and isotropic nature.[4] The microstructure of cement paste can be considered to satisfy those stereological conditions. The imaging technique for BSE images is a method for quantitatively evaluating the actual 3D microstructure by analysing information on 2D cross sections.

The pioneering work with the imaging technique for revealing local variations of microstructure, e.g. porosity gradients in concretes were made by Scrivener and Pratt.[5,6] At early stages in the development of this technique, the area fraction of measurable image parameters was a key quantity to understand features of microstructure in concrete. However, recently, the imaging technique for concrete has been extended to extract not only the total quantity, but also geometric features of phases of interest.[7,8] In view of recent advanced technology of diagnostic imaging system in medical engineering, geography etc., it is expected to establish such a system in concrete science and engineering.

Four phases in cement paste, i.e. unhydrated cement particles, calcium hydroxide crystals, CSH and capillary pores, are resolved by the imaging technique. Usually, only the total amount of a specific phase has been focused upon. However, few studies on the evolution of microstructure with an emphasis on quantitative interrelations between phases by image information are found. If the quantitative balance between phases is properly related to a reliable model for the hydration of cement, the image-based operations on the hydration of cement can give quantitative and/or qualitative information on the evolution of microstructure including more refined features than the resolution of BSE images.

The object of this study is to quantitatively pursue changes in the amount of constituent phases with time in cement pastes and mortars by the BSE image analysis. Two contrast phases of pores and unhydrated cement grains in gray levels are examined to quantify their balance based on their volume changes with time during the hydration of cement. It was assumed that variations in area fractions of these two phases can be directly related to each other by means of the degree of hydration of cement determined by the image analysis. In the analysis, the results of image-based operations for apparent variations of the two phases are complemented for the presence of fine pores less than the resolution. In the complement for the presence of fine capillary pores, the Powers model[9] for the hydration of cement was used. Effects of water/binder ratio and the addition of mineral admixtures on the microstructure of cement pastes and mortars were discussed in terms of the combination of image analysis and the Powers model. The validity of the dependence of the compressive strength on the gel/space ratio obtained by the BSE image analysis was also discussed.

2 EXPERIMENTAL

2.1 Materials and Mix Proportion of Cement Pastes and Mortars

Table 1 *Physical properties of admixtures*

	Silica Fume	Fly Ash
Density (g/cm^3)	2.20	2.19
Specific Surface Area (m^2/g)	20.0	0.345
Ignition Loss (%)	1.20	0.70
SiO$_2$ (%)	90.8	70.9

Table 2 *Mix proportions of cement pastes and mortars*

		Water/Binder Ratio	Replacement Ratio (%)	Binder□Sand
Cement Paste	Admixture-free	0.25	0	1:0
		0.4	0	1:0
		0.6	0	1:0
	Silica Fume paste	0.4	10	1:0
	Fly ash paste	0.4	15	1:0
Mortar	Admixture-free	0.25	0	1:2
		0.4	0	1:2
		0.6	0	1:2
	Silica Fume mortar	0.4	10	1:2
	Fly ash mortar	0.4	15	1:2

The cement used was an ordinary Portland cement produced in Japan. A silica fume and a fly ash with a specific surface area of 20m^2/g and 0.345m^2/g were used as mineral admixtures (Table 1). The replacement of silica fume and fly ash were 10% and 15%, respectively. The fine aggregate used was a natural river sand with F.M. of 2.59. A polycalboxylic acid type superplasticizer was used in cement pastes with silica fume. Their water/binder ratios were 0.25, 0.40 and 0.60. Mix proportions of the cement pastes and mortars are given in Table 2.

2.2 BSE image analysis

Cylinders of 100mm in length and 50mm in diameter were produced. They were demolded at 24 hours after casting, and then cured in water at 20□. At the prescribed ages, slices about 10mm in thickness were cut from cylinders for the BSE image analysis. They were dried by ethanol replacement and in the vacuum drying, and then impregnated with a low viscosity epoxy resin. After the resin hardened at room temperature, the slices were finely polished with SiC papers. The polished surfaces were meticulously finished with diamond slurry for a short time.

Samples were examined using the SEM equipped with a quadruple backscatter detector. BSE images were acquired at a magnification of 500×. In order to avoid influences of interfacial transition zones around fine aggregate particles on results, regions of interest for acquiring images in mortars were sufficiently away from surfaces of sand particles. Taking account of statistical variations in the results of image analysis, ten fields in each specimen were randomly chosen and analyzed.[6,10] Each BSE image (e.g. Fig.1) consists of 1148×1000 pixels. The size of one pixel is about 0.22×0.22μm. The dynamic thresholding method was used to make binary segmentation based on the grey level histogram. Pixels for unhydrated cement particles and pores were tallied so as to obtain area fractions of the phases. The stereology principles state that area fractions in 2D cross sections are equal to 3D volume fractions. It was assumed that volume fractions of hydration products (i.e. CSH and calcium hydroxide crystals) were obtained by subtracting the volume of unhydrated cement particles and capillary pores from the volume of a sample.

The degree of hydration of cement (α) was calculated by Equation (1) .[6, 11]

$$\alpha = 1 - \frac{UH_i}{UH_0} \qquad (1)$$

where:
UH_i: area fraction of unhydrated cement particles at the age of t_i
UH_0: initial area fraction of unhydrated cement particles (i.e. $t_i=0$)

It was confirmed in advance that the volume fraction of cement in a mixture of cement and epoxy resin was closely equal to the volume fraction determined by the image analysis.

Representative BSE micrographs of admixture-free and fly ash-containing cement pastes are given in Fig. 1. As found in Fig.1(b), brightness of fly ash particles were sufficiently different from that of unhydrated cement particles. Therefore, in the calculation of the degree of hydration of cement, it was easy to determine the threshold value in the gray level between unhydrated cement and fly ash particles. Silica fume particles cannot be resolved in the SEM-BSE examination. However, the degree of hydration of cement in silica fume and fly ash-containing cement pastes was able to be calculated based on the same procedure as for the pastes without admixture.

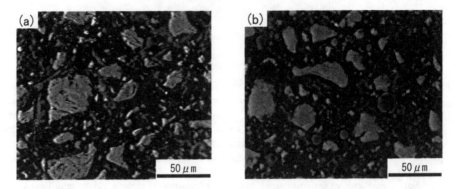

Figure 1 *BSE image micrographs of cement pastes (water/binder ratio of 0.4) at 1 day: (a)Admixture free (b) Containing fly ash.*

2.3 Compressive strength tests

Cylinder specimens of 50mm in diameter and 100mm in height were prepared according to JIS R 5201 and JSCE-F506. The specimens were cured under the same condition as those used in the SEM examinations. Compressive strength tests were conducted at the age of 1,3,7,28 and 91days.

3 RESULTS

3.1 Constituent phase fractions and the degree of hydration in cement pastes

Fig. 2 shows the volume fractions of constituent phases in cement pastes at various ages. The volume fractions of unhydrated cement remarkably decreased for the initial hydration of 24 hours, but subsequent reduction in the volume fraction of unhydrated cement in the cement paste with a water/cement ratio of 0.25 was not so conspicuous as in other cement pastes. The volume of coarse capillary pores also decreased with time in concert with decrease in the volume of unhydrated cement. As expected, the higher the water/cement ratio, the greater the volume fraction of coarse capillary pores in the cement pastes.

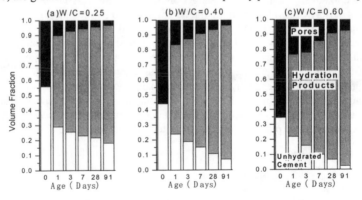

Figure 2 *Measured volume fractions of constituent phases in cement pastes*

Fig. 3 shows the degrees of hydration of cement which were calculated with Equation (1) using volume fractions of unhydrated cement measured by the BSE image analysis (Fig. 2). About a half of cement has reacted for the first 24 hours in the cement pastes with water/cement ratios of 0.25 and 0.40. The higher the water/cement ratio of cement pastes, the lower the degree of hydration up to the age of 3 days. However, thereafter, the degree of hydration rapidly increased with increasing water/cement ratio so that the cement paste with a water/cement ratio of 0.6 achieved the highest value of about 90% at 91 days. It should be noted that the degree of hydration in the cement paste with a water/cement ratio of 0.25 little changed approaching the ultimate degree of hydration of 0.69 (in water) which was estimated based on the Powers model.

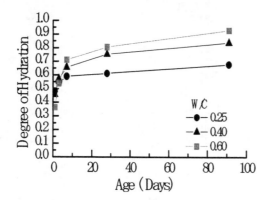

Figure 3 *Degrees of hydration in cement pasts without admixture.*

3.2 Changes in phase fractions with the degree of hydration

The volume of hydration products can be calculated with a model for the hydration of cement. In this study, the Powers model was applied to the results of image analysis. In the calculation, the volume of cement gel produced by the hydration of $1cm^3$ dry cement was assumed to be $2.1cm^3$. The non-evaporable water content in the reacted cement is about 23% by mass. Chemical shrinkage was also assumed to be 0.254 of the volume of non-evaporable water. The porosity of cement gel used in the calculation was 28%; gel pores saturated with gel water. The volume of cement gel was estimated using the degree of hydration determined by Equation (1). The volume of capillary pore was obtained by subtracting the volume of unhydrated cement and cement gel from the initial volume of the mixture. Thus, differences in the volume fraction between the capillary pore volumes calculated based on the Powers model and the coarse pore volumes obtained by the image analysis represent the volume fractions of fine pores of which diameters were less than the resolution of the image analysis ($0.2\mu m$ in this study).

Assuming that excess water was supplied from outside, the volume fractions of constituent phases in cement pastes with various water/cement ratios are given in Fig. 4. The volume fractions of capillary pores estimated based on the Powers model (the sum of coarse and fine pores in Fig. 4) were greater than those of coarse pores($>0.2\mu m$) obtained by the image analysis. These results indicate that considerable amounts of fine capillary pores not counted in the image analysis exist. It has been pointed out that the BSE image analysis could detect large capillary pores not measured by the mercury intrusion porosimetry.[8] A noteworthy result obtained in this study is that amounts of small and medium capillary pores (50nm~2.6nm) and parts of large capillary pores (0.2μm~0.05μm) not counted by the BSE image analysis were extremely small in the cement paste with a low water/cement ratio of 0.25, compared with the cement pastes with water/cement ratio of 0.40 and 0.60. In 90 days old cement pastes with a low water/cement ratio of 0.25, the medium-sized pores (0.2μm~2.6nm) have disappeared, but some amounts of larger capillary pores than 0.2μm existed.

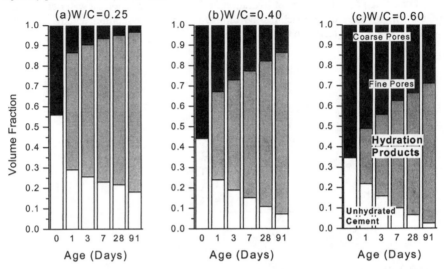

Figure 4 *Volume fractions of constituent phases in cement pastes (Coarse pores : >0.2μm Fine pores:0.2μm-2.6nm)*

It is also found from Fig.4 that an increase in water/cement ratio results in a decrease in the volume of solid phases (hydration products + unhydrated cement). However, differences in the degree of hydration reflecting the volume of unhydrated cement between different water/cement ratios were not so great as differences in the solid volume in cement pastes with various water/cement ratios because of lack in parts of fine pores (Fig.4). It is found from these results that the image analysis gave more quantitative evaluation for capillary pore structures. Furthermore, it should be noted that volume fractions of fine capillary pores at a low water/cement ratio of 0.25 were far less than those in cement pastes with water/cement ratios of 0.40 and 0.60. Especially, the application of the Powers model to the results obtained by the image analysis drew a conclusion that at 91 days, such fine capillary pores were not present. As for cement pastes with higher water/cement ratios of 0.4 and 0.6, the volume fractions of fine pores little changed with time. Lack in capillary pores ranging from 0.2μm to 2.6nm and the presence of considerable amounts of pores larger than 0.2μm in 91 days old cement pastes with a low water/cement ratio of 0.25 means that pore size distributions in the cement pastes may be discontinuous. The gap-graded coarse pores are found to be embedded in the C-S-H gel matrix in cement pastes with an extremely low water/cement ratio of 0.25. The characteristic of pore structure in cement pastes with an extremely low water/cement ratio is considered to be important in the mechanism of concrete deterioration in a freeze-thaw environment.

The solids and pore phase constitutions shown in Fig. 4 also represent the characteristics of deposition of hydration products under various water/cement ratios. At high water/cement ratios, enough water and spaces are available for the formation of hydration products. As hydration of cement proceeds, refinement of pores may occur in all the ranges of capillary pore sizes. As already described, reductions in large pores were accompanied by the formation of fine capillary pores and increases in solid phase. The translation of large pores to fine pores was also found in Fig.4. However, at an extremely low water/cement ratio of 0.25, amounts of free water may not be large enough for the progress of hydration. In such a cement paste, C-S-H gel was produced, but large pores were not sufficiently filled with hydration products so that gap-graded large pores might be left in the cement paste.

3.3 Effects of mineral admixtures on the degree of hydration of cement

The degrees of hydration in cement pastes with mineral admixtures are presented in Fig. 5. It is found from Fig. 5 that the degree of hydration of cement in pastes with the mineral admixtures was higher than that in admixture-free pastes at early ages. As reported by some workers,[12,13] the acceleration of hydration of cement by the addition of fly ash and silica fume was also indicated in this study. On the assumption that the volume fraction of silica fume and fly ash are assumed not to change with time, volume fractions of the phases in these cement pastes were calculated (Fig. 6). As shown in Fig. 6, differences in the fractions of hydration products between the cement pastes with and without admixture were small. However, it should be noted that the cement pastes with fly ash contained more coarse capillary pores than others at the age of 3, 7 and 28 days. However, the total amounts of coarse and fine capillary pores were almost the same in all the cement pastes. This result agrees with a generally accepted concept that low-calcium fly ashes do not react appreciably at early ages.[14] The volume fractions of phases in mortars with admixtures are presented in Fig. 7. There are no differences in the proportions of phases between pastes and mortars. The addition of sand grains did not affect phase constituents in cement paste.

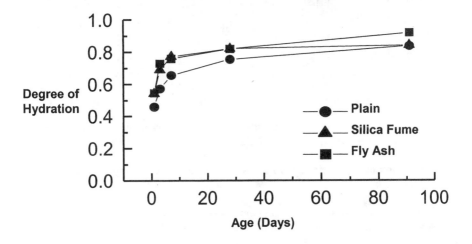

Figure 5 *Degrees of hydration in cement pastes with a mineral admixture*

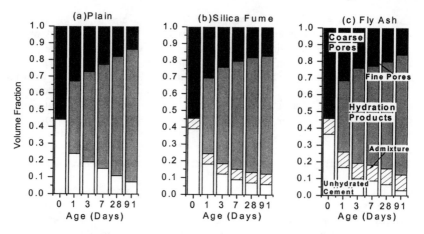

Figure 6   *Volume fractions of constituent phases in cement pastes with a mineral admixture (Coarse pores:>0.2µm Fine pores:0.2µm-2.6nm)*

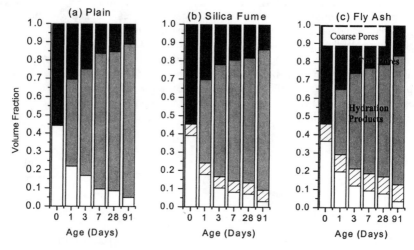

Figure 7   *Volume fractions of constituent phases in cement mortars with a mineral admixture (Coarse pores:>0.2µm Fine pores:0.2µm-2.6nm)*

3.4 Correlation of capillary pore volumes with compressive strength of cement pastes and mortars

Fig. 8 shows the relationship between compressive strength (σ_c) and coarse capillary pore porosity (x) for cement pastes and mortars. The relationship between both can be written as Equation (2).

$$\sigma_c = C_0 e^{nx} \qquad (2)$$

where, C_0 is a constant representing the strength of the matrix containing capillary pores smaller than 0.2µm. The value of n was -16.7 in this case. Correlation between both was good. However, examining the plotted data in detail, the dependence of compressive strength on the coarse capillary porosity is found to be different for different water/cement ratios. As shown by broken lines of regression in Fig. 8(b), the gradient of each regression line decreased with increasing water/cement ratio. These results suggest that the relationships between the strength and the porosity of coarse pores detected by the image analysis are liner, but that the slope of a straight line increases with decreasing water/cement ratio of cement pastes and mortars. It is also found from Fig. 8(b) that data for both cement pastes and mortars are plotted along a straight regression line.

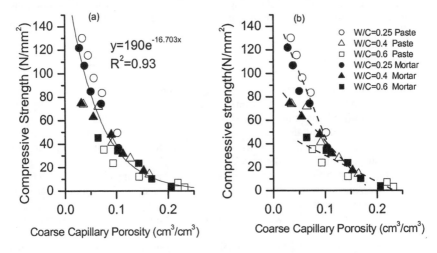

Figure 8 *Correlation between compressive strength and coarse capillary porosity (a)Non linear regression for all data (b)Linear regression for various water/cement ratios*

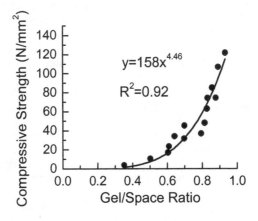

Figure 9 *Compressive strength vs. gel/space ratio for mortars without admixtures*

The gel/space ratio in the Powers-Brownyard expression[15] for estimating compressive strength of mortars can be calculated using the degree of hydration obtained by Equation (1). The relationship between compressive strength and the gel/space ratio determined by the BSE image analysis is presented in Fig. 9. It is found that the relationship between compressive strength and the gel/space ratio can be written by the same expression as Powers and Prownyard proposed.[15] However, the compressive strength of mortars at a gel/space ratio in this study was considerably smaller than that estimated by the Powers-Brownyard equation. A part of the reduction in compressive strength in mortars in this study may be due to the difference in the shape and size of specimen between both studies. Furthermore, the degree of hydration determined by the image analysis on the volume basis may give slight different values than those based on the mass proportions.[11] As suggested in Figs. 2 and 4, if the image analysis overestimates the gel/space ratio, the compressive strength at the gel/space ratio of 1.0 in the regression line may be reduced.

4 DISCUSSION

As mentioned above, the result of the image analysis for cement pastes and mortars appears to be consistent with the Powers model on the hydration of cement. Furthermore, the combination of the image analysis with the hydration model was useful for understanding significant differences in capillary pore structures in cement pastes and mortars between various water/cement ratios. Particularly, it should be noted that the cement paste with a low water/cement ratio of 0.25 had a discontinuous pore size distribution in which most of the coarse capillary pores were directly connected to gel pores. At long ages, fine capillary pores, which are defined in this study, disappeared so that relatively large pores appeared as if isolated at long ages (Fig. 4). However, in cement pastes with a relatively high water/cement ratio, considerable amounts of fine capillary pores existed even at long ages.

The presence of gap-graded large capillary pores in cement pastes with an extremely low water/cement ratio may be related to the moisture conditions in the cement pastes at early ages. The hydration of cement under such an insufficient water content as in cement pastes with an extremely low water/cement ratio leads to the reduction in relative humidity within cement paste specimens even at early ages. Water meniscus was generated in coarse capillary pores so that large empty pores must have been formed in the pastes. Hydration products by subsequent hydration of cement can grow in finer pores containing liquid water, but not in larger empty pores. The characteristic of volume fractions of phases in the cement paste with a water/cement ratio of 0.25 (Fig. 4) may suggest the occurrence of self-desiccation in the process of hydration of cement at early ages.

Effects of water/cement ratio and curing time on the pore size distributions in cement pastes have been extensively investigated in terms of the MIP method.[1,16,17,18] It has been reported that the total porosity and the threshold diameter of pores decreased in terms of cumulative curves of pore volume. As pore size distributions move toward smaller sizes, greater strengths are achieved. The same trend in changes of pore structure with time was observed in the results of the BSE image analysis. However, it should be noted that large capillary pores were still left even in the mature cement pastes with an extremely low water/cement ratio.

As shown in Fig. 4, the volume fraction of pores greater than the size of pixel ($0.2\mu m$) in the cement paste with a water/cement ratio of 0.25 was not different from that of the cement paste with a water/cement ratio of 0.40 at 91 days. However, there were great differences in strength between the cement paste with a water/cement ratio of 0.25 and 0.40 at 91 days (Fig. 8). These results show that the densification in the range of fine capillary pores greatly contributed to the increase in strength. The MIP method cannot evaluate variations in large capillary porosity. The densifications in microstructure of cement pastes in the MIP method have been misunderstood as a simple decrease of threshold diameter with time. Since the large capillary porosity do not play a significant role in differentiating pore structures in the mature cement pastes, the MIP has qualitatively succeeded in giving a reasonable interpretation for pore refinement in the cement systems.

By using the gel/space ratio calculated from the result of image analysis, the compressive strength of cement pastes and mortars could be expressed as the same function of gel/space ratio as Powers and Brownyard proposed.[15] As mentioned above, Figs.4 and 9 also appear to indicate that parts of the increase in strength resulted from the densification of microstructure in the range of fine capillary pores, in particular at long ages. This is to say, the reduction of porosity in the medium-sized pores was accompanied by the increase in the gel/space ratio. The presence of gap-graded coarse capillary pores at a water/cement ratio of 0.25 was not so crucial in the development of strength as far as those coarse pores are properly dispersed so as to be isolated. This fact suggests that spatial distribution and connectivity of large pores are important in the development of strength in concretes, but that the elimination of the gap-graded coarse capillary pores lead to an increase in the strength of the ultra-high strength concrete.

5 CONCLUSIONS

Effects of water/cement ratio and mineral admixtures on the microstructure of cement pastes and mortars were discussed by the combination of the BSE image analysis with the Powers model. Major results obtained are as follows;

(1) Considerable amounts of large capillary pores which are not correctly measured in the MIP method exist in cement pastes. These large pores decreased with time and with decreasing water/cement ratio.

(2) The degrees of hydration determined by the BSE image analysis were consistent with the Powers model for hydration of cement. The presence of fine capillary pores (0.2μm-2.6nm) predicted on the basis of the Powers model characterized the capillary pore structures. The development of strength in cement pastes appears to result from the elimination of these fine pores, leading to the increase in hydration products.

(3) At an extremely low water/cement ratio of 0.25, there were little fine capillary pores at long ages. The gap-graded coarse capillary pores in the cement paste with a low water/cement ratio of 0.25 were supposed to be connected by gel pores. At a high water/cement ratio, coarse pores were subdivided into finer capillary pores. The coarse pores left in the cement paste may be still interconnected by fine capillary pores at long ages.

(4) The presence of the gap-graded large capillary pores may impede the development of compressive strength in cement pastes with a low water/cement ratio at long ages.

(5) Good correlation between the compressive strength and the volume fractions of coarse capillary porosity was found. However, the sensitivity of the strength to the volume fractions of coarse capillary porosity depends on water/cement ratio.

(6) The relationship between compressive strength and the gel/space ratio calculated based on the results obtained by the BSE image analysis in mortars could be expressed by the same function as proposed by Powers and Brownyard.

References

1   Diamond, S., Cement and Concrete Research, 2000, **30**, 10, 1517.

2   Scrivener, K.L., Papers to be presented, but not included in the Proceedings of the 10th International Congress on the Chemistry of Cement, Gothenburg, Sweden, 1997

3   Kjellsen, K.O., Detwiler, R.J. and Gjorv, O.E., Cement and Concrete Research, 1990, **20**, 308.

4   Russ, J.C. and Dehoff, R.T., Practical Stereology, Second Edition, 2000, Kluwer Academic Press, New York, USA.

5   Scrivener, K.L. and Pratt, P.L., Proc. 6th International Conference on Cement Microscopy, New Mexico, 1984, 145.

6   Scrivener, K.L., Patel, H.H., Pratt, P.L. and Parrott, L.J., MRS Symposium Proceedings, 1987, **85**, 67.

7   Lange, D.A., Jennings, H.M. and Shah, S.P., Cement and Concrete Research, 1994, **24**, 5, 841.

8   Diamond, S. and Leeman, M.E., MRS Symposium Proceedings, 1995, **370**, 217.

9   Powers, T.C., Proceedings of the 4th International Symposium on the Chemistry of Cement, 1960, **1**, 577.

10 Kjellsen, N.O., Detwiler, R.J. and Gjorv, O.E., Cement and Concrete Research, 1991, **21**, 388.

11 Kjellsen, N.O. and Fjallberg, L., Proc. Workshop on Water in Cement and Concrete Hydration and Pore Structure, Skagen Denmark, The Nordic Concrete Federation, 1999, 85.

12 Taylor, H.F.W.: Cement Chemistry, Thomas Telford, London, 1997.

13 Wu, Z.-Q. and Young, J.F., Journal of Materials Science, 1984, **19**, 11, 3477.

14 Diamond, S., Proc. Symp. Annual Meeting of MRS, 1981, 12.

15 Powers, T.C. and Brownyard, T.L., Journal of American Concrete Institute, 1946, 43.

16 Winslow, D.N. and Diamond, S., Journal of Materials, 1970, **5**, 3, 564.

17 Cook, R.A. and Hover, K.C., Cement and Concrete Research, 1999, **29**, 6, 933.

18 Uchikawa, H., Part4, Journal of Cement and Concrete, 1987, 488, 33 (in Japanese).

MODIFICATION OF CEMENT PASTE WITH SILICA FUME—A NMR STUDY

B. Lagerblad[1], H.M. Jennings[2,3] and J.J.Chen[3]

[1]Swedish Cement and Concrete research Institute,
10044 Stockholm, Sweden.
[2]Department of Civil and Environmental Engineering,
[3]Department of Materials Science and Engineering,
Northwestern University, Evanston IL. USA

1 INTRODUCTION

1.1 Background

Cement based materials used in repositories for spent nuclear fuel. In some cases, it is desirable to add silica fume to the cement mix. This can lead to higher strengths, lower permeability, and hence greater resistance to leaching by percolating water. The main products of cement hydration, calcium silicate hydrate (C–S–H) and calcium hydroxide (CH), are significantly influenced by the presence of silica fume, especially at high dosages. However, the long-term stability and behaviour of these phases have not been fully studied. Here we investigate the structure of the C–S–H as a function of time and silica fume content using ^{29}Si Magic-Angle Spinning (MAS) NMR. Aged mortar and concrete specimens, the oldest being nearly 100 years old, were also analysed.

1.2 Cement paste

Cement paste, which is the binder of concrete and other cementitious products, is comprised of ordinary Portland cement (OPC) mixed with water. Cement paste is dominated by (C–S–H) but also contains calcium hydroxide (CH), ettringite (AFt), monosulphate (AFm) and some minor amounts of other compounds like hydrogarnet. The relative amount of each phase depends on the composition of the cement and the degree of hydration. Moreover, the cement paste is porous and contains a pore solution rich in alkalis, calcium, and hydroxide ions. The C–S–H is cryptocrystalline with a variable structure and composition, while the other components are generally crystalline. The amount of AFt and AFm is controlled by the amount of aluminate, ferrite and sulphate in the cement. A majority of the Ca is bound in CH and C–S–H, while nearly all of the Si is bound in C–S–H.

Hydration of cement occurs through a series of complex, kinetically controlled chemical reactions. Cement hydration is exothermic reaction initiating when the cement grains come in contact with water. The C–S–H and CH phases are derived from dicalcium silicate (C_2S, belite) and tricalcium silicate (C_3S, alite). During this reaction calcium, silicate, and hydroxide ions are released into the water-filled pores. During the early stages of hydration, C–S–H and CH precipitate rapidly at or close to the hydrating cement grains. A slower and different reaction follows, and continues for decades. Thus, the formation and structure of the C–S–H must be considered as a function of time. Eventually, however, the most stable C–S–H will form in equilibrium with the aqueous phase.

Silica fume (SF) is a by-product from the production of silicon or silicon alloys. When it comes from the electric furnace it is a gaseous silicon suboxide, which in contact with air oxidises and condenses to very small particles (around 100 nm) of amorphous silica. Due to its high reactivity the SF interacts with the early hydration processes. SF reacts with calcium ions provided by CH or C–S–H; at a sufficient concentration of SF, approximately 20–25%, CH can be completely consumed. The basic reactions are as follows:

$Ca(OH)_2 + SiO_2 + H_2O = C–S–H$

C–S–H (1) + SiO_2 = C–S–H (2) where the Ca/Si ratio is lower in (2)

To elucidate the long-term performance and chemical characteristics we must understand the effects of SF on the structure of the C–S–H as a function of time. Thus, we prepared a series of mortar mixes with different amounts of SF. These mixes were investigated at different ages by SEM/EDS on flat polished surfaces, X-ray diffraction and by ^{29}Si MAS NMR. Scanning electron microscope (SEM) gives textural details, and energy dispersive spectroscopy (EDS) gives chemical analysis of paste phases. X-ray diffraction identifies crystalline phases. Nuclear magnetic resonance (NMR) yields quantitative information on the local Si structure of C–S–H from which an average silicate chain length can be computed.

In addition to the above laboratory samples we have analysed edge beams made with silica fume taken from bridges that are more than 20 years old. These are compared with aged normal concretes. In an earlier work[1] we analysed concretes from old water basins by ^{29}Si MAS NMR. These concrete samples, between 40 and 90 years old, have been submerged in rainwater throughout their life. These samples have been more fully examined here.

2 MATERIALS

White cement from Aalborg Portland A/S was used for our laboratory mixes. This white cement has the lowest content of alkalis available on the commercial market and thus gives a lower pH in the pore solutions than normal OPC. Moreover, it does not contain any ferrite, which makes the NMR analysis simpler. The SF was uncompacted SF (grade 983) from ELKEM A/S.

Mortar was made from 1 kg of binder, 0.5 kg of water, and 2 kg of fine (0–0.5 mm) limestone. In the mixes we replaced 5, 10, 15, 20, 25, 30, and 40 wt. % of the OPC with SF. Higher amounts of SF demanded increased amounts of superplasticizer to get a proper flowability.

The concrete from the 20+ year old side beams were normal concretes containing granitoid aggregates with a maximum diameter of 18 mm. The SF came from ELKEM A/S.

The oldest water basins, built in 1910, 1916, 1927, were made with a fairly porous concrete (compacted with tamping) covered by a dense mortar concrete to give protection against leaching by fresh water. The younger basins (from 1944 and 1960) were built with cast concrete of good quality. The binder contains pure OPC and the aggregates were of granitoid origin. The concretes have been submerged except for one day a year when the filter sand was exchanged.

3 RESULTS

3.1 Powder X-ray diffraction analysis

Examination of the different mortars was done by powder X-ray diffraction (XRD) after 2 and 60 weeks (Table 1). The samples were cured at 20 °C under humid conditions. The intensities of the peak from CH, C_3S, and C_2S were measured to get an estimate of the relative amounts of these phases in the paste.

Table 1 *X-ray diffraction of mortars with different amounts of SF. Intensities from X-ray peaks of (CH) and (C_2S, C_3S) are shown. Values give an estimate to the relative amounts of these phases.*

Intensity of peaks after 2 weeks of humid curing at 20 °C.

Amount of SF wt. %.	5 %	10 %	15 %	20 %	25 %	30 %	40 %
CH	581	428	202	199	98	81	38
C_3S, C_2S	206	189	175	205	157	185	155

Intensity of peaks after 60 weeks of humid curing at 20 °C

Amount of SF wt. %	5 %	10 %	15 %	20 %	25 %	30 %	40 %
CH	484	586	125	12	0	0	0
C_3S, C_2S	76	188	173	79	98	72	69

These results (Table 1) shows that after 2 weeks CH was found in all samples, while the 60-week analysis showed that the CH had disappeared from all samples that contained more than 20 wt. % SF. This shows that CH forms during early ages and then reacts to form more C–S–H. Later, hydration of the remaining cement gives more CH and C–S–H.

3.2 SEM/EDS analysis

The SEM/EDS analyses were performed on carbon coated flat-polished samples. Between 10 and 20 spot analysis of C–S–H were made on homogenous cement gel in each sample. The Ca and Si measurements have been calibrated against a wollastonite standard and the values were regularly controlled against alite and belite crystals. The analyses show that the gel is inhomogeneous. Thus, it is difficult to get an accurate value. We have tried to find homogeneous C–S–H but instead found that the composition varies depending on whether it is an inner or outer gel, and on the distance from cement grains.

Analyses on the test specimen have been made after 1, 4, and 14 month (same mortar, different sample). The Ca/Si remain fairly constant through time. As expected the Ca/Si decreases with increasing amount of silica fume.

The edge beams have somewhat higher Ca/Si ratio than the test samples with similar amount of silica fume. The concrete of the edge beams is, however, fairly badly mixed with distinct inhomogenities.

In the samples from the old basin the cement paste is more homogeneous in composition.[1] This is probably because the cement paste has been water saturated for decades. The side beams are in air and thus less saturated, which slows the homogenisation process. In all the old concretes the Ca/Si of the C–S–H is somewhere near 1.6–1.8.

Table 2   *SEM–EDS analyses on C–S–H from the test specimen and the side beams. The analyses have been done on flat-polished samples cured 1, 4, and 14 months. The analyses are normalised to 100-wt % oxide, and the ratio calculated from this value.*

1 month

Cement/ SF	95 5	90 10	85 15	80 20	75 25	70 30	60 40
CaO/SiO_2	1.8	1.7	1.6	1.5	1.3	1.2	0.9

4 month

Cement/ SF	95 5	90 10	85 15	80 20	75 25	70 30	60 40
CaO/SiO_2	2.0	2.0	1.7	1.6	1.5	1.2	0.9

14 month

Cement/ SF	95 5	90 10	85 15	80 20	75 25	70 30	60 40
CaO/SiO_2	1.8	1.7	1.6	1.4	1.1	1.0	0.8

20+ year concretes

Sample	5E2	1B5	2E19	2C25
Cement/ SF	50 50	80 20	80 20	83 17
CaO/SiO_2	1.0	1.6	1.7	1.8

3.3 ^{29}Si MAS NMR Spectroscopy

^{29}Si MAS NMR (Magic Angle Spinning Nuclear Magnetic Resonance) gives information about the local silicate structure of C–S–H. The relative proportions of Q^0 (isolated tetrahedra), Q^1 (end groups or dimers), Q^2 (middle-groups), Q^3 (cross-linking), and Q^4 (network) Si sites can be determined quantitatively. Since C–S–H is based on a single chain structure, it shows only Q^1 and Q^2 sites; Q^3 sites only arise at Ca/Si ratios near 0.7–0.8[2]. Q^0 sites are present in the anhydrous silicate phases in cement.

In an earlier work one of us reported NMR analysis of concrete from water basins built in 1910.[1] Data and analytical method can be found in another study[3]. These samples were analysed again in this study. The surface of the concrete from these basins has been leached while the interior has slowly densified by the prolonged curing in water. The analysis from the porous interior (unleached, without any remaining cement, no Q^0) concrete shows that C–S–H is dominated by Q^2, dimer, and is more polymerised than the denser cover mortar (with Q^0). The outermost layer of the cover (not shown) contains C–S–H that is strongly decalcified (Ca/Si < 1) and polymerised, showing Q^3 peaks[1].

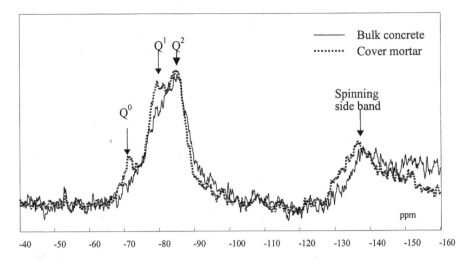

Figure 1 *^{29}Si NMR spectra of 1910 cement paste from Norsborg water basins. The basins are made of porous concrete covered by a dense mortar.*

The ^{29}Si solid state MAS NMR spectra were collected on a Varian VXR300 (7 T) spectrometer operating with a magnetic field of 59.578 MHz and a spinning rate of 4 kHz. A 5-mm zirconia rotor with Aurum caps was used for all experiments. Samples were collected for 800 scans, each scan consisting of a single pulse of width 4.7 μs followed by a delay of 10 seconds. Chemical shifts are reported relative to tetramethylsilane (TMS). It is assumed that Q^3 is not present in the C–S–H of the above preparations; hence, the mean chain length is calculated as $2(Q^1+Q^2)/Q^1$.

Mortar and concrete samples were lightly crushed into a fine powder before testing by ^{29}Si MAS NMR. To remove some of the larger aggregate particles in the concrete samples, several crushed samples (1B5, 2E19, 2C25) were passed through a 600 μm sieve before testing. The concrete samples cast in 1910, 1916, 1927, 1944, and 1960 were passed through a 75μm sieve.

The results from the laboratory mortar samples (Table 3, Fig. 2) show that the amount of Q^2 and mean chain length increases with the amount of SF. In the concretes with 5 and 15 wt % SF the mean chain length is similar at 40 days and 470 days, while the samples with more SF and no CH exhibited a distinctly increasing mean chain length with time. The degree of hydration increases with the amount of SF. This confirms the result from the semi-quantitative X-ray diffraction (Table 1).

The old concrete from the side beams with SF has similar or somewhat longer chain lengths compared to the laboratory mortars. The sample with 50 wt % SF, sample 5E2, is highly polymerised. Analyses from the ordinary concrete in the old water basins show that the mean chain length of the C–S–H ranges from 4.3 to 5.4 (Table 3), suggesting an average near 5. This indicates that C–S–H in ordinary paste polymerises with time.

Table 3 *Summary of ^{29}Si MAS NMR results. Degree of cement hydration comes from $(Q^1+Q^2)/(Q^0+Q^1+Q^2)$. Since Q^0 reflects the proportion of unhydrated cement, a low value indicates a low degree of hydration while a value of 1 indicates full hydration. The mean chain length is calculated from $2(Q^1+Q^2)/Q^1$. The uncertainty is +/- 6 %.*

Laboratory specimen

Silica fume (%)	Age days	Degree of hyd.	Mean chain length	Age days	Degree of hydr.	Mean chain length	Age days	Degree of hyd.	Mean chain length
5	40	0.70	3.5	167	0.79	3.7	470	0.81	3.2
15	83	0.76	3.5	167	0.78	3.9	470	0.83	3.6
25	85	0.80	3.8	167	0.79	4.6	470	0.89	5.2
40	40	0.82	7.1	172	0.89	12.4	470	0.92	9.3

Old concrete from water dams and 20+ years old side beams

Cased year 100% OPC	Degree of hyd.	Mean chain length	Side-beams. Cast around 1980	Degree of hyd.	Mean chain length
1910	1.0	5.3	5E2, 50% SF	1.0	Long, no Q^1
1916	1.0	4.3	1B5, 20% SF	0.78	4.7
1927	1.0	5.4	2E19, 20% SF	0.82	4.6
1944	1.0	5.2	2C25, 17% SF	0.80	5.3
1960	1.0	4.7			

4 DISCUSSION

Silica fume modifies the hydration of ordinary Portland cement and particularly effects the composition and structure of C–S–H. As expected, this NMR study showed that increasing amounts of SF decreases the Ca/Si of C–S–H and increases the mean silicate chain length. This trend between Ca/Si ratio and mean chain length is a general rule for C–S–H.[2,4,5] Additional new insights, described below, are also drawn from the data.

The data in Table 3 show that the rate of polymerisation is also affected by the content of SF in the cement paste. For example, the mortar samples containing low amounts of SF, the 5 and 15% samples, show only small increases in mean silicate chain length between 40 and 470 days. However, the samples containing high amounts of SF, the 25 and 40% SF samples, show an appreciable increase in mean chain length during this time. Hence, the mechanism of silicate polymerisation in C–S–H appears to be facilitated at high SF contents, and thus lower Ca/Si ratios. The presence of CH in the 5 and 15% SF samples may also play a role in these trends.

The NMR data on the old concrete and mortar specimens in this study are valuable not only because data on such samples are rare, but also because information of this kind can give insight into the silicate structure of C–S–H. All the data in Table 3 for samples containing less than 20% SF (i.e. those that presumably still contain CH) suggest that the mean silicate chain length in C–S–H is limited to a maximum of about 5. This value agrees with data on old C_3S pastes. Rodger et al.[6], for instance, show by ^{29}Si MAS NMR that the mean chain of a 26 year old C_3S paste is 4.8. Mohan and Taylor[7] show by trimethylsilation

(TMS) that the linear pentameric species accounts for a majority of the silicate polymers in C_3S pastes aged 22–30 years.

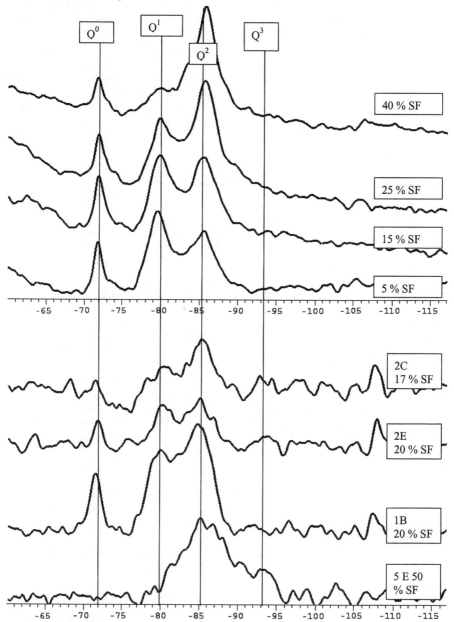

Figure 2 ^{29}Si *MAS NMR spectra for 6-month mortar (Tab. 3) and for 20+ years old concrete from bridge side beams.*

These findings suggest that the mode of silicate polymerisation in concrete is not significantly different from that in C_3S. Moreover, contrary to intuition, the mean chain length in C–S–H does not approach an infinite length even after nearly 100 years of hydration. All available evidence, including the SEM-EDS data in Table 2, also suggests that the Ca/Si ratio in C–S–H remains constant at 1.7–1.8 with time.

From the above evidence, there appears to be two ways by which polymerisation can occur in C–S–H. From an energetic standpoint, the easier way seems to be by removing Ca from C–S–H; this can occur through addition of SF, as in the present case, or by leaching with water or acid.[2,5] The Ca/Si ratio of C–S–H decreases by this method. In the other method, the Ca/Si ratio remains constant as the mean chain length increases. This mechanism, which occurs in the old concrete and mortar samples, presumably involves more complicated rearrangements of the silicate chains, thereby producing Ca–OH bonds in C–S–H, and a presumed transition to a jennite-like structure.[5,8] The maximum mean chain length of 5 seen in the old samples appears to be explained by the fact that, at a Ca/Si ratio of 1.8, a jennite-like C–S–H can only have a mean chain length of. 5.[9] With high amounts of silica fume the Ca do not hinder further polymerisation. With more than 40 wt. % silica fume in the mix long chains dominated by Q^2 appear. This C-S-H has a Ca/Si ratio of close to 1 and probably has a structure similar to a defect 1.4 nmTobermorite.[2]

5 CONCLUSIONS

Silica fume (SF) has a significant influence on the Ca/Si ratio and silicate structure of C–S–H. With increasing concentration of SF, the Ca/Si ratio of C–S–H decreases and the mean chain length increases. At concentrations above 20–25% SF (i.e. when CH is fully consumed), the rate of polymerisation also increases. From ^{29}Si MAS NMR studies of old concrete specimens, the earliest dating back to 1910, it is shown that the maximum mean chain length in C–S–H in equilibrium with CH is about 5. This finding is similar to results of old C_3S pastes, and was interpreted as reflecting the intrinsic tendency for C–S–H of high Ca/Si to move towards a jennite-like structure. With larger amounts of silica fume the lower Ca/Si ratio in the C-S-H the will allow a more polymerised silica chain and a structure closer to a defect 1.4 nm Tobermorite crystal. Thus concrete with 25-50 wt % of the binder replaced by silica fume will give a C-S-H with long-term stability.

Acknowledgement

This work based on material from a project named "Qualification of Low Alkali Cementitious Products in the Deep Repository" co-financed by the Nuclear Waste Management Organisations of Finland (POSIVA), Japan (NUMO) and Sweden (SKB). We also want to thank the Danish Road Authorities, COWI Consulting and ELKEM for the access to the samples from the bridge side beams and Stockholm Water Works for the access to the old water basins.

References

1   B. Lagerblad, Swedish Nuclear Fuel and Waste Management Co. SKB TR-**01-27**, 2001.
2   X. Cong R., J Kirkpatrick,. Adv. Cem. Based Mat., 1996, **3**, 144.
3   C. Porteneuve, PhD Thesis, Universite Paris VI, Sciences des Matériaux, 2001.
4   H. Matsuyama, J. F. Young, Adv. Cem. Res., 2000, **12**, No. 1, 29.

5   J.J. Chen, J.J. Thomas, H.F.W. Taylor, and H.M. Jennings, submitted to J. Phys. Chem. B, 2003.

6   S.A. Rodger, G.W. Groves, N.J. Clayden, and C.M. Dobson, Mater. Res. Soc. Symp. Proc., 1987, **85**, 13.

7   K. Mohan and H.F.W. Taylor, Cem. Concr. Res., 1982, **12**, 25.

8   H.F.W. Taylor, Cement Chemistry, 2nd Ed., Thomas Telford, London, 1997.

9   H.F.W. Taylor, J. Am. Ceram. Soc., 1986, **69**, 464.

Part 3 : Modelling

MODELLING AND TEMPERATURE DEPENDENCE OF MICROSTRUCTURE FORMATION IN CEMENT-BASED MATERIALS

T. Kishi and K. Ito

Institute of Industrial Science, The University of Tokyo, Komaba 4-6-1, Meguro-ku, Tokyo 153-8505, JAPAN

1 INTRODUCTION

In the early age of hardening concrete hydration of cement, microstructure formation and moisture state change and movement simultaneously take place under their mutual dependencies. To predict property development of concrete and establish a life span simulation scheme for concrete structures the modelling of hardening process of concrete is important as a core platform. Especially, micro pore structure which varies from nm to mm orders seems to be the most substantial feature of cement-based materials in a view point of nanoscience. In this paper an integrated computational model for simulating property development of young concrete is briefly introduced in the former part. To enhance the applicability and reliability of this technology a wider versatility of materials, such as various cement-based binders and mix proportions of concrete should be coped with and the accuracy of prediction should be guaranteed not only at standard curing condition but also at high temperature and/or dried condition. Regarding to mix proportion its applicability to concrete of low water to cement ratio (W/C, hereafter), such as high strength concrete and self-compacting concrete is especially important from the recent issue of application. Thus, in this paper a feature of microstructure formation of low W/C concrete is investigated and a rational way for further modelling is discussed.

2 AN INTEGRATED COMPUTATIOAL PROGRAM - DuCOM -[1]

To simulate varying concrete performance based on microscopic modelling of cement hydration, microstructure formation, and mass transport the integrated computational program so-called DuCOM has been established and it is still being developed. Since details of this technology are introduced in a reference book[1] only its outline is introduced in this section.

2.1 Multi-component Hydration Heat Model

To cope with versatile type of Portland cement and powder admixtures the multi-component modelling is adopted for hydration heat of cement-based materials. A modified Arrhenius' law is applied to calculation of heat generation rate of each compound. The

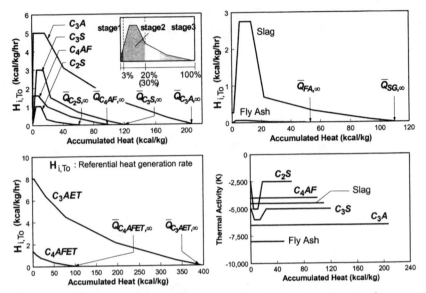

Figure 1 *Assumed reference heat generation rates and thermal activities of cement clinkers and pozzolans.[1] The reference heat rate is set at 293°K. Ettringite reaction that precedes main hydrations is also implemented.*

heat generation rate of each component is expressed by equation (1) in which several functions for the mutual interactions are implemented.

$$H_i = \gamma \cdot \beta_i \cdot \lambda \cdot \mu \cdot s_i \cdot H_{i,T_0}(Q_i) \exp\left\{-\frac{E_i}{R}\left(\frac{1}{T} - \frac{1}{T_0}\right)\right\} \qquad (1)$$

Where Q_i is the accumulated heat and E_i is the activation energy of component i; R is the gas constant; $H_{i,T0}$ is the reference heat generation rate of component i at constant reference temperature T_0 and is a function of the accumulated heat; Coefficients, γ, β_i, λ, μ and s_i are coefficients expressing mutual interactions among compounds and so on (β_i is a coefficient expressing the reduction in heat generation rate due to the reduced availability of free water (and precipitation space) which is discussed later). $-E_i/R$ is defined as thermal activity. The reference heat rate and the thermal activity are set as shown in Figure 1.

This model can be applied for binary and ternary blending binders among Portland cement, blast furnace slag and fly ash. In this modelling the amount of calcium hydroxide is adopted as an important state variable, which is produced by cement hydration and react with pozzolanic admixtures as a catalyser. For inert powder admixtures like limestone powder the filler effect, which accelerates cement hydration is also taken into account. As the micro filler effect a thickness of diffusion layer formed around remaining unhydrated part of cement is mitigated since the surface of inert powders plays roll as additional precipitation site for a part of hydrates. For versatile types of Portland cement such as Early-hardening, Ordinary, Moderate heat and Low heat cements the reference heat rate of clinker mineral is modified by the ratio of mass fraction between C_3S and C_2S.

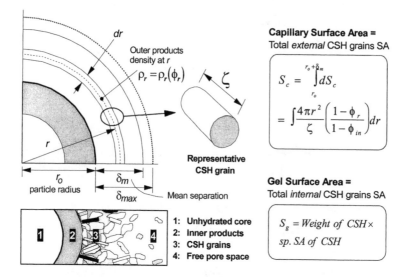

Figure 2 *A reactant particle at arbitrary stage of hydration.[1] The surface area to volume ratio of gel grain ζ is assumed and then total surface area of capillary pores is computed. ϕ_{in} is the characteristic porosity of the inner layer set constant.*

2.2 Microstructure Formation Model and Moisture Movement

Characteristic of microstructure in cement-based materials is very important for not only moisture conductivity but also moisture capacity. In this system three categories of pores, such as capillary, gel and interlayer pores are taken into account. The most important feature of this modelling is to compute not only those porosities but also total surface area of pores, so that the pore size distribution and its representative size are obtained. Figure 2 shows the schematic representation of microstructure formation at inner and outer layers around the original cement particle.

Based on the pore size distribution moisture capacity can be computed in accordance with thermodynamic equilibrium under arbitrary relative humidity (RH, hereafter). The amount of water condensed in pore structures due to thermodynamic equilibrium is available for further hydration as free water. A phenomenon of self-desiccation, which would be a driving force for autogeneous shrinkage, can be automatically simulated by taking the change of specific gravity of chemically combined water into account. Further, hysteresis phenomenon of saturation in wetting and drying processes is expressed by taking the ink-bottle effect into account based on a simple assumption of pores connectivity.

Based on the pore size distribution moisture conductivities both in vapour and liquid phases in the porous media are estimated and the pressure based formulation is derived for moisture transport analysis. Chemically combined water, which should be excluded from the mass conservation of moisture as a sink term, is computed from chemical equation and degree of hydration of each reaction.

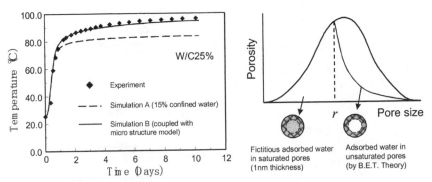

Figure 3 *Comparison between simulations with experimental result of adiabatic temperature rise of low W/C concrete (Left) and tentative definition of confined water in simulation B (Right).*[3]

3 FEATURES OF LOW W/C CONCRETE IN ADIABATIC TEMPERATURE RISE AND DISCUSSIONS

The computational program DuCOM is continuously extended to integrate other physico-chemical phenomena into the scheme, such as chloride ion transport, carbonation and corrosion of reinforcement with chemical ion equilibriums in the solution.[2] Simultaneously it is necessary to pursue more rational modelling of core platform for microstructure formation and hydration to enhance its accuracy and applicability. Recently, concrete of low W/C mix proportion, such as high strength concrete and self-compacting concrete are being widely used. An application of this model to those low W/C concrete is one of important issues. Here, this model is applied to simulate the adiabatic temperature rise of low W/C concrete and then the rationality of modelling is discussed from the microscopic point of view. Though the model is expected to cover a wide range of temperature it is not yet verified, especially in low W/C ratio case.

An adiabatic temperature rise of concrete of 25% W/C ratio was measured by a very accurate apparatus. In Figure 3 the experimental result and two simulations are shown. The difference of two simulations is due to the estimation of the physically confined water which is not able to be used for further hydration. In the simulation the amount of free water is being monitored according to hydration, which is computed by subtracting chemically combined water for hydration and the physically confined water in pore structure from the initial mixed water. Then, the stagnation of hydration rate due to shortage of remaining free water is simulated with the following equation,

$$\beta_i = 1 - \exp\left\{-r\left\{\left(\frac{\omega_{free}}{100 \cdot \eta_i}\right)\middle/ s_i^{\frac{1}{2}}\right\}^s\right\}\qquad(2)$$

Where, r and s are material constants common for all mineral compounds. The coefficient β_i represents the reduction of heat rate and is simply formulated in terms of both amount of free water and the thickness of internal hydrates layer. The function adopted varies from 0

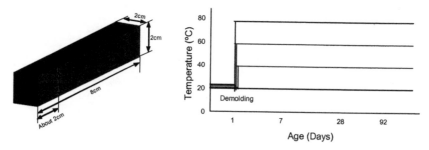

Figure 4 *Size of cement paste specimen (Left) and Curing temperature patterns (Right) for Mercury Intrusion Prosimetry. Formwork is removed at 1 day and specimen is wrapped by aluminium adhesive sheet. Then sealed curing is applied.*

to 1 and it sharply reduces the heat rate when the amount of free water is reduced and unhydrated particles are covered by thick clusters of hydrates.

In Simulation A, which underestimates the ultimate temperature of the adiabatic temperature rise, the constant 15% of confined water to hydrated compounds is assumed. Though enough accuracy of this simulation has been verified for the conventional W/C concrete it is impossible to adjust the computational result to the experimental one by modifying the function shown in equation (2). That is because the amount of free water is substantially not enough at the ultimate stage in this simulation. In a case of adiabatic temperature rise of low W/C concrete it seems that 15% confined water to hydrated compounds is overestimated and it is not rational to give constant ratio.

In the other Simulation B, whose analytical result coincides with the experimental one, the confined water is estimated by a tentative definition based on the pore structure computed. As an attempt, a) water within 1nm thickness on the surface of hydration products in saturated pores, b) adsorbed water on the surface of unsaturated pores and c) water trapped in interlayer pores are assumed as confined water. This definition of the confined water in Simulation B is also illustrated in Figure 3. Thus, the amount of free water is given by subtracting aforementioned confined water a) from thermodynamically condensed water in pore structure. The feature of this treatment, which differs from the former simulation, is that the amount of confined water is computed in accordance with porosity change and thermodynamic equilibrium. As a result a part of water treated as confined water in Simulation A is released and then can be consumed for additional hydration in Simulation B. With these treatments in the model the extremely slow but continuous temperature rise and the ultimate temperature rise of low W/C concrete could be satisfactorily reproduced though the adequacy of this assumption is not yet verified.[3]

4 TEMPERATURE DEPENDENCE OF PORE STRUCTURE FORMATION

The pore structure formation of low W/C concrete, which shows a very unique continuous adiabatic temperature rise as shown in Figure 3 and then be cured at high temperature, is thought to be very interesting. Then, the pore structure formation and the degree of hydration of low W/C paste cured at various different temperatures with sealed condition are systematically investigated in this study. Test pieces are subjected to oven dry and microstructure is measured by the mercury intrusion porosimetry. Ordinary Portland

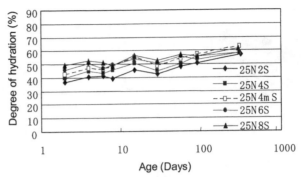

Figure 5 *Degree of hydration of cement paste specimen at various curing temperatures.*
Symbol "25N2S", for example, indicates that "25": W/C 25%, "N": Normal
ordinary Portland cement used, "2": Curing temperature 20°C and "S":
Sealed curing condition.

cement is used and water to cement ratio is set at 25%. Curing temperature patterns and
specimen size are shown in Figure 4 and sealed curing is applied.

4.1 Degree of Hydration with Time at Various Curing Temperatures

Figure 5 shows that degree of hydration, which is computed from weight loss due to water
evaporation in oven drying, continuously increases up to over 300 days at each curing
temperatures. The temperature dependence on degree of hydration is verified. Hydration of
cement consistently continues without remarkable stagnation even at room temperature
though its rate is not so fast. Further, though it is recognized that once concrete experiences
curing at high temperature the following hydration is stagnated, this kind of tendency is not
observed when continuous high temperature curing is applied. From this result it is
confirmed that degrees of hydration are not so different between concrete cured at room
temperature in long time range and concrete cured at high temperature for several weeks.

4.2 Features of Microstructure Development at Various Curing Temperatures

The cumulated porosity and pore size distribution of hardened cement paste cured at
various temperatures in terms of the elapsed time are shown in Figure 6. Though the
mercury intrusion porosimetry can not perfectly capture the actual pore structure due to
geometrical connectivity of pores, some remarkable tendencies can be obtained. In this
figure the following clear features about the pore structure formation are seen in terms of
curing temperature.

In 20 and 40°C curing cases the reflecting point of pore size distribution reaches to
around 50nm diameter at 1 day before temperature is elevated (shown as -0day in Figure
6). Then, the peak of pore size is fixed at same position and does not become smaller any
more even though the degree of hydration is increased. This feature of microstructure
formation at room temperature seems to be very unique since it is natural to suppose that
the representative pore size continuously becomes smaller according to degree of
hydration. On the other hand, in 60 and 80°C curing cases the reflecting point of pore size
distribution continues to become smaller up to around 10nm diameter. Here, it must be
reminded that the degrees of hydration in those cases are not so different from those of 20

and 40°C curing cases as stated before. Thus, it must be recognized that there is a clear and characteristic temperature dependence of microstructure formation of cement paste matrix. In other words it can be said that the ultimate pore size, to which capillary pores can reach has a clear temperature dependence. Further, it is noted that this change of pore size

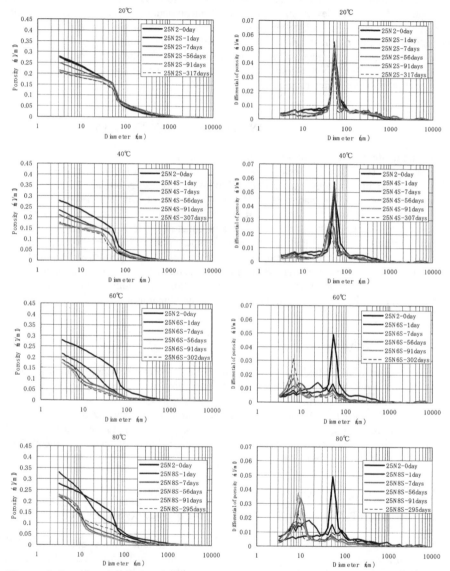

Figure 6 *Porosity of cement paste specimen cured at various curing temperatures at the stipulated ages up to 300 days (Left) and Its pore size distribution (Right). "-*days"in symbol means age after the elevated temperature curing starts. Thus, "0day" is corresponding to 1day after casting.*

It is supposed that this temperature dependence of pore structure formation is related to size of hydration products or surface energy of pores, which may govern formation's possibility of hydrates in ultimate fine pores. This continuous micro structure formation at high temperature, which reaches to around 10 nm peak size of pore diameter, seems to be associated with the continuous adiabatic temperature rise with excess increment as shown in Figure 3. For further development this kind of phenomenon should be implemented into the model to make it faithfully applicable to versatile mix proportion and temperature condition. Additionally, it is also observed at 80 °C curing condition that the porosity of finer pores is inversely increased according to the elapsed time. This may indicate the existence of another phenomenon, which makes microstructure loser at the higher temperature in the long time range.

5 DISCUSSION ABOUT INTERNAL RH IN MICROSTRUCTURE

In the model the pore structure formation is simulated and water state equilibrium is taken into account. Thus, the RH change is also computed according to hydration. In the simulation of adiabatic temperature rise of low W/C concrete stated at Chapter 3 RH is drastically reduced lower than 50% at last. If a situation of low W/C concrete at room temperature is computed the internal RH is similarly reduced in long term range at last. However, it is reported that the internal RH in concrete of low W/C around 25% still remains around 75% even if it is cured at room temperature for 450 days.[4] Though there is no report about RH in adiabatic temperature rise of low W/C concrete, an inconsistency seems to exist about internal RH between current computation and measurement at this moment. Thus, further study is requested on the relationship among degree of hydration, microstructure formation, state equilibrium of water and the internal RH.

6 CONCLUSIONS

1) To simulate a featured adiabatic temperature rise of Low W/C concrete the estimation of water availability based on the microstructure development and water phase equilibrium is important.
2) Low W/C concrete shows quite different pore structure formations between at room temperature curing and high temperature one. It seems to be necessary that temperature dependence of microstructure formation should be taken into account for more versatile modelling.

References

1 K. Maekawa, R.P. Chaube and T. Kishi, *Modelling of concrete performance*, E&FN SPON, 1999.
2 K. Maekawa and T. Ishida, Modeling of structural performances under coupled environmental and weather actions, Materials and Structures, 35, 591-602, 2002.
3 T. Kishi, T. Ishida and K. Maekawa, Interaction of hydration heat of low water to cement ratio concrete with moisture equilibrium in solid micro-pore, Proc. of Japan Society of Civil Engineers, No.690/V-53, pp.45-54, 2001. (in Japanese)
4 B. Persson: Moisture in concrete subjected to different kinds of curing, Materials and Structures, Vol.30, pp.533-544, 1997.

NUMERICAL MODELLING OF VOLUME CHANGES IN CEMENT-BASED SYSTEMS AT EARLY AGES

K. van Breugel, Ye Guang, E.A.B. Koenders

Delft University of Technology, Faculty of Civil Engineering and Geosciences, Stevinweg 1, 2600 GA Delft, The Netherlands

1 INTRODUCTION

The Greek philosopher Democritus (465-370), who is often considered as one of the precursors of modern chemistry, assumed that attempts to understand the basic properties of matter would finally lead us to adoption of inseparable entities. He called these entities atoms. At that time he did speculate about the size of those atoms. His assumptions show, however, that throughout the centuries people have been looking for a reference from which on they could explain phenomena observed in the real world. For building materials, like concrete, the reference scale used by concrete technologists ranges between tenth of millimetres, the typical size of the smallest sand particles, and centimetres, the typical size of aggregate particles. Structural engineers, on their turn, compose high-rise buildings of hundreds of meters with concrete elements with cross sections in the centimetre range.

Decades of joint research of concrete technologists and structural engineers have finally made it possible to determine the structural response of an high rise building to any type of loading on the basis of the knowledge of the concrete properties and the underlying mixture composition. With these analyses concrete technologists and structural designers span a length scale from tenth of millimetres to hundreds of meters, i.e. six orders of magnitude. Scaling down another six orders of magnitude brings us from the millimetre scale to the nanoscale. Although the laws that are in force at the nanoscale are different from those, which act on the millimetre scale, there is similarity in the challenge to bridge the gap of six orders of magnitude from the nanoscale to the millimetre scale, like civil engineers span the six orders of magnitude when designing their buildings.

In fact most of the comprehensive experimental en theoretical research on cement-based systems performed by Powers[1] in the forties and fifties of last century was related to mechanisms and processes that are operational on the nano- and microscale. Also with the models for cement-based materials proposed by Feldmann et.al[2] and Wittmann[3] the nanoscale is addressed. The nanoscale is considered explicitly in models for explaining volume changes in hardening and hardened cement-pastes[4]. However, the mere fact that in past research on cement-based systems the nanoscale has been addressed frequently has never been a reason to call this type of research nanotechnology. It seems that the fact that today nanotechnology has become a solid vehicle to generate research money, and probably also an ultimate attempt to get rid of the low-tech image of research on building materials, are the main reasons to join the nanotechnology hype. But this is not all that can be said. On second sight there is more than just joining a modish trend. The challenge of nanotechnology is to intervene in and to manipulate the nanostructure of cement-based materials in order to improve the performance of traditional materials and to develop new advanced materials. The techniques required to reach this aim have to cover a range of length scales of six orders of magnitude. Multi-scale modelling seems the appropriate approach the span this range of length scales. In this contribution the focus of attention will be on modelling of the microstructure and volume changes of cement-based systems during the hydration process. Where appropriate reference will also be made to the potential, or even the need, for modelling at the nanoscale.

2 VOLUME CHANGES IN CEMENT-BASED SYSTEMS AT EARLY AGE

2.1 Types of volume changes at early age

The increasing use of low water-binder ratio concrete mixtures has generated a revival of the interest of researchers and practitioners in volume changes of hardening cement-based systems. The primary and inherent cause of shrinkage of hardening cement pastes is chemical shrinkage. This type of shrinkage is well known, but not much "feared" by structural engineers. Strongly related to chemical shrinkage is the autogenous shrinkage. This type of shrinkage is typical for low water-binder ratio mixtures. Apart from shrinkage some mixtures exhibit swelling at early ages. The simultaneously operating mechanisms responsible for shrinkage and swelling complicate the quantification of the individual phenomena and hence also the modelling of volume changes at early ages

2.1.1 Chemical shrinkage. Chemical shrinkage is the volume reduction observed when cement reacts with water. The volume of the hydration products is about 7% less than that of the cement and water of which these products are formed. This volume reduction is considered proportional to the degree of hydration and can even be used as a measure for the latter. Only in the very early stage of hydration the external volume change will correspond to chemical shrinkage. When the spatial network of hydration products has got some strength, the chemical shrinkage will result in capillary pores, partly filled with water and partly empty.

2.1.2 Autogenous shrinkage. With progress of the hydration process the pore systems gradually dries out. The remaining water will accumulate in the narrow pores and is also partly adsorbed at the surface of the empty pores (Figure 1). The thickness of the adsorption layer depends on the relative humidity (RH) in the pores system[5]. Going from 0% to 100% RH the thickness of the adsorption layer increases from 0.3 nm (mono-molecular water film) to 18 nm. In low water-binder ratio mixtures the relative humidity can reach values as low as 60% – 70%[4]. The drop in relative humidity is accompanied by capillary tension in the pore water. The tensile stresses are equilibrated by compressive stresses in the solid skeleton of hydration products[1]. The generated compressive stresses in this skeleton are accompanied by deformations of it. These deformations may consist of an elastic and time-dependent part[1,6,7].

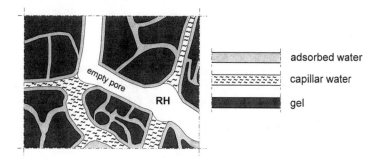

Figure 1 *Schematic representation of pore structure and state of water*

2.1.3 Swelling. Swelling at very early ages has been observed and studied by, among others, Steinour[8], Budnikov[9] and Tezuka et al[10]. For cement pastes with a water-cement ratio of about 0.55 an expansion before final set of 0.09...2.1% has been observed. In low water-cement ratio pastes the volume expansion might reach up to 2.7%[8]. According to Budnikov the conversion of a cement particle in thousands of new hydration particles through topochemical reactions would create a "volume hydration effect". This effect refers to the phenomenon that the agglomeration of small hydration products occupies a larger volume than that of the original cement particle from which the hydration products were formed. According to Budnikov this hydration dispersion pressure had not been investigated quantitatively very much. What is known for sure is that macroscopic expansion of hydrating pastes at early ages has been observed indeed, whereby the expansion appears to depend very much on the type of cement, the water-cement ratio and the presence of admixtures. According to Tezuka et al. ettringite formation would cause early expansion of pastes made with cement and calcium sulphate. The more ettringite was observed in the paste, the larger the expansion.

2.2 Quantitative models for early-age shrinkage

In quantitative models dealing with volume changes at early ages, swelling at very early ages is generally not considered. A major reason for this is that this expansion occurs in the stage that the stiffness is still very low. Large stresses are, therefore, not to be expected. Autogenous shrinkage is considered to be of much greater importance and will be focused on in more detail here. The starting point for modelling of volume changes in *hardening* cement-based systems is the extensive research done in the past on volume changes caused by moisture changes in *hardened* cement pastes by Powers[1]. He suggested three mean mechanisms that could explain these volume changes, viz. changes in the surface tension of the solid, disjoining pressure and hydrostatic tension.

2.2.1 Changes in surface tension. Due to the presence of surface tension γ_s a solid colloidal particle with radius r will experience a compressive stress P [MPa]:

$$P = \frac{2\,\gamma_s}{r} \tag{1}$$

Adsorption of a water layer will change the surface tension in the solid particle by an amount $\Delta\gamma$. According to Gibbs[11] the change in surface tension $\Delta\gamma$ is related to the change in vapour pressure. For a change in relative vapour pressure from h_1 to h_2 it holds:

$$\Delta\gamma = RT \int_{h_1}^{h_2} \frac{w_a}{V_s} \, d \ln h \tag{2}$$

where R is the universal gas constant, T the temperature in [K], w_a the amount of water adsorbed at the solid particles, V_s the volume to the adsorbent and h the relative vapour pressure. For the volume change of a solid particle the following equation was derived:

$$\frac{\Delta V}{V} = \frac{2}{3} K \frac{RT}{M} \int_{h_1}^{h_2} \frac{w_a}{V_s} \, d \ln h \tag{3}$$

where K is compressibility coefficient of the solid particle, M the molecular weight of the adsorbate, i.e. the water.

It has been illustrated[12] that changes in the RH from 1% to 100% can result in an elastic expansion of gel particles with a radius of 10 μm of about 0.23 10^{-3}. Wittmann[3] emphasised that the correlation between length changes on the one hand and changes of the surface tension on the other hand is of a semi-phenomenological type. These length changes could be approximated with Bangham's equation[13]:

$$\frac{\Delta L}{L} = \lambda \, \Delta\gamma \tag{4}$$

with the proportionality factor λ after Hiller[14]:

$$\lambda = \frac{\Sigma \rho}{3 E} \tag{5}$$

In this equation Σ the internal surface area of the material, ρ the specific mass and E the modulus of elasticity of the material.

2.2.2 Disjoining pressure. A second mechanism that could cause volume changes is the disjoining pressure. This disjoining pressure is active in areas of hindered adsorption between gel particles. The disjoining pressure has its maximum at full saturation of the system ($h_0 = 1$). When the humidity drops to $h < h_0$ the change in disjoining pressure ΔP_d is calculated with:

$$\Delta P_d = \frac{R T}{M \upsilon_w} \ln h \tag{6}$$

with υ_w the specific volume of the adsorbed water. For the volume changes caused by disjoining pressure for a drop in the relative vapour pressure from a maximum value $h_0 = 1$ to $h < h_0$ Powers proposed the equation:

$$\left(\frac{\Delta V}{V}\right)_d = \beta \, f(w_a) \frac{R\,T}{M\,\upsilon_w} \ln h$$

(7)

In this equation β is the compressibility coefficient of the specimen, $f(w_a)$ the fraction of internal surface area over which the disjoining pressure is acting and is considered a function of the adsorbed water content w_a.

By assuming that the simple relationship $f(w_a) = k(w_a/V_s)$ would hold, where k is a proportionality constant depending on the internal geometry of the system, equation (7) can be combined with equation (3). For the humidity range from 0 to h the volume change can then be represented with the equation:

$$\frac{\Delta V}{V} = \lambda \int_0^h \frac{w_a}{V_s} \, d \ln h$$

(8)

with:

$$\lambda = \left(\frac{2}{3}K + \frac{k\,\beta}{\upsilon_w}\right) \frac{R\,T}{M}$$

(9)

2.2.3 Hydrostatic tension in capillary water. For the volume changes caused by hydrostatic tension in the capillary water Powers proposed the following equation:

$$\frac{\Delta V}{V} = \beta \left(\frac{R\,T}{M\,\upsilon_w} \ln h\right)$$

(10)

where β is the compressibility coefficient of the specimen. This equation is similar to equation (8) derived for quantification of the volume changes caused by the combined effect of changes in surface tension and disjoining pressure.

2.2.4 Evaluation of proposed models in view of modelling. A comparison of the equations for volume and length changes reveals that all these equations have a similar shape. With the Gibbs equation (2) the equations (8) and (9) can be rewritten as:

$$\frac{\Delta V}{V} = \lambda^* \, \Delta\gamma$$

(11)

with:

$$\lambda^* = \left(\frac{2}{3}K + \frac{k\,\beta}{v_w}\right) \frac{1}{M}$$

(12)

In the latter expression the factor k was defined as a proportionality constant depending on the internal geometry of the system and β as the compressibility coefficient of the system as a whole. Both k and β can be written as functions of the degree of hydration α.

Particularly the factor $k\{\alpha\}$ is of great interest, since this factor is strongly correlated with the properties of the microstructure. Apart from a function of the degree of hydration, this factor also depends on the water-cement ratio, the particle size distribution of the cement and the curing temperature. Hence, $k = k\{w/c, T, psd, \alpha\}$. In equation (2) also the factors w_a and h will change with progress of the hydration process. From these observations it can be inferred that for modelling of the volume changes in hardening cement-based systems quantification of the evolution of the microstructure, and even the nanostructure, for example when the effect of adsorption layers on volume changes has to be considered, with progress of the hydration process is essential.

3 NUMERICAL MODELLING OF HYDRATION AND MICROSTRUCTURE

3.1 The concept of the HYMOSTRUC model

In the past models for hydrating cement-based systems often focused on either the hydration kinetics or the microstructure. Kinetic hydration models were very much chemistry-oriented. Microstructural models, on their turn, were dominated by physical and stereological considerations. In fact hydration kinetics cannot be considered indepen-dently from the formation of the microstructure[16]. The formation of interparticle contacts, which is the basis of the development of the microstructure, affects the rate of hydration of individual cement particles, particularly in the diffusion-controlled stage of the hydration process.

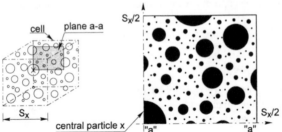

Figure 2 *Cell concept adopted in simulation program HYMOSTRUC[16,17]*

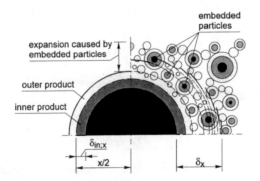

Figure 3 *Growth mechanism for central particles with encapsulating smaller particles in the outer shell of the central particle [17].*

In the last two decades a small number of numerical simulation programs have been proposed for quantification of the formation of interparticle contacts[15]. One of these models is the HYMOSTRUC model, developed at TU Delft [16]. This program is based on the cell-concept and an associated growth mechanism. In the model a cement particle of a certain size is defined as the central particle of a cubic cell (Figure 2). This central particle, located in the centre of the cube, is the biggest particle in the cell. Going from the surface of the central particle in outward direction small cement particles will be found. When the central particle starts to react and reaction products are formed, this particle will "grow" in outward direction. When growing, small particles will become encapsulated in the growing outer shell (Figure 3). The encapsulated particles may be partly hydrated. If we assume a constant density of the reaction products, the encapsulation of small particles in the outer shell of the growing central particle will create another additional outward growth of the outer shell. This additional growth, on its turn, will result in encapsulating more cement particles and hence in additional growth of the outer shell, etc.. The growth of the central particle, whereby the encapsulation of adjacent particles in the growing outer shell is explicitly taken into account, can be developed in a geometrical series[16]. For each particle in the system the growth process can be determined numerically. With further mathematical manipulations information can be deduced about the amount of cement that is embedded in the outer shell of the central particle. Some of these particles, particularly the bigger ones with a diameter larger than the thickness of the outer shell, will only partly be embedded in the outer shell of central particle. In principle these particles can become encapsulated in the outer shell of another nearby growing particle as well. In that case these particles form bridges between growing clusters and can give the whole complex of growing clusters its stiffness and strength.

The thus created microstructure, consisting of clusters and connecting bridge particles, constitute the basis for the determination of the pore structure. At the same time information can be deduced about the self-desiccation process and the relative humidity in the pore system.

3.2 Pore structure analysis

From the generated three-dimensional microstructure, pore structure data can be determined. This data can be obtained in a direct way and an indirect way. The direct way takes the generated microstructure as starting point. By leaving out the solid matter, i.e. the clusters and the bridge particles, the remaining volume resembles the capillary pore structure. An example is shown in Figure 4. The figure shows the simulated capillary pore network of a cement paste with water-cement ratio 0.3 at a degree of hydration 30% and 90%, respectively. Some typical pore structure features can be observed. First of all it is obvious that, at least in the model, the shape of the capillary pores is irregular. It is, therefore, not appropriate to consider these pores as cylindrical pores as assumed in a mercury intrusion porosity test. In Figure 4b typical pore structure features are indicated, such as isolated pores and dead-end pores. These pores will not contribute to the transport properties. Another important feature shown by the dots in the Figure 4b is the existence of the necks or throats. These necks are the critical links for liquid transport.

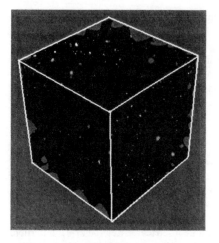
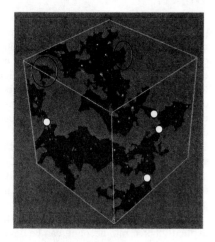

a). Degree of hydration 30%   b). Degree of hydration 90%

Figure 4 *Simulated capillary pore network of paste with water-cement ratio 0.3* [18]

A challenging and most tedious problem still ahead concerns the simulation of the very small pores. In fact the model is able to produce very narrow pores, namely when two growing spheres reach each other at a very close distance. However, when the distance of two growing spheres is of the order of magnitude of the size of the hydration products, i.e. 2 to 100 nm[19], the pore size is more likely to be related to the size and geometrical composition of the reaction products than to the calculated distance between growing spheres (Figure 5). At this point the concept of growing spheres is too crude to generate the information that is needed for simulating nanoscale features of the microstructure of cement paste. This calls for modelling at a lower length scale, at which not the growing cement particles, particle clusters, bridge particles or fillers are the building blocks, but the hydration products themselves.

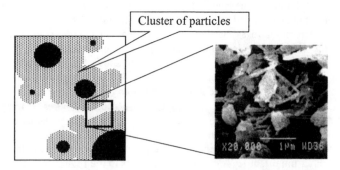

Cluster of particles

Figure 5   *Schematic representation of an arbitrary stage of the formation of the virtual microstructure with "clusters" at close distance of about 1 μm (left) and SEM picture of microstructure (right)*

The indirect method for determination of the pore structure consists of the experimentally determined correlation between the pore size distribution and the degree of hydration. Lots of data have been generated in the past that can serve as a database, which can be consulted if information about the pore size distribution and its evolution is required. For the use of this indirect method one needs a reliable estimate of the degree of hydration. In fact, this estimate can be obtained with less complicated predictive models as described in the foregoing section. A drawback of these simpler models is, however, that they do not give information about some specific features of the microstructure and are, for that reason, less appropriated for in-depth research of cement-based systems. The advantage of these models is that they can easily address the range of small pores. These small pores are of great interest for modelling the volume changes in the low RH range, up to RH = 50%.

4. NUMERICAL MODELLING OF VOLUME CHANGES AT EARLY AGES

Volume changes in hydrating pastes are strongly related to changes in the state of water in the pore system and the interaction between the pore water and the solid. A schematic representation of the cumulative pore size distribution and the state of (capillary) water is shown in Figure 6. The total capillary pore volume is denoted as $v_{por}(\alpha)$. This pore volume consists of pores in the range from the gel pores ϕ_0 (about 2 nm) to the biggest pores $\phi_{por}(\alpha)$. The capillary water is concentrated in the smaller pores. The volume of the capillary water is $v_{wat}(\alpha)$. The biggest pore that is still filled with water is denoted $\phi_{wat}(\alpha)$. With the help of the Kelvin equation the RH in the empty pores can be calculated as a function of the pore diameter $\phi_{wat}(\alpha)$ [16,17]. In this way the evolution of the RH is obtained

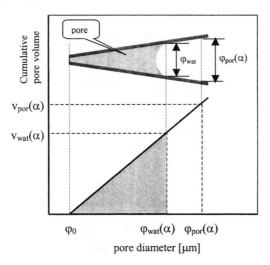

Figure 6   *Schematic representation of cumulative pore size distribution and distribution of capillary water*

as a function of the degree of hydration. From the equations in chapter 2 it can be seen that the volume changes, whether they occur at low, medium or high relative humidity, can all be related to the relative humidity in the pore system. The mechanism through which the total system would experience volume changes could differ from changes in the surface tension of the solid matter, disjoining pressure to hydrostatic tension in the capillary water. Koenders[17] has proposed model based on changes in surface tension of the solid due to adsorption of additional water layers. The thickness of the adsorption layer was correlated to the relative humidity as proposed by Hagymassy[5]. He succeeded in satisfactory simulations of the autogenous shrinkage of isothermally cured low water-binder ratio cement pastes and concretes. Figure 7 shows the predicted and measured autogenous shrinkage of cement paste with water-cement ratios 0.3, 0.4 and 0.5 as a function of time and of the degree of hydration.

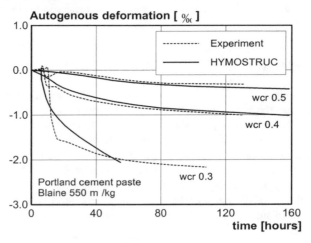

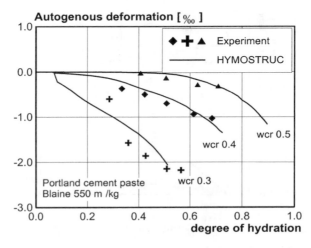

Figure 7   *Predicted and measured autogenous shrinkage of cement pastes with water-cement ratios 0.3, 0.4 and 0.5 as function of time and degree of hydration[17]*

Hua et al.[20] took the hydrostatic tension as the driving force for autogenous shrinkage. Also Lura [7] concentrated on changes in the hydrostatic tension as starting point for the quantitative analysis of length changes in the early stage of hardening. Both Hua and Lura calculated the hydrostatic tension as function of the relative humidity. Lura's results showed a good correlation between measured and simulated autogenous shrinkage during a certain period. At later ages an underestimation of the observed shrinkage pointed to additional shrinkage mechanisms, probably creep. The effect of creep, suppose it was creep, was different for different types of cement [7].

5 DISCUSSION AND CONCLUSIONS

Volume changes in hardening cement-based systems are the result of a number to simultaneously operating processes and mechanisms that are operational at the micro- and nanoscale. The interaction between the water and the solid substance seems to play a crucial role in this respect. Apart from the three shrinkage mechanisms briefly addressed in this paper additional mechanisms have to be considered, e.g. swelling mechanisms with a distinct geometrical background as well as the formation electrical double-layers close to the surface of the gel particles. The latter are typical nanoscale phenomena, which certainly have to be addressed in comprehensive models for volume changes of hardening systems. Most probably we have to consider combinations of all these mechanisms. The complexity of these mutually interacting mechanisms has been indicated clearly by Wittmann[21] when he stated, speaking about shrinkage phenomena: "…. we have to keep in mind that we do not yet fully understand the complex interaction of water with pore walls in fine pores". A challenging task that follows from this statement is that, in view of numerical modelling of volume changes at early age, the modelling of the microstructure, i.e. the pore structure, should be given high priority. With this task a lot of complex nano- and microscale issues have to be considered. In this respect the studies of Sellevold[22] and Radjy[23] on a phenomenological theory of pore structure analysis are very illustrative. In one study they assumed slit-shape pores, whereas in the second study cylindrical pores were considered. The authors found that essentially all the results developed for the case slit-shaped pores were assumed had to be modified in case cylindrical pores had been adopted.

The foregoing clearly illustrates that modelling of the pore structure, as the basis for quantitative modelling of the water-solid interactions and resulting volume changes, is full of pitfalls. On the other hand it can be said that the power of numerical models for predicting pore structures of cement-based systems has increased significantly in the last decade. Today the reliability of simulations of capillary pore structures is quite promising. Reaching the same reliability for the pore structure at the nanoscale is still a challenge but, as we believe, a matter of time. There are, however, still a lot of challenges ahead. For example, the modelling of the occurrence of microcracks and of time dependent effects needs to be improved. Both microcracking and creep are expected to affect the volume changes of hardening systems. Another influencing factor is temperature, which parameter significantly affects the correlation between volume changes and relative humidity[24]. All these complicating factors should make us reluctant to focus too much on good correlations between measurements and predictions in case the latter have been obtained with very simple models.

For real mature models co-operation of different disciplines with the aim to developed multidisciplinary models seems to be a prerequisite. Researchers from the field of (colloid) chemistry, physics, stereology, fracture mechanics and (numerical) modellers have to co-operate, thus bridging the gap between these disciplines as well as the gap between different length scales, i.e. the nano-, micro-, meso- and macroscale. Scepticism, as if it would be just a modish trend to join the nanotechnology movement, should make place for serious attempts to bridge these gaps. Today it is not difficult anymore to show how the knowledge of concrete technologies can be applied for modifying the behaviour of real concrete structures under external loads. In that way concrete technologists and structural engineers span six orders of magnitude. Similarly "genetic manipulation" of the nano- and microstructure of cement-based systems can be expected to result in a pre-defined performance of cement-based materials. If there has ever been a time to get rid of the low-tech image of research on building materials, it is now!

References

1   T.C. Powers, Int. Symp. *Cem. & Concr. Ass.,* London, 1965, 319.
2   R.F. Feldmann, P.J. Sereda, *Materials and Structures*, **1**, 1968, 509.
3   F.H. Wittmann, *Conference on 'Hydraulic cement paste: Their structure and properties'*, Sheffield, 1976, 96.
4   V. Baroghel-Bouny, *Proc. Int. Workshop on 'Self-desiccation and its importance in concrete technology'*, Lund, 1997, 72.
5   J. Hagymassy, S. Brunauer, R.Sh. Mikhail, *J. Coll. and Interface Science*, **29** (3), 1969, 321.
6   B. Person, *Proc. 5th Int. Symp. on Utilization of high strength/high performance concrete*, Sandefjord, 1999, 1272.
7   P. Lura, *PhD thesis,* Delft University of Technology, 2003.
8   H.H. Steinour, *Cement & Concrete Research*, **12** (1), 1982, 33.
9   P.P. Budnikov, M.I. Strelkov, *ACI Special Report 90*, 1966, 447.
10   Y. Tezuka, et al. *Proc. 8th Int. Congress on Chemistry of Cement*, Rio de Janeiro, **II**, 1986, 323.
11   J.W. Gibbs, *'Collected Works'*, New Haven, Yale Univeristy Press, 1957.
12   K. van Breugel, *Shrinkage 2000,* Ed. Baroughel Bouny, RILEM, 2000.
13   D.H. Bangham, N. Fakhoury, *Proc. Royal Society of London*, 1931, 81.
14   K.H. Hiller, *J. Applied Physics*, **35,** 1964, 1622.
15   D.P. Bentz, *J. Am. Ceram. Soc.*, **80**, (1), 1997, 3.
16   K. van Breugel, PhD Thesis, Delft, Sec. Ed.,1991, 305.
17   E.A.B. Koenders, *PhD Thesis*, TU Delft, 1996, 171
18   Ye Guang, K.van Breugel, A.L.A. Fraaij. *Cem.Concr.Res.*, 2003, **33** (2), 215.
19   A. Grudemo, *Proc. Swedish Cem. Concr. Inst.*, Stockholm, (26), 1955.
20   C. Hua, P. Acker, A. Ehrlacher, *Cem. And Concr. Res.*, **27** (2), 1997, 245
21   Z.P. Bazant, F.H. Wittmann, *John Wiley & Sons,* New York, 1982
22   E.J. Sellevold, F. Radjy, *J. of Colloid and Interface Science*, **39** (2), 1972, 379.
23   F. Radjy, E.J. Sellevold, *J. of Colloid and Interface Science*, **39** (2), 1972, 367.
24   O.M. Jensen, *Dept. of Structural Eng. & Mat., DTU*, Lyngby (Dk), Series R, (47), 1998.

NUMERICAL MODELLING AND EXPERIMENTAL OBSERVATIONS OF THE PORE STRUCTURE OF CEMENT-BASED MATERIALS

G. Ye[1] and K. van Breugel[1]

[1]Faculty of Civil Engineering and Geosciences, Delft University of Technology P.O. Box 5048, 2600 GA Delft, the Netherlands

1 INTRODUCTION

In recent years, it has been realized that nanotechnology plays an important role in construction materials. In particular, when we talk about water and ion migration through concrete cover, the pores in the range from 1 nm to 1 mm are involved. However, because of the complexity of the pore structure of cementitious materials, for example, the range of pore sizes varies over 6 orders of magnitude[1]. Therefore, measuring the total porosity, pore size distribution, and especially the connectivity of the pore structure, are difficult.

In general, several methods have been used for the determination of the pore structure in the past decades. Among these methods are Mercury Intrusion Porosimetry [MIP], Nitrogen Sorption, Backscattered Image Analysis (BSE) and Nuclear Magnetic Resonance [NMR]. MIP is so far the most widely used method for determining the pore structure of cement-based materials, not because it "fits" the system best, but because it is the only available procedure that purports to cover nearly the whole range of sizes that must be tallied[2]. Several investigations[3, 4] have drawn attention to the fact that the results of MIP measurements have to be interpreted carefully considering the pore shape, ink-bottle effect and high pressure damage to the pore structure. Compared to the MIP technique, the use of backscattered electron (BSE) imaging to study the microstructure of cement paste is a comparatively new technique. This technique allows large cross-sectional areas of a sample to be studied over a wide range of magnifications.[4] From these images, the amounts and distributions of the microstructural constituents can be quantified by different image analysis techniques.[5-7] With the BSE technique, the pore geometry can be explicitly revealed, and the algorithm used to "size" a pore of complex geometry can be stated explicitly.[2] Once combining mercury porosimetry and BSE image analyses, the main characterization of a pore structure can be captured by careful interpretation of the experimental data.

Due to the limitation of experimental techniques, it is difficult to determine the 3D structure of pore sizes in a sufficient resolution. However, with the rapid development of computer technology, a number of experimental complications can be circumvented by using computer-based simulation models. The numerically simulated pore structures have the advantage to give more detailed information on the evolution of microstructure, special on the connectivity of pore structures in 3D. The connectivity of pore structures is the base for making a good permeability model.

During the last two decades a few fundamental pore structure models [8-11] have been developed that can be used, in principle, for predicting the transport properties of cement-based materials. These models are based on numerical simulation techniques and on a digital-image processing approach. Pioneering work in this field dates back to Jennings and Johnson [8], who first proposed a cement hydration model. In their model, the cement particles were modelled as growing spheres. Another advanced model has been accomplished with the digital-image-based microstructure, which was developed by Bentz and Garboczi.[9] Their model allows the direct representation of multiphase and non-spherical cement particles. Van Breugel's model HYMOSTRUC[10] had the advantage of considering the individual particles interaction and reaction kinetics. These models vary widely in concept, accuracy, required computation time and "degree of reality". It is still very difficult for the present generation models to mimic the pore structure over the whole range of pore sizes satisfactory, i.e. from nano- to micro- scale.

In the present paper, MIP and ESEM were utilized to observe the pore structure of cement paste. The features of the cement hydration model HYMOSTRUC and reproduced microstructure are described. Cement pastes with different water/cement ratio, i.e. 0.4 to 0.6, have been simulated and the microstructures have been analysed. Both results from experiments and numerical simulation are compared with each other.

2 3D PORE STRUCTURE SIMULATION

3D pore structures have been simulated by the HYMOSTRUC model as shown in Figure 1. In the HYMOSTRUC model, the rate of cement hydration and the formation of inter-particle contacts are modeled as a function of the particle size distribution, the chemical composition of the cement, the w/c ratio and the reaction temperature. The cement particles are modeled as spheres. Particles of the same size are considered to hydrate at the same rate. In the beginning, hydration reactions are considered as phase-boundary reactions. At later stages, when the shell of reaction products has reached a certain predefined thickness, the reactions become diffusion controlled. The simulation starts from a random distribution of cement particles in a cubic volume. During hydration, the cement grains gradually dissolve and a porous shell of hydration products is formed around the grain. With progress of the hydration process, the growing particles become more and more connected. Thus a porous structure is formed.

In this model, the hydration products are considered as solid phases with inherent gel porosity. The capillary pores are considered as the space between the growing spheres. Using a serial sectioning algorithm combined with an overlapping criterion, the effective pores, dead-end pores and isolated pores can be distiguished and the connectivity of pores can be derived from the pore network structure[12]. The pore size distribution curve can be determined by calculating the sub-volumes of the pore space that are accessible to "testing spheres" of different radii r[13].

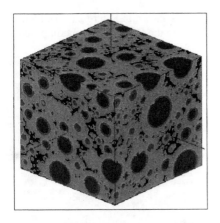 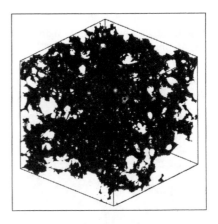

Figure 1 *3D simulated cement paste (left) and its pore structure (right) (w/c=0.3, degree of hydration α=0.75)[13]*

3 EXPERIMENTAL OBSERVATIN IF PORE STRUCTURE

3.1 Mercury intrusion porosimetry

With MIP, porous samples are introduced into a chamber. The chamber is evacuated and then the samples are surrounded by mercury. The pressure on the mercury is gradually increased. With the increase of pressure, mercury is forced into the pores that are connected to the surface of the sample. If the pore system is continuous, a pressure may be achieved at which mercury can penetrate the smallest pore necks of the pore system and penetrate the bulk sample volume. If the pore system is not continuous, mercury may penetrate the sample volume by breaking through pore walls. By tracking pressures and intrusion volumes during the experiment, it is possible to get a measure of the connecting pore necks. The relation between pressure P and the pore diameter D is described by the Washburn equation (1) based on a model of cylindrical pores:

$$P = -\frac{4 \ \sigma \times cos \ \theta}{D} \tag{1}$$

where σ is the surface tension of the mercury and θ is the contact angle between the mercury and the pore surface of the cement. The surface tension value of 480.0 dyn/cm and the contact angle of 130 were suggested by Diamond[2].

3.2 Backscatter Image Analysis

In order to obtain reliable information of the pore structure of cement paste with BSE techniques, a number of details have to be addressed. The BSE technique includes sample preparation, image acquisition and image analysis. Proper sample preparation includes casting of cement paste, drying specimen, epoxy impregnation and cutting, grinding and polishing. Image acquisition was done by a Philips-XL30-ESEM. In order to strictly compare each other, the technique parameters, such as accelerating voltage and spot size were ultimately kept identical. The image analysis includes binary image segmentation and pore size distribution. A typical grey level BSE image is shown in Figure 2 (left). Figure 2 (right) shows a histogram of distribution of grey levels in the BSE image superimposed on the original image.

Figure 2   *Left: A typical BSE image. Right: Threshold of grey level to original image.*

The threshold value of the pores corresponds to the left peak in the histogram. The image shown in Figure 3 (left) was obtained after applied threshold of grey level on original BSE image. After the binary segmentation, the median 7×7 filter was applied to remove the isolated pixels due to the noise in the image acquisition process. Figure 3 (right) is the final binary image used for further quantitative measurements. In order to compare the pore size distribution obtained by numerical simulation with the result from image analysis, the pore area fractions are obtained by sorting the pores by area A and plotting the pore area against the pore diameter d. The equivalent diameter of a circle with an area equal to that of the feature was calculated. The cumulative pore area fraction against equivalent pore diameter was obtained.

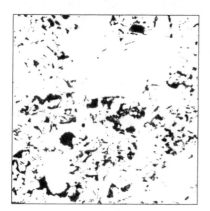
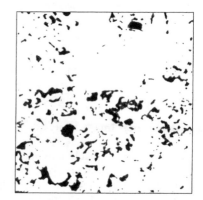

Figure 3   *Left: Pores obtained after applied threshold on original BSE image. Right: pores obtained after applied median 7×7 filter on left image to remove noise.*

4 RESULT AND DISCUSION

4.1 The geometry of the capillary porosity

The simulated pore structures are illustrated in Figure 4 for the sample with w/c ratio 0.3, 0.4 and 0.5, at a degree of hydration of 0.75. It has to been noted that the minimum size of cement particles used in the simulation was 2 μm. The chemical composition of cement was that of CEMI 32.5R. It is obvious that at the same degree of hydration, the sample with the lower water/cement ratio shows less porosity than the sample with a higher water/cement ratio. The calculated result revealed a capillary porosity of 6%, 19%, 28% for w/c 0.3, 0.4 and 0.5 respectively.

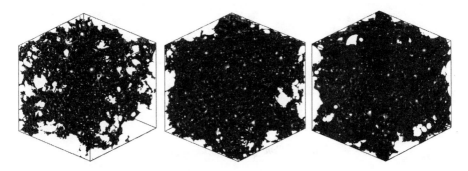

Figure 4 *Simulated pore structure. Left: w/c ratio 0.3, ρ=6%, middle: w/c ratio 0.4, ρ=19%, right: w/c ratio 0.5, ρ=28%, at degree of hydration 0.75.*

4.2 Connectivity of the capillary porosity

The connectivity of capillary pores, in terms of connected porosity fraction, is illustrated in Figure 5, as function of the degree of hydration. It is noticed that, for all samples, the capillary pores are almost always connected with each other even at ultimate hydration. No de-percolation of capillary porosity can be found. These results are quite different from the result of the NIST model[14], where a de-percolation threshold of the capillary porosity of 12% was found in cement paste with w/c ratio 0.4. On the other hand, the present results are quite similar to the results previously calculated by Navi et.al[15]: when a resolution of 2μm/pixel was used, their results revealed that almost all pores were connected with each other. For an explanation of these differences, several aspects have to be considered. First, as discussed previously, the percolation threshold of the capillary porosity is largely dependent on the digital resolution. In the present simulation, the digital resolution was 250 nm/pixel; at this high digital resolution, even very small pathways were detected, leading to a very low percolation threshold of the capillary porosity or no threshold at all. Another important factor, which influences the connectivity of the capillary porosity, is the fundamental concept of the different numerical models. As discussed earlier[14], the NIST model is a pixel-based model, where the different cement components are simulated as different sets of digital pixels and no shape constraint exist. In other models, such as HYMOSTRUC or Navi's model, the cement particles and hydration products are simulated as spheres. The growing spheres are overlapping when cement hydration take place. The pore space percolation threshold of the spherical based models has been numerically found always around a few percent porosity[16]. However, theoretically, in the cement pastes with w/c ratio from 0.3 to 0.6, even at the ultimate hydration stage, the capillary porosity will not be less than 5% (Figure 5). This explains why no de-percolation of porosity can be found in a spherical based cement hydration model.

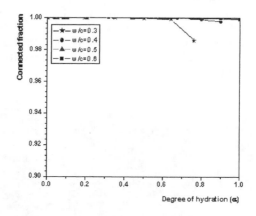

Figure 5 *Connected fraction of porosity*

4.3 BSE image and HYMOSTRUC 2D section

The cement paste and its pore structure are shown in Figure 6, taken from backscatter image, and Figure 7 simulated with the HYMOSTRUC model, both with w/c ratio 0.3 at a age of 14 days, degree of hydration 0.63.

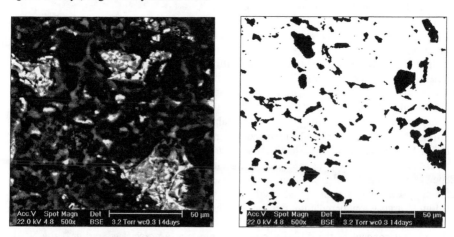

Figure 6 *BSE image of cement paste with w/c=0.3 at age 14 days. Left: original grey image. Right: after binary segmentation and noise removal.*

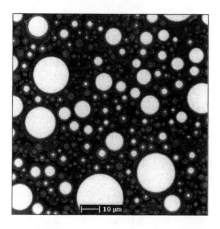
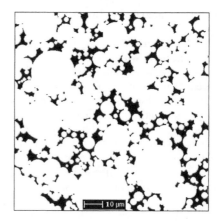

Figure 7 *Simulated cement paste with w/c=0.3 at degree of hydration 0.63. Left image grey level from white to dark show unhydrated cement core, inner products, outer products and capillary pores. Right image shows capillary pore only.*

Figure 8, left, presents the results from the image analysis and the HYMOSTRUC simulation. The equivalent pore diameter versus pore area fraction represents the two-dimensional pore size distribution. For 5 samples, taken from arbitrary positions both from BSE image and simulations, the regression curves are plotted in Figure 8 (right). The R-squared value of a simulated 2D section and a BSE image are 0.97 and 0.99, respectively. The regression curve could represent the experiments very well. These figures show that there is good agreement between the simulated pore structure and BSE image analysis. The simulated pores are slightly bigger than deduced from the BSE image in the big pore region only. The size of almost all pores is less than 20 µm, both in the experiments and in the simulations.

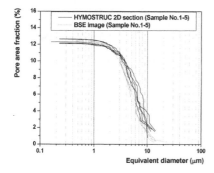
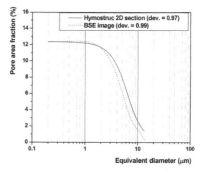

Figure 8 *Left: 2D "pore size" distribution of simulated cement paste and BSE image analysis for the w/c 0.3 cement paste. Right: Regression curve calculated from both 5 samples.*

4.4 MIP comparison between experiments and simulations

Figure 9, right provides a comparison of MIP by experiment and simulation for a 14-day old cement paste with w/c ratio 0.30. Obviously, the pore size distribution measured by experiment show a smaller pore volume in a range of pore size from 0.03 μm ~ 2μm. This difference might have to be attributed to the limitations of MIP test method. As we know, the MIP test method "suffers" the "ink bottle" effect, in which a large pore are preceded in the intrusion path of the mercury by a smaller neck. This problem produces pore size distribution curves with erroneously high volumes of smaller pores and erroneously small volumes of larger pores[17, 18]. On the other hand, in the simulation, the pore size distribution curve was determined by calculating the sub-volumes of the pore space that are accessible to testing spheres of different radii r. This method avoids "ink bottle" effect. It takes all pores into account in the "right" size.

Willis et al[17] extensively investigated the "ink bottle" effect using Wood's metal intrusion porosimetry (WMIP) and image analysis. They concluded that pore size distributions obtained from MIP are shifted more towards smaller pores than those obtained from WMIP and image analysis. A comparison between present simulation and image analysis size distribution of WMIP by Willis et al is shown in Figure 9, left. In his original paper, Willis et al presented their plots as volume versus pore area at a pressure of 5000 psi. Diamond recalculated the data indicated as the common volume versus pore diameter[18]. From Figure 9, right, the pore size distribution from MIP simulation with HYMOSTRUC model is more close to the Wood's metal image analysis, rather than mercury intrusion porosimetry test as shown in Figure 9, left.

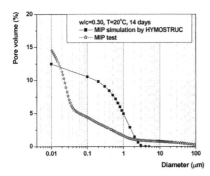 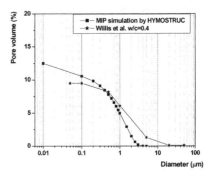

Figure 9 *Left: Comparison of MIP curve by experiment and simulation for 14-day-old w/c 0.3 cement pastes. Right: Comparison of simulation and Wood's metal intrusion porosimetry test by Willis et al (After Diamond)*

5. CONCLUSIONS

The numerical cement hydration model HYMOSTRUC provides a description of the evolution of cement paste microstructure, including the pore microstructure. With the 3D analysis program, which is based on a serial sectioning algorithm and is integrated into HYMOSTRUC3D, both the geometrical properties of the solid and pore phase can be characterized quite well. Obviously, the model provides a reliable representation of the real-time development of cement microstructure, which can be visualized. The connectivity of porosity and the volume fraction of solid, as well as the internal surface area of each phase, can easily be calculated with the model.

The quantitative comparisons between the simulated pore structure and experiments both from 2D backscatter image and 3D MIP are presented. The findings show that the pore area percentage versus equivalent diameter from simulated 2D slice is in good agreement with backscatter image analysis. However, due to the limitation of MIP test, the pore volumes from experiments are much lower in the range of 0.03μm ~ 10 μm compared to the numerical simulation with the HYMOSTRUC model. The numerical result fits quite well to the pore size distribution obtained from Wood's metal intrusion porosity test.

The good correlation between the measured and simulated capillary pore structures shows that the model can serve as a starting point for more advanced simulation models, in which the nano-pores are addressed in a more realistic way. This is what future research and modelling activities should focus on.

References

1   A. M. Neville, *Properties of Concrete*, 4nd edn., Addison Wesley Longman Limited, Edinburgh Gate, 1996, Chapter 10, p. 483.
2   S. Diamond, 'Methoddologies of PSD Measurements in HCP: Postulates, Peculiarities, and Problems' in *Mat. Res. Soc. Symp. Proc.* eds., L.R. Roberts and J.P. Skalny, 1989, Vol, 137, pp. 83-89.
3   S. Diamond, *Cem. Concr. Res.,* 1971, 1, pp. 531.
4   R.A. Cook and K.C. Hover, *ACI Mater.,* 1993 **90** (2) pp. 152.
5   K.L. Scrivener, 'The Use of Backscattered Electron Microscopy and Image Analysis to Study the Porosity of Cement Paste' in *Mat. Res. Soc. Symp. Proc.* eds., L.R. Roberts and J.P. Skalny, 1989, Vol 137, pp. 129-140.
6   S. Diamond and M. Leeman, 'Pore Size Distribution in Hardened Cement paste by SEM Image Analysis' in *Mat. Res. Soc. Symp. Proc.* eds., S. Diamond, S. Mindess and F.P. Glasser, 1995, Vol 370, pp. 217-226.
7   D.A. Lange, H.M. Jennings, and S.P. Shah, *Cem. Concr. Res.,* 1994, **24**, 841.
8   H.M. Jennings and S.K. Johnson, *J. American Ceramic Soc*, 1986, **69**, p.790.
9   D.P Bentz and E.J. Garboczi, 'A digitised simulation model for microstructural development', in *Advances in Cemetious Materials*, edited by S. Mindess. Westerville, Ohio, American Ceramic Society, 1989, p.211.
10 van Breugel, K. *Simulation of Hydration and Formation of Structure in Hardening Cement-based Materials*, PhD thesis, Delft University of Technology, 1991.
11 P. Navi and C. Pignat, *Cem. Concr. Res.,* 1999 **29** (4), p.507.
12 G. Ye, K. van Breugel and A.L.A. Fraaij, *Cem. Concr. Res.,* 2003, **33** (2), p.215.
13 G. Ye, J. Hu, K. van Breugel and P. Sreoeven, *Mats and Strucs.,* 2002, **35**, p. 603.
14 E.J. Garboczi and D.P Bentz, *Cem. Concr. Res.,* 2001, **31**, p.1501.
15 P. Navi and C. Pignat, *Advanced Cement Based Material*, 1996, **4**, p.58.
16 W.T. Elam, A.R. Kerstein, and J.J. Rehr, *Physical Reviewer Letters*, 1984, **52**, p.1516.
17 K.L. Willis, A.B. Abell and D.A. Lange, *Cem Concr Res* 1998 **28** (12) p.1695.
18 S. Diamond, *Cem. Concr. Res.,* 2000 **30** (10), p.1517.

VIRTUAL CONCRETE: WORKING AT THE NANOMETER SCALE

E.J. Garboczi[a] and D.A. Neumann[b]

[a]Materials and Construction Research Division
[b]Center for Neutron Research
National Institute of Standards and Technology
Gaithersburg, Maryland USA

EXTENDED ABSTRACT

Concrete, a complex composite material, is made even more complicated by the time dependence of its properties and most especially by its multi-scale nature. Aggregates are at the millimeter scale, the cement paste matrix is at the micrometer scale (according to the average size of the cement grains), and the main hydration product, calcium-silicate-hydrate or C-S-H, has important pore and layer structure at the nanometer scale. All these length scales work together. For example, it is well known that concrete is a viscoelastic material. The only phase in this composite material that is itself viscoelastic is the C-S-H phase. The processes that make C-S-H viscoelastic take place at the nano-scale. Therefore, to fully understand why concrete creeps at the millimeter or meter scale, one has to investigate nano-scale processes. This is true for many other properties of concrete as well, since they are dominated by the nano-scale "glue", C-S-H, which binds concrete together.

The Virtual Cement and Concrete Testing Laboratory (VCCTL) is a NIST-led industrial consortium of nine major companies and industrial groups that is developing the VCCTL software to accelerate the industrial R&D process by reducing physical testing and therefore optimizing the use of concrete in the built infrastructure[1] (Ref. [2] provides more technical details of some of the science of these models). This software has been designed to realistically simulate, based on fundamental materials science, the micrometer to millimeter size range of concrete. To predict and optimize those properties of concrete (with or without waste-stream mineral admixtures) whose properties are dominated by its nanoscale structure, requires that the VCCTL software be quantitatively extended to the nanometer-scale. To accomplish this model extension, we intend to combine molecular dynamics (MD) simulations of C-S-H with neutron scattering measurements, using direct computation of experimental quantities in order to obtain a much more complete understanding of the critical nanoscale structure and dynamics of C-S-H.

Molecular dynamics (MD) simulation methods have been employed to study materials for many years. In these computer calculations, atoms (or molecules) are positioned in a box in accordance with known structural information, and are allowed to move according to parameterized phenomenological forces. The calculations give both spatial and temporal information on the locations of the atoms and molecules, thereby allowing one to easily visualize the structure and dynamics of a material on the atomic scale. This powerful technique has not yet been brought fully to bear on the atomic scale properties of cement, primarily due to the disordered atomic scale structure and the nanoporous nature of the C-S-H gel, which makes it difficult to find an adequate starting configuration. Moreover, these structural properties necessitate that any MD simulation of C-S-H requires thousands of atoms and molecules to fully capture the important nanoscale features. Only in the last few years has computing speed advanced to the point where such simulations are feasible. Some promising MD work on cements has begun, showing the power of such techniques[3].

Neutron scattering methods are the perfect experimental complement to MD simulations because they measure the Fourier transform of correlation functions that are based on the positions of the atoms and molecules. Thus one can simply and directly compare the results of a scattering measurement with the results of an MD simulation. This allows one to adjust the initial conditions and/or the phenomenological forces in the calculation in order to better describe the data. Moreover, the simulations allow one to interpret the features seen in the scattering from complex composite materials such as concrete. This symbiotic relationship allows an understanding of materials to be obtained that is more complete than either method could yield on its own.

Three types of scattering measurements are particularly relevant for understanding the structure of a disordered, nanoporous material such as the C-S-H gel phase. The first is small-angle neutron scattering (SANS), which provides a quantitative characterization of solid/pore nanostructures down to a length of 1 nm [4]. One of the most difficult aspects of the MD simulation will be setting up realistic nanopores. In fact, this is a quite general problem for many other functional nanomaterials. The procedure for accomplishing this task, which will be developed as part of this work, will be transferable to a wide range of other problems. Small angle scattering results will be crucial for determining if the nanopores in the simulations are representative of real C-S-H.

Another important scattering technique is typically termed "total scattering" [5]. This type of measurement yields information on the local structure of a material via the atomic radial distribution function. This information will allow us to verify that the local atomic arrangement produced in the MD simulation is representative of that in real C-S-H.

Finally, neutron spectroscopic techniques allow one to probe the motions of atoms and molecules on time scales ranging from fractions of a picosecond all the way to a nanosecond, a range accessible by MD simulations. Because inelastic neutron scattering is extremely sensitive to hydrogen atoms, these methods will be most useful for looking at the motion of the water molecules. In particular, quasielastic neutron scattering studies[6] will provide information on the local physical environment of the water molecules, which can be directly and quantitatively compared to the environment experienced by the water molecules in the MD simulation.

Based on neutron scattering and other relevant experimental results that probe the nano-scale, we plan to validate and incorporate nano-scale models into the VCCTL suite of models, extending its range across all the length scales of importance to concrete.

References

1   http://vcctl.cbt.nist.gov/, http://www.bfrl.nist.gov/862/vcctl/

2   http://ciks.cbt.nist.gov/monograph/;   http://math.nist.gov/mcsd/savg/vis/concrete/; http://math.nist.gov/mcsd/savg/parallel/epc/.

3   A.G. Kalinichev and R.J. Kirkpatrick, Molecular dynamics modeling of chloride binding to the surfaces of Ca-hydroxide, hydrated Ca-aluminate and Ca-silicate phases, *Chemistry of Materials*, **14**, 3539-3549 (2002).

4   J.J. Thomas, H.M. Jennings, and A.J. Allen, The surface area of cement paste as measured by neutron scattering: evidence for two C-S-H morphologies, Cem. Conc. Res. **28**, 897-905 (1998).

5   H. Brequel, S. Enzo, F. Babonneau, and P.G. Radaelli, Neutron diffraction study of nanocrystalline oxycarbide glasses prepared by sol-gel, Mater. Sci. Forum, **386**, 275-280 (2002).

6   J.J. Thomas, D.A. Neumann, S.A. FitzGerald, and R.A. Livingston, The state of water in hydrating tricalcium silicate and portland cement pastes as measured by quasielastic neutron scattering, J. Amer. Ceram. Soc., **84**, 1811-1816 (2001).

EVALUATION OF THEORETICAL MODELS FOR ASSESSING INTERFACIAL PROPERTIES IN AGED GRC USING FIBRE PUSH-IN TEST

J.J. Gaitero[1], W. Zhu[2] and P.J.M. Bartos[2]

[1]Labein, Cuesta de Olaveaga 16, 48013 Bilbao, Spain
[2]Advanced Concrete and Masonry Centre, University of Paisley, Paisley, PA1 2BE, UK

1 INTRODUCTION

Glass fibre reinforced cement (GRC) has been used as a construction material for many years. In such a composite material, the basic reinforcing element is a fibre strand or bundle, instead of a single fibre. Typically, the number of fibre filaments per bundle varies from 20 to 400, and the diameter of the filament ranges from 8 to 20 μm. An important feature associated with the bundle structure of the reinforcement is the continuous change of interfacial microstructure with age when exposed to wet environment. It has been widely recognised that the ageing performance of GRC and the reinforcing efficiency of the fibre bundle has a strong dependence on the characteristics/properties of the fibre-matrix and fibre-fibre interfaces as well as of the fibre bundle structure.[1-5] Study of the interfacial properties in GRC, particularly in the aged composites, however, has proved to be difficult due to the very complex nature of the fibre reinforcement.

A novel nanotechnology based indentation method was adopted previously in a fibre push-in/push-out test with limited success.[4, 6] Progress has since been made in the nano-indentation equipment which offers significant improvement in sample positioning accuracy and load-displacement resolutions. The overall objective of this study is to study the interfacial properties in aged GRC, and particularly to extract quantitative results from experimental fibre push-in test data by using suitable models.

2 METHODS FOR ASSESSING INTERFACIAL BOND PROPERTIES IN GRC

To study the interfacial bond properties of GRC, many different techniques/methods have been tried. These include SIC (Strand in Cement) test, composite direct tensile test, multiple or single fibre bundle/rod pullout test,[7] and more recently, fibre push-in or push-through tests. The main shortcomings of common tests, such as pullout test and direct tensile test, are their inability to evaluate bond properties in aged composites due to the fact that a substantial portion of the fibres ruptures rather than being pulled out during the test. Furthermore, practical difficulties/complexities in specimen preparation for pullout test make it impossible to study the interfacial properties at the fibre-fibre interface within a fibre bundle.

To overcome the above-mentioned difficulties, fibre push-in and push-through tests based on a depth-sensing nanoindentation apparatus were developed.[4, 6] The fibre push-through and push-in tests, as shown in Figures 1a and 1b respectively, differ from the fibre pull-out test mainly in that the fibre is pushed through or into the matrix rather than pulled out. A clear advantage for the fibre-pushing method is that the specimen can be cut from a practical, commercially produced composite that has been subjected to different ageing/service conditions.

As demonstrated in Figure 1, the application of load (by very small indenter) usually leads to a progressive debonding that starts at the top end of the fibre and a sliding of the fibre in the debonded region against frictional resistance at the interface. In the case of the fibre push-through test (Figure 1(a)), a very thin specimen (e.g. 50 – 300 μm thick) is used and the applied load is increased until complete debonding is detected and the fibre pushed through. Though relatively simple to extract bond properties from the load-displacement curve in such a case, preparation of good thin, polished specimens proved to be a difficult task. To avoid this problem, the fibre push-in test (Figure 1(b)) can be used instead. The fibre push-in method allows the use of normal specimens (e.g. 10 – 50 mm thick), but during the test only partial fibre debonding and frictional sliding in the debonded region is possible. As a result, the extraction of interfacial properties relies on fitting the fibre push-in test data with a suitable theoretical model. A few such models which have been proposed/used previously for studying ceramic, metallic and polymer composites,[8-10] will be evaluated in this study for test on aged GRC.

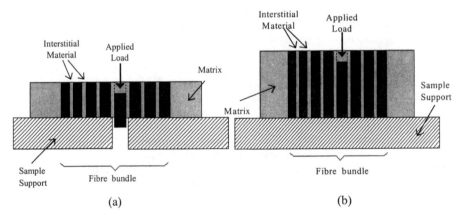

Figure 1 *Test set-up for fibre push-through test (a) and fibre push-in test (b).*

3 METHODOLOGY OF THE FIBRE PUSH-IN TEST

3.1 Test Apparatus and Procedures

The equipment used for the fibre push-in test is a depth-sensing nanoindentation apparatus, Nanoindenter XP (MTS), which continuously monitors the displacement of a probe/indenter in contact with a fibre surface during a programmed load cycle. The operating principle of the test apparatus was given previously.[4] Instead of using a normal sharp Vickers or Berkovich diamond indenter, a specially produced 90 degrees conical diamond indenter with a rounded tip of 3 μm radius was used in this study for pushing the fibre. Such an arrangement greatly reduces the risk of damaging/splitting the fibre end during the test.

A typical test result of the fibre push-in test is shown in Figure 2. The test procedure usually consists of two main steps: 1) first, the indenter tip is aligned with the centre of the fibre to be tested and then the indenter is slowly moved towards the fibre; 2) once contact between indenter tip and the fibre end surface is detected, a pre-programmed loading-unloading scheme starts and the load-displacement data are obtained (as in Figure 2). Usually only one loading-unloading cycle is carried out, as the data required for extraction of the interfacial properties are those from the 1st loading-unloading cycle. However, data from the 2nd loading-unloading cycle (i.e. reloading in Figure 2) can provide an indication of the extent of frictional sliding of the tested fibre that occurred during the test. This is because in the case of no frictional sliding (e.g. when the maximum applied load is not enough to initiate partial debonding), the reloading (i.e. the 2nd loading) curve will follow the same trace of the unloading curve of the 1st cycle (i.e. no hysteresis between the unloading and reloading curve).[8] In this study, the maximum load levels were adjusted so as to produce sufficient frictional fibre sliding displacement and at the same time to avoid damage/split of the fibre end (a problem associated with high maximum load).

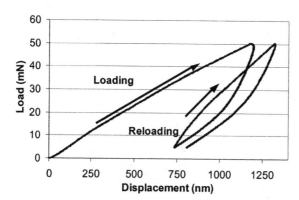

Figure 2   *Typical result of a fibre push-in test with two load-unload cycles*

3.2 Acquiring Net Load-Displacement Data in the Fibre Push-in Test

Generally, the load applied in the fibre push-in test not only leads to frictional sliding in part of the fibre length, but also generates a certain indentation penetration/impression into the fibre end surface, as shown in Figure 3(a). Therefore, the displacement obtained from a fibre push-in test is the total of the penetration depth of the indenter and the net displacement of the tested fibre, as shown in Figure 3(b).

Under the test conditions in this study, the penetration depth was found to be comparable to the net sliding displacement of the test fibre. As a result, ignoring the contribution of indenter penetration to the fibre push-in load-displacement data would lead to erroneous results of the interfacial properties. Therefore, it is necessary to acquire net Load-Displacement data by eliminating the indenter penetration from the original fibre push-in test data. To do this, an indentation test with the same programmed loading-unloading cycle as that used in the fibre push-in test was carried out on a bulk glass

specimen of similar chemical compositions to the glass fibre.   This   supplementary
test produced the penetration-only load-displacement curve.   The net sliding load-
displacement curve can then be acquired by subtracting the penetration-only data from the
original total displacement data, as illustrated in Figure 4.

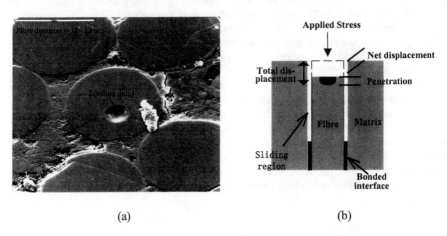

(a)   (b)

Figure 3   *(a) SEM photograph of a tested fibre, (b) displacements generated in fibre*

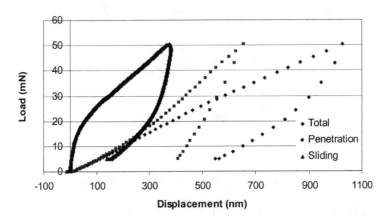

Figure 4   *Acquiring the net sliding load-displacement curve by subtracting the*
                    penetration data from the total displacement data

4 MODELS STUDIED FOR THE FIBRE PUSH-IN TEST

4.1 Marshall and Oliver (1987)

One of the earliest models proposed for fibre push-in test was very simple, assuming that
the only interaction between the interface and the tested fibre is by means of a constant

frictional shear stress τ that extends all along the sliding region.[8] Thus, the load-sliding displacement relationship is

$$u = \frac{1}{4\pi^2 R^3 \tau E_f} F^2 \qquad (1)$$

where u is the sliding displacement, F is the load applied by the indenter tip, R and E_f are the radii and Young's modulus of the tested fibre.

To account for the possible presence of a chemical bond between the fibre and the matrix that had to be broken to allow fibre sliding, a new term could be added to Equation (1).[8] The following equations (henceforth mentioned as Marshall-Oliver 1987) were thus derived,

$$u = \frac{1}{4\pi^2 R^3 \tau E_f} F^2 - \frac{2\Gamma}{\tau} \qquad \text{for loading} \qquad (2)$$

$$\text{or,} \qquad u = \frac{1}{4\tau E_f \pi^2 R^3}\left(F^2 - F_d^2\right) \qquad \text{for loading} \qquad (2a)$$

$$u = u_m - \frac{(F_m - F)^2}{8\tau_u E_f \pi^2 R^3} \qquad \text{for unloading} \qquad (3)$$

$$u = \frac{u_m}{2}\left[1 + \frac{F}{F_m}\right] \qquad \text{for reloading} \qquad (4)$$

where Γ represents a fracture surface energy which is related to debonding load F_d, u_m and F_m the maximum displacement and load respectively, and τ_u the constant frictional shear stress for the unloading segment. Equation (2a) is equivalent to Equation (2), with the difference being that it is load based instead of energy based.

4.2 Marshall and Oliver (1990)

Marshall and Oliver extended their models further to cover the effect of axial pre-stress in the fibre that could be present due to sample preparation and matrix shrinkage.[9] This was not an easy task because there are two different cases depending on whether this pre-stress is compressive or tensile, and it is necessary to add some restrictions to avoid a theoretical spontaneous movement of the fibre.[9] This led them to the development of the following equations:

$$u = \frac{1}{4\tau E_f \pi^2 R^3}\left(F^2 - 2FF_r - F_d^2 + 2F_d F_r\right) \qquad \text{for loading} \qquad (5)$$

$$u = u_m - \frac{(F_m - F)^2}{8\tau_u E_f \pi^2 R^3} \qquad \text{for unloading} \qquad (6)$$

where F_r is the load associated with the pre-stress in the fibre. The unloading part in this model (Equation (6)) is exactly the same as in the previous model (Equation (3)) because the residual stress does not have any effect on the unloading curve.

4.3 Hsueh (1990)

Using a different approach, a model was proposed by Hsueh[10] to account for the clamping stress, the Poisson effect, and the deformation of the matrix. The model was derived by considering a single fibre of radii a embedded in a cylindrical composite sample of external radii b. This gave a very complicated load-sliding relationship,

$$u = \frac{1}{E_f}\left\{\frac{A_3}{A_2}\left[l + \frac{1-\exp(m_3 l)}{m_2}\right] + X_1\left[\frac{\exp(m_1 l)-1}{m_1} - \frac{\exp(m_2 l)-1}{m_2}\right]\right\} \tag{7}$$

$$l = \frac{1}{m_1}\ln\left[(m_1 - m_2)\left(\frac{F}{\pi a^2} - \frac{A_3}{A_2}\right)\right] \bigg/ \left(\frac{A_3}{A_2}m_2 + \sigma_c\frac{2\mu}{a}\right) \tag{8}$$

where A_1, A_2, m_1, and m_2 are parameters related to geometric dimensions, modulus and Poisson's ratios of both the fibre and matrix, and the friction coefficient μ; A_3 and X_1 are also a function of the applied load F, the clamping stress σ_c; while l is an estimate of the length of the debonded region. Detailed equations can be found in the reference.[10] More complex equations were also derived for the case of bonded interface,[11] but to solve them requires special mathematical software packages. Therefore, only the unbonded case (i.e. the Hsueh 1990 model) was attempted in this study.

5   EVALUATION OF MODELS AND DISCUSSIONS

5.1   Specimens Used

Two types of aged GRC specimens were tested.
- Specimens A - were extracted from GRC samples prepared by hand in laboratory using a Portland cement (Blue Circle, class 42.5), an alkali-resistant glass fibre (NEG) and a w/c ratio of 0.35. The samples had undergone an accelerated ageing in water at 60°C for 6 weeks before specimen preparation.
- Specimens B - were extracted from an external GRC cladding panel of a hotel in Glasgow, which was produced commercially by a spray process and installed over ten years ago. One side of the panel (~ 50 mm thick) was exposed to natural weathering while the other (inner side) was attached to insulation materials.

The procedures for specimens preparation included: precision cutting, embedding in resin, grinding, polishing and ultrasonic cleaning.

5.2   Results of Specimens A

A large number of fibres in the specimens were examined. These included: outer fibre filaments of a fibre bundle in close contact with the matrix, inner fibre filaments within a fibre bundle, and single isolated fibre filaments separated from a fibre bundle. The maximum load applied to the fibre in the fibre push-in test was adjusted for the different fibre filaments so that the fibre sliding length produced was greater than 200 nm and no damage/split occurred at the fibre end. The damage/split of fibre end could be identified by a sudden jump of displacement in the load-displacement curve obtained and further confirmed by observations under SEM.

Typical fibre push-in test curves for the three different types of fibre filaments examined and the theoretical fits using the models described are presented in Figures 5-7. The results in Figures 5 and 6 appeared to indicate that there was not much difference between the load-sliding displacement curves for the outer and inner filaments of a fibre bundle. As for the model fits, Figures 5 and 6 show that except for some small deviations in the initial part of the loading curve, all three models can produce nearly perfect fits to the experimental data for both the outer and inner fibre filaments.

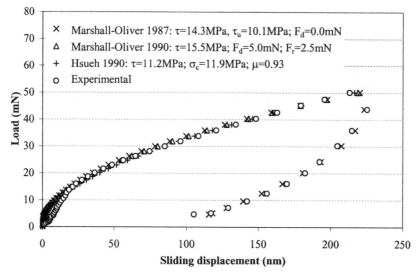

Figure 5 *Typical fibre push-in test results and model fits for Inner filaments of a fibre*

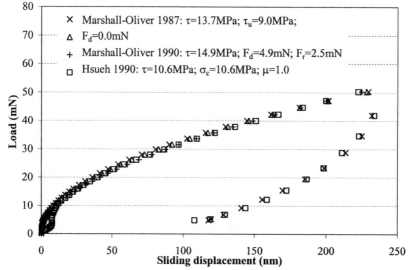

Figure 6 *Typical fibre push-in test results and model fits for Outer filaments*

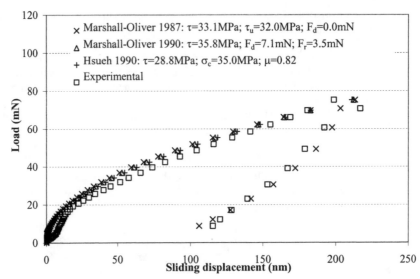

Figure 7   *Typical push-in test results and model fits for Isolated filaments separated from a fibre bundle*

For the isolated fibre filaments, however, the results were quite different. As shown in Figure 7, to generate the same level of sliding displacement the maximum loads applied were approximately 50% higher for the isolated fibre than for the outer or inner filaments. This suggested that the bond between isolated fibres and the matrix was much greater than that associated with fibres in a bundle. The model fits for the isolated filaments were also found to be reasonable, but not as good as those for the outer/inner filaments of a bundle.

To evaluate the three theoretical models used, careful examination of the model fits for the early part of the load-displacement curves were made. It appeared to reveal that using Marshall & Oliver 1987 model and Hsueh 1990 model produced very similar curves, and the fitting was slightly better than using Marshall & Oliver 1990 model. It is not clear whether the poor fits of all three models for the initial part of the loading curve were due to the theoretical models used or errors in the experimental data. It was observed that the fibre end and the matrix surface were not always at the same level in the polished specimens. Due to preferential polishing of the matrix in the interfacial area, the matrix surface was usually slightly lower than the fibre end.

The debonding loads F_d extracted from the model fits using Marshall & Oliver 1990 (Figures 5-7) would suggest a debonding stress in a range of 40 – 60 MPa, which seemed to be too high as tensile strength of the matrix is likely to be smaller than 10 MPa. The shear stress τ extracted from the model fits using all three models appeared to be similar and in a reasonable range. Also, as expected the shear stress for unloading (reverse sliding) τ_u was slightly lower than that for loading, due to wearing at the sliding interface. The values of kinetic frictional coefficient μ extracted from the Hsueh 1990 model were also too high for such a material combination.

Judging by the level of complexity and the range of interfacial parameters extracted, the Marshall & Oliver 1987 model appears to be the most suitable model among the three models examined for the test system.

5.3 Results of Specimens B

Specimens B were extracted from a commercial GRC panel manufactured over 10 years ago by a spray process. Compared to specimens A, the fibre content was much higher (5% by volume and the fibre strands were randomly distributed in specimens B. Microscopic observations revealed that there were large areas of fibre strands entangled together with thin layers of porous matrix present between the fibre strands, while in some patches a small number of fibre strands were surrounded by dense matrix. These structure features made it difficult to obtain a good polished specimen for the fibre push-in testing, since the inner fibres or fibre strands surrounded by porous matrix were pulled out or damaged during the polishing process. As a result, testing could only be carried out on selected fibre filaments in fibre bundles that bonded well to the matrix.

Typical fibre push-in test results and model fits are presented in Figure 8. The results presented were only part of the loading curve as damage/split of the fibre end was often detected/observed before the maximum load of 75 mN was reached. The damage/split of the fibre end at a lower load seemed to suggest the fibres in specimens B were more brittle, compared to the fibres in specimens A.

The model fits were again similar for all the three models used. The general fitting of the models to the experimental data was not good for the initial part of the loading curve. The interfacial parameters extracted from the model fits indicated that the bond associated with the fibres selected in specimens B was greater than that associated with outer/inner fibres in specimens A.

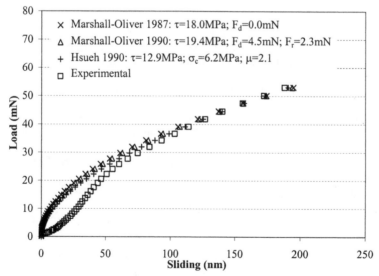

Figure 8 *Typical push-in test result for outer/inner fibre filaments in well bonded fibre strands and model fits*

6 CONCLUSIONS AND FURTHER WORK

The results of this preliminary study appear to indicate that all three theoretical models selected work reasonably well in most cases, except a small deviation in the early part of the loading curve. Among the three models used, the constant shear stress model (Marshall-Oliver 1987) is the simplest to use and the interfacial parameters extracted from it were within a reasonable range. Therefore, with the use of a suitable model, the fibre push-in test proves to be a valuable method for studying interfacial properties (or fibre-matrix interaction) in aged GRC composites.

For commercially produced GRC samples, sample preparation, especially the polishing process to acquire a good surface finish, remains a critical problem. Better specimen preparation techniques and equipment may be necessary to overcome this problem. There could be experimental errors in the initial part of the loading-sliding displacement curve, due to preferential polishing of the matrix in the area surrounding the fibres. It may be possible to take this (i.e. a free fibre length above the matrix) into account by modification of the existing models.

Subtraction of the penetration depth of indenter into fibre surface from the total displacement data could also generate errors in the experimental data. This is because the reference glass plate used as a reference would not always match the chemical compositions, ageing process and mechanical response of the glass fibres tested. An alternative approach could be to prepare a specimen from the same sample with fibres/strands horizontal to the surface, so that the indentation load applied to the fibre (perpendicularly) generates pure penetration. Work is currently in progress in this aspect.

References

1   M.S. Stucke and A.J. Majumdar, *Journal of Materials Science*, 1976, **11**, 1019.

2   A. Bentur, *Proceedings of the Durability of GFRC Symposium*, S. Diamond (ed.) 1985, 109.

3   P. Bartos, *Proceedings of the Durability of GFRC Symposium*, S. Diamond (ed.) 1985, 136.

4   W. Zhu and P.J.M. Bartos, *Cement and Concrete Research*, 1997, 27 (11), 1701.

5   W. Zhu, P, Trtik and P.J.M. Bartos, *Proceedings of the 12th GRCA Congress*, 2001, 103.

6   P. Trtik and P.J.M. Bartos, *Proceedings of 24th Conference on Our World in Concrete & Structures*, 1999, 415.

7   A. Bentur and S. Mindess, Fibre Reinforced Cementitious Composites, Elsevier Applied Science, London and New York, 1990.

8   D.B. Marshall and W.C. Oliver, *Journal of the American Ceramic Society*, 1987, **70** (8), 542.

9   D.B. Marshall and W.C. Oliver, *Materials Science and Engineering*, 1990, **A126**, 95.

10 C.H. Hsueh, *Journal of Materials Science*, 1990, **25**, 818.

11 C.H. Hsueh, *Journal of Materials Science*, 1990, **25**, 4080.

MOVING-WINDOW REPRESENTATION OF INTERFACIAL DEBONDING IN CONCRETE

L.L. Graham-Brady and D.J. Corr

Civil Engineering Department, The Johns Hopkins University, Baltimore MD 21218, USA

1 EXTENDED ABSTRACT

The analysis of concrete behavior at the structural scale has progressed in large part due to the development of empirical knowledge about the bulk mechanical properties, based on field experience and experimental research. These bulk mechanical properties can be used to accurately predict global structural behavior under many loading conditions; however, they provide little information about fracture and failure of concrete, processes that are dominated by micro- and even nano-scale random heterogeneities in the concrete microstructure. This paper presents a technique for stochastic simulation of concrete mechanical properties and the associated local stresses that occur under given loading conditions. These simulations are based on statistical descriptors of concrete mechanical and damage properties, which are obtained from analysis of digital concrete microstructural images applied in a moving-window context.

2 MOVING-WINDOW GMC

The composite description of concrete is widely considered to contain three components: aggregate, hydrated cement paste, and the interfacial transition zone[1]. The interfacial transition zone (ITZ) is a region of cement paste that surrounds large aggregate particles, typically 20-50μm thick. Because the aggregates in concrete are 3-4 orders of magnitude larger than the ITZ, the microstructure of concrete exhibits critical mechanical behavior at multiple scales. Accurate models of concrete behavior must therefore be able to capture important phenomena at all these scales. The ITZ is particularly important to the analysis of concrete, as it is characterized by higher porosity and lower stiffness and strength than the bulk hydrated cement paste[2,3]. Thus, the ITZ acts as a "weakest link" in concrete, and is responsible for many failure characteristics of concrete, because this region is where the first microcracks form during mechanical loading.

At the core of the concrete micromechanics model used here is the generalized method of cells (GMC). GMC is a composite mechanics model developed by Aboudi[4], and later generalized by Paley and Aboudi[5], which allows consideration of debonding between different phases of the composite microstructure[6]. In GMC, a repeating unit cell is identified from a periodic composite microstructure. Within the unit cell, the microstructure is divided into "subcells," with each subcell containing only one material of known constitutive properties. The mechanical properties and microstructural configuration of the subcells are used to determine the effective stress-strain curve of the repeating unit cell. In the context of a random concrete microstructure, the local constitutive behavior is anisotropic (due to the locally heterogeneous microstructure) and nonlinear (due to the debonding that occurs between aggregates and cement paste as a result of ITZ failure). Both the anisotropy and the nonlinearity due to debonding of any one portion of a concrete microstructure is modelled accurately using GMC.

Applying GMC in a moving-window context to digital microstructural images allows for mechanical characterization of heterogeneous composites[7]. With this technique, a window of fixed pixel size (e.g. a 10x10 pixel square) is scanned across a digital microstructural image, with the window moving one pixel at a time. At each window stop, the microstructure is extracted and analyzed with GMC, treating each pixel as a subcell, and each window as the unit cell in the GMC formulation. The resulting constitutive properties of the unit cell are mapped to an equivalent microstructure (location at the center of the original window) and stored for subsequent analysis.

This moving-window GMC technique allows for efficient analysis of a heterogeneous microstructure, without making assumptions about the size or shape of inclusions in the material. All that is required for this analysis is a microstructural image in which the phases of the microstructure can be visually distinguished, the constitutive properties of the component materials, and appropriate parameters describing the debonding between component materials. It should be noted that this analysis is an approximation, because GMC assumes periodic boundary conditions to describe a random microstructure; however, Graham et al.[8] demonstrate that the errors associated with moving-window GMC are less that 1% compared with an "exact" finite-element solution. The moving-window GMC process has been applied specifically to concrete in later work by the authors of this paper[9].

The field of local constitutive properties obtained from moving-window GMC are inserted into a finite element model that represents the concrete microstructure. Using this finite element model, the variation in local stresses and strains due to an applied load are obtained. This affords an accurate representation of how that particular sample microstructure would respond to mechanical loading.

3 STOCHASTIC SIMULATION

Perhaps of more interest than the response of a single sample of concrete would be a statistical description of the failure behavior in an ensemble of random concrete microstructures. This type of analysis would yield more information regarding the reliability of a given concrete under specified loading conditions. Since any one analysis of a physical sample involves sectioning, imaging and analyzing a specimen, it would be quite time consuming to develop meaningful ensemble results based solely on moving-window GMC analysis of separate specimens.

In order to curtail the difficulties associated with gathering multiple physical samples, stochastic simulation techniques are explored in this paper as a way to analyze many statistically equivalent microstructures based solely on computational modelling. One option is to simulate the microstructure explicitly. Concrete contains irregular shaped inclusions, however, so the only techniques available to perform these simulations are extremely costly[10]. The other option is to simulate the constitutive property fields obtained from the moving-window GMC analysis[11]. The constitutive properties are non-Gaussian and exhibit differing degrees of correlation with respect to each other, but stochastic simulation techniques for generating such fields are well-documented in the literature[12,13,14,15]. Based on each sample of the constitutive properties simulated, a finite element model is used to obtain the local stresses corresponding to that sample. Based on the ensemble of sample local stresses, the statistics on the local stresses is examined.

References

1   A.U. Nilsen and P.J.M. Monteiro, "Concrete: A Three Phase Material," *Cement and Concrete Research*, 1993, **23**, 147.

2   A.M. Neville, "Aggregate Bond and Modulus of Elasticity of Concrete." *ACI Materials Journal*, 1997, **94**(1), 71.

3   K. Mitsui, Z. Li, D.A. Lange, and S.P. Shah, "Relationship between Microstructure and Mechanical Properties of the Paste-Aggregate Interface," *ACI Materials Journal*, 1994, **91**(1), 30.

4   J. Aboudi, "Micromechanical Analysis of Composites by the Method of Cells," *Applied Mechanics Reviews*, 1989, **42**(7), 193.

5   M. Paley and J. Aboudi, "Micromechanical Analysis of Composites by the Generalized Cells Model," *Mechanics of Materials*, 1992, **14**, 127.

6   A.C. Sankurathri, S.C. Baxter, and M-J Pindera, "The Effect of Fiber Architecture on the Inelastic Response of Metal Matrix Composites with Interfacial and Fiber Damage." in *Damage and Interfacial Debonding in Composites*, eds., G.Z. Voyiadjis and D.H. Allen, Elsevier Science, The Netherlands, 1996, pp. 235-257.

7   S.C. Baxter and L.L. Graham, "Characterization of Random Composites Using Moving-Window Technique," *Journal of Engineering Mechanics*, ASCE, 2000, **126**(4), 389.

8   L.L. Graham, E.F. Siragy and S.C. Baxter, "Analysis of heterogeneous composites based on moving-window techniques," *Journal of Engineering Mechanics*, ASCE, in press.

9   D.J. Corr and L.L Graham, "Mechanical Analysis of Concrete with the Moving-Window Generalized Method of Cells," *ACI Materials Journal*, in press.

10   D. Cule and S. Torquato, "Generating Random Media from Limited Microstructural Information," *Journal of Applied Physics*, 1999, **8**(6), 3428.

11   L.L. Graham and S.C. Baxter, "Simulation of Local Material Properties Based on Moving-Window GMC," *Probabilistic Engineering Mechanics*, 2001, **16**, 295.

12   G. Deodatis and R.C. Micaletti (2001). "Simulation of Highly Skewed Non-Gaussian Stochastic processes," *Journal of Engineering Mechanics*, ASCE, 2001, **127**(12), 1284.

13   M. Grigoriu, "Simulation for Stationary Non-Gaussian Translation Processes," *Journal of Engineering Mechanics*, ASCE, 1998, **124**(2), 121.

14   K. Gurley and A. Kareem, "Analysis, interpretation, modeling and simulation of unsteady wind and pressure data," *Journal of Wind Engineering and Industrial Aerodynamics*, 1997, **69-71**, 657.

15   F. Yamazaki and M. Shinozuka, "Digital Simulation of Non-Gaussian Stochastic Fields," *Journal of Engineering Mechanics*, ASCE, 1988, **114**(7), 1183.

MOLECULAR MODELING OF CONFINED FLUIDS AND SOLID-FLUID INTERFACES IN PORTLAND CEMENT AND RELATED MATERIALS

R. James Kirkpatrick, Andrey Kalinichev, Jianwei Wang

Department of Geology and ACBM Center, University of Illinois, Urbana, Il 61801, USA.

EXTENDED ABSTRACT

The interaction of pore fluid with silicate and aluminate cement and cement hydration products plays a central role in controlling most of the important chemical reactions and mechanical properties of concrete, including strength development, rheology, durability, creep and cracking. Many of these interactions occur at the fluid-solid interface or in confined volumes with characteristic dimensions from 1 to 100 nm. Investigation of these nano-scale interactions on a molecular scale is difficult because of the chemical and microstrucutral complexity of cement-based materials. Even data from advanced experimental spectroscopic and scattering methods is often difficult to interpret unambiguously. Molecular modeling techniques, especially molecular dynamics (MD) methods have advanced substantially in recent years and can now provide an effective bridge between theory and experiment that yields both useful molecular and nano-scale insight and guidance for design of future experimental research. Classical MD methods are highly developed theoretically and in practice, and software is widely available. In this presentation we will focus on the fundamental aspects of MD modeling, the range of information it can provide, and current applications to Portland cement materials.

In a typical simulation, a three-dimensional system of atoms, ions and molecules containing as many as 100,000 particles and with lengths of the order of a few nm is constructed in the computer, and the interactions among the particles is calculated using a set of pre-determined interatomic potential (the "force field") among the different types of particles. The particles are given initial positions and velocities, and in classical MD their motion is calculated by numerical integration of Newton's equations over time for all particles. Simulations are carried out for periods of 10^{-10} - 10^{-8} s, which is quite long compared to, e.g., the time scales of molecular vibrations. Individual calculations often take several CPU days on fast multi-processor computers. The simulated system can consist of fluid, solid, or both. Much of our work has focused on fluid-solid interfaces and fluid confined in interlayers and other nano-spaces.

Successful molecular modeling depends critically on the interatomic potentials used, and there has been much effort in recent years to develop effective set of potentials for silicate, aluminate, and other metal oxide and hydroxide systems. Our simulations have been carried out using the newly developed CLAYFF force field that is based on quantum chemical calculations of partial atomic charges but is empirical optimized based on the structures of simple oxide and hydroxide materials. This is a relatively simple forcefield, but has the significant advantage that it can be readily used for materials that are chemically complex, disordered and poorly characterized on the molecular scale. It also allows proper formulation of the energy and momentum transfer between fluid and solid, which is essential to effective modeling of fluid-solid interfaces. Hydrogen bonding is key to understanding the pore fluids and fluid-solid interactions in cement-based materials, and CLAYFF uses the well-tested flexible SPC model for water molecules. We and others have tested this force field extensively, and the results show that it is highly effective in reproducing the known structures of a wide range of well understood crystalline compounds that were not used in its development. It also successfully reproduces many dynamical features, including molecular rotations, site hopping frequencies and residence times, and diffusion coefficients. Recently, we have used it to successfully model the complex cooperative molecular motions of interlayer and surface water and anions that are responsible for the experimentally observed low frequency (50 - 180 cm^{-1}) bands in far infrared spectra of metal hydroxides. We now have high confidence in the ability of MD methods using this forcefield to capture the essential features of the molecular and nano-scale structure and dynamics of a wide range of complex fluid-solid systems, including cements and concretes. The result of an MD simulation is 10^4 - 10^6 instantaneous particle configurations that represent the structure of the system at each time step of the calculation. This record of the trajectories of the particles (i.e., their positions and velocities, and if they are molecules their orientations and angular velocities) is a complete description of the system in a classical mechanical sense. With the help of statistical mechanics, the thermodynamic, structural, spectroscopic, and transport properties of the system can then be calculated from the time averages of these parameters. It is generally assumed that the total potential energy of the system is the sum of the energies of all interactions among all particles. Most commonly these are taken pair-wise, but many body terms can be included. It is possible, for instance, to calculate pressure, temperature, internal energy, heat capacity, interatomic distances, coordination numbers, radial distribution functions, diffusion coefficients, residence times of individual species on particular structural sites, and the power spectrum (atomic density of states) that can be related to observed vibrational spectra.

We have undertaken successful MD computer models of the structures of representatives of the most important groups of phases present in hydrated cement paste, including Ca-hydroxide (portlandite), the hydrated Ca-aluminate phases ettringite (AFt) and AFm, and tobermorite (model C-S-H) and the interaction of aqueous Cl^- and associated cations (Na^+, K^+, Cs^+) with their surfaces. The crystal structures are very similar to those observed experimentally and the results of the surface interaction studies provide significant insight into the controls exerted by the structure and composition of the solid phase on the structural environments of the surface-associated species, the structure of the solution near the interface, and the translational, librational, and vibrational dynamics of the surface species including effective diffusion coefficients and approximate surface site lifetimes. Ongoing work is investigating the effects of surface composition and structure on the structure and dynamical behaviour of near-surface fluid.

DENSITY FUNCTIONAL CALCULATION OF ELASTIC PROPERTIES OF PORTLANDITE AND FOSHAGITE

J. L. Laugesen

Department of Civil Engineering, Technical University of Denmark, 2800 Lyngby

1   INTRODUCTION

The structure and properties of calcium-silicate-hydroxide (CSH) has been investigated by chemists and physicists for about half a century. Many methods, nano-, micro- and macroscopic methods have been applied to study in particular the structure. The crystal structure is very important since this is the key to understand crystal interaction with environmental materials. Portlandite is the second most abundant product in hydrated cement pastes. It is always present as crystals of typically a few micrometers. Its influence on mechanical properties of cements is believed to be of minor importance, but a final conclusion have not been achieved.

CSH in the foshagite phase only occurs at elevated temperatures. However, the structural similarity with other phases, as for instance tobermorite, makes it reasonable to study this phase.

The primary goal of this paper is to confirm and determine the structure of the minerals foshagite and portlandite, of which the first one is a member of the CSH group. Portlandite is perhaps the most simple crystal appearing in Portland cement and the properties are well-known from experiments. The elastic properties has as far as the author knows not yet been determined theoretically. However, experimental studies of portlandite have been done[11] and experimental studies are being done by the author for foshagite at the moment of writing this paper. In this paper the elastic properties are calculated using density functional theory.

2   PRINCIPLES OF DENSITY FUNCTIONAL THEORY

Density functional theory (DFT) provides a way to determine the total energy of a system as a functional of the electron density, for detailed information see the pioneer work of Hohenberg and Kohn[1] and Kohn and Sham[2]. An introduction to DFT have been written by W. Koch and M. C. Holthausen[3].

In this study a plane wave pseudopotential software package, called DACAPO[4], has been used. In the plane wave approximation, the plane waves only describe the chemically active electrons (the valence electrons). The effects of the nucleus and the core electrons (the strongly bonded electrons) are replaced by a pseudopotential. Thus, the valence electrons are moving in a potential due to all non-chemically active charges. The basic problem is then to obtain a set of plane waves that describe the valence electrons in a given potential.

The pseudopotentials used are generated by use of the generalized gradient approximation (GGA) developed by Perdew and Wang (PW91). These potentials are included in the software package. It should be noted that since the number of equations to solve grows by the cube of the number of electrons, it is extremely beneficial to apply pseudopotentials. To this may be added that to describe the innermost electrons a very large number of plane waves are often required due to fast variation of the nuclear potential, which further increases the computation times.

2   THE STRUCTURE OF PORTLANDITE AND FOSHAGITE

2.1 Portlandite

The structural formula of portlandite is $Ca(OH)_2$. Its structure has been experimentally determined by X-ray[5,6] and Neutron[7] diffraction spectroscopy. It is a layered hexagonal lattice with space group symbol $P\bar{3}m1$. The lattice parameters are $a = b = 3.593\text{Å}$, $c = 4.909\text{Å}$ and $\gamma = 120^0$, $Z = 1$. The Ca^{2+} ions are octahedrally coordinated by oxygen and the O^{2-} ions are tetrahedrally coordinated by three calcium and one hydrogen ion.

2.2 Foshagite

Foshagite has together with wollastonite the lowest number of atoms per unit cell of all crystals in the CSH group. It has been shown by Taylor and Gard[8] that the composition is $Ca_4Si_3O_9(OH)_2$. By use of X-ray spectroscopy the unit cell was later[9] shown to be monoclinic with parameters $a = 10.32\text{Å}$, $b = 7.36\text{Å}$, $c = 7.04\text{Å}$, $\beta = 106.4^0$ and $Z = 2$ with space group symbol $P\bar{1}$. The unit cell contains 40 atoms of which 20 have independent positions. The space group $P\bar{1}$ contains an inversion symmetry from which the remaining positions is determined. The Si^{4+} ions are tetrahedrally coordinated by oxygen and the tetrahedra are linked in chains to form dreierketten. As typical for CSH, the Ca^{2+} ions are octahedrally coordinated by oxygen or hydroxyl. The calcium is approximately positioned in a plane parallel to (101).

Since the X-ray spectroscopy method is practically incapable of determining the positions of light atoms only the positions of the heavy atoms has for the time being been obtained experimentally. However, qualitative positions can be argued. Since OH^- is a dipole and coordinated towards Ca^{2+}, the H^+ must be placed away from the positive calcium plane towards oxygen in the SiO_4^{4-}-chains.

2 STRAIN, STRESS AND ELASTICITY

1.1 Stress and strain, Hooke's law

A deformation is defined by the relative displacements respectively to the axes, $\varepsilon_{ij} = \partial u_i / \partial x_j$, which defines the strain tensor. It is sometimes convenient to use matrix notation instead. The strain matrix relates to the strain tensor by

$$e_i = \begin{pmatrix} e_1 & e_6 & e_5 \\ e_6 & e_2 & e_4 \\ e_5 & e_4 & e_3 \end{pmatrix} = \begin{pmatrix} \epsilon_{11} & 2\epsilon_{12} & 2\epsilon_{31} \\ 2\epsilon_{12} & \epsilon_{22} & 2\epsilon_{23} \\ 2\epsilon_{31} & 2\epsilon_{23} & \epsilon_{33} \end{pmatrix} \tag{1}$$

The stress is the force per unit area corresponding to the response of the material to a strain. The stress is described by a stress tensor or stress matrix and is given by

$$\sigma_i = \begin{pmatrix} \sigma_1 & \sigma_6 & \sigma_5 \\ \sigma_6 & \sigma_2 & \sigma_4 \\ \sigma_5 & \sigma_4 & \sigma_3 \end{pmatrix} = \begin{pmatrix} \sigma_{11} & \sigma_{12} & \sigma_{31} \\ \sigma_{12} & \sigma_{22} & \sigma_{23} \\ \sigma_{31} & \sigma_{23} & \sigma_{33} \end{pmatrix} \tag{2}$$

In linear elasticity stress relates to strain by

$$\sigma_i = c_{ij} e_j \tag{3}$$

where c_{ij} is the stiffness matrix. The relation between a unit cell U_{ij} and the unit cell U'_{ij} subject to a strain, ε_{ij} is

$$U'_{ij} = (\delta_{ik} + \epsilon_{ik}) U_{kj} \tag{4}$$

where δ_{ij} is Kroneckers delta. The strain tensor may be replaced by the strain matrix e_i, according to (eq. 1).

In practice there will always be a small residual stress, σ_i^r, after a refinement of the lattice parameters. As long as this residual stress is within the linear region, this stress can be accounted for. Assume that U_{ij}^0 is the exact Bravais lattice and U_{ij}^r is the lattice subjected to the residual strain ε_i^r, then according to the above

$$U_{ij}^r = (\delta_{ik} + \epsilon_{ik}^r) U_{kj}^0 \quad \text{and} \quad \sigma_i^r = c_{ij} e_j^r \tag{5}$$

The exact stress is given by

$$\sigma_i = c_{ij} (\epsilon_{ik} + \epsilon_{ik}^r) \tag{6}$$

This leads directly to the corrected Hooke's law

$$(\sigma_i - \sigma_i^r) = c_{ij}(\epsilon_j - \varepsilon_j^r) \tag{7}$$

where $(\varepsilon_j - \varepsilon_j^r)$ is the external strain applied. This relation holds as long as the stresses are within the linear region. The residual stress should of course be as small as possible, not only to minimize the error, but also in order to make sure that the system is not in some metastable state.

2 RESULTS

1.1 The calculated equilibrium states of the crystals.

Before calculation of the stresses in the strained crystals, the equilibrium state of portlandite and foshagite must be found in the frame of DFT. A cut off energy of 700eV and 340eV, respectively, have been used for portlandite and foshagite. The number of kpoints within the first Brillouin zone is 108 for portlandite and 2 for foshagite. The kpoints are distributed along the axes (6,6,6) for portlandite and (1,2,2) for foshagite. In the case of portlandite the interaction between layers along the c-axis is very weak. It has been shown[10] that the energy minimum depends highly on the spin polarization of the interaction between the layers. Hence, for portlandite the calculation is performed spin polarized.

The relaxation of the system is performed iteratively. The ionic structure is relaxed such that the maximum force is below 0.01 eV/Å. The lattice parameters are relaxed to a state where the maximum residual stress component is lower than $2 \cdot 10^{-4}$ ev/Å^4.

The equilibrium lattice parameters are listed in table 1 together with the order of magnitude of the residual stress. Apart from the c-axis of portlandite the values are seen to be within 0.5% of the experimental values. The c-axis of portlandite is about 1% too small compared to the experimental value.

Table 1   *Calculated and experimental lattice parameters for portlandite and foshagite. Also shown is the residual stress at equilibrium.*

Portlandite	Calculated	Experimental
a / Å	3.609	3.593
c / Å	4.864	4.909
Res.stress / Gpa	0.032	
Foshagite		
a / Å	10.31	10.32
b / Å	7.32	7.36
c / Å	7.07	7.04
γ	106.76	106.4
Res.stress / Gpa	0.032	

1.1 The hydrogen ions in foshagite

In order to determine the positions of the protons in foshagite a number of different orientations of the hydroxyl groups has been tried. It is assumed that the inversion symmetry also holds for the positions of the hydrogen positions. However, this needs not to be true, since the X-ray study[9] is unable to give information of the non-heavy hydrogen. Thus, this is only an assumption and should be studied further.

The configuration that gave the lowest energy was found when the positions of the hydrogen ions were about one OH bond length off from the oxygen position and almost on the straight path towards to nearest oxygen atom in the tetrahedra chains. After an iterative relaxation of the ions and the unit cell the atoms converged to their equilibrium positions, see table 2.

Table 2   *The positions of the hydrogen atoms in scaled coordinates. The inversion symmetry determines the positions of the remaining two hydrogen atoms.*

Hydrogen	x/a	y/b	z/c
H1	0.3700	0.6275	0.0085
H2	0.3523	0.6275	0.2498

The bond lengths in the OH groups are 0.988Å. The hydrogen labelled H1 is in a distance of 1.876Å from the closest oxygen atom in the SiO_4^{4-} tetrahedra, while the H2 is 3.09Å from the closest oxygen atom in the tetrahedra chains. Since the bond lengths between H2 and the oxygen atom in the OH1 hydroxyl group and the bond lengths between the hydroxyl groups and the calcium ions approximately are of the same size as the corresponding bond lengths in portlandite this indicates that the hydroxyl groups are more coordinated towards the calcium layer than to the SiO_4^{4-} tetrahedra.

Figure 1 shows the unit cell of foshagite together with the plane (010). This plane intersects 2 hydroxyl groups, 2 calcium ions and a part of a SiO_4^{4-} tetrahedra. Figure 2 shows the electron density in the (010) plane. This further illustrates the coordination of the calcium ions by the hydroxyl groups as described above. The SiO_4^{4-} tetrahedra is clearly recognized and is indicated by a triangle. A straight line connects the two calcium ions. The contour curves in the region of the hydroxyl group in the right of the figure are clearly non-spherical and is seen to interact with the closest oxygen in the SiO_4^{4-} tetrahedra.

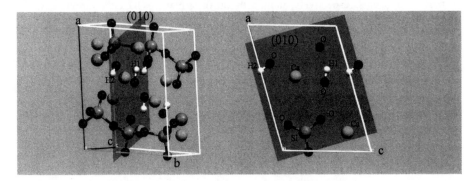

Figure 1   *The unit cell of foshagite shown with the plane (010) intersecting the hydroxyl groups. To the right the unit cell is shown perpendicular to the (010) plane and only the atoms intersecting the plane are shown.*

Figure 2   *The electron density in the plane (010). The endpoints of the straight line indicates the positions of Ca^{2+} ions in the plane and the triangle in the lower left corner illustrates the SiO$_4^{4-}$ tetrahedra.*

1.1 Stress-strain calculations

The applied strains are all uniaxial, meaning that only one of the strain components are non-zero. In this way a whole column of the stiffness matrix may in principle be calculated with only one non-zero strain component. However, the uniaxial strains are not independent from the shear strains and a least square method is used to fit the linear dependence.

It may be shown[10] that a hexagonal and a monoclinic lattice has 5 and 13 independent stiffness components, respectively, and that the stiffness matrices have the forms

$$
c_{ij}^{hexa} =
\begin{pmatrix}
c_{11} & c_{12} & c_{13} & 0 & 0 & 0 \\
 & c_{11} & c_{13} & 0 & 0 & 0 \\
 & & c_{33} & 0 & 0 & 0 \\
 & & & c_{44} & 0 & 0 \\
 & & & & c_{44} & 0 \\
 & & & & & c_{66}
\end{pmatrix}
\qquad
c_{ij}^{mono} =
\begin{pmatrix}
c_{11} & c_{12} & c_{13} & 0 & c_{15} & 0 \\
 & c_{22} & c_{23} & 0 & c_{25} & 0 \\
 & & c_{33} & 0 & c_{35} & 0 \\
 & & & c_{44} & 0 & c_{46} \\
 & & & & c_{55} & 0 \\
 & & & & & c_{66}
\end{pmatrix}
$$

$$(8)$$

where the component $c_{66} = \frac{1}{2}(c_{11} - c_{12})$ in the hexagonal case. The strains applied to the unit cells of portlandite and foshagite are all chosen to be ±0.01. Since all components of the stress tensor are returned by the code a large number of calculations are saved.

The calculated stiffness matrix for portlandite is

$$
\mathbf{c}_{ij}^{portlandite} =
\begin{pmatrix}
98.51 & 29.02 & 8.58 & 0 & 0 & 0 \\
 & 98.51 & 8.58 & 0 & 0 & 0 \\
 & & 39.73 & 0 & 0 & 0 \\
 & & & 12.86 & 0 & 0 \\
 & & & & 12.86 & 0 \\
 & & & & & 34.75
\end{pmatrix} GPa
\tag{9}
$$

The elastic constants are relatively large in the (001) plane both regarding tensile and shear components, while the elastic constants are a factor 2.5-3 smaller when they includes the c-axis. The values of the components are in reasonable agreement with the experimental results of Monteiro and Chang[11], who report the constants to be $c_{11} = 99.28 GPa$, $c_{12} = 36.18 GPa$, $c_{13} = 29.65 GPa$, $c_{33} = 32.6 GPa$, $c_{44} = 9.846 GPa$, $c_{66} = 31.55 GPa$. There is, however, a factor 3.4 in difference on the c_{13} component, a deviation which is at present unexplained. The remaining components differ by less than 25%.

The calculated stiffness matrix for foshagite is

$$
\mathbf{c}_{ij}^{foshagite} =
\begin{pmatrix}
114.28 & 59.04 & 40.69 & 0 & 1.93 & 0 \\
 & 189.89 & 43.72 & 0 & -7.12 & 0 \\
 & & 123.33 & 0 & -19.26 & 0 \\
 & & & 39.05 & 0 & 1.95 \\
 & & & & 31.41 & 0 \\
 & & & & & 56.94
\end{pmatrix} GPa
\tag{10}
$$

The tensile components for foshagite are all of the same order. Only some of the shear moduli are small and some of them even negative. It should here be noted that negative elastic coefficients are allowed as long as the strain energy $W = \frac{1}{2} c_{ij} e_i e_j$ is positive.

The definition of the uniaxial Young's modulus is the inverse of the compliance matrix elements along the crystallographic axes. The values of the special moduli are listed in table 3.

Table 3 *Young's moduli along the crystallographic axes for portlandite and foshagite.*

	[100]/GPa	[010]/GPa	[001]/GPa
Portlandite	89.02	89.02	38.58
Foshagite	86.98	154.09	94.21

2 CONCLUSION

The equilibrium states of portlandite and foshagite, which are of the highest importance for the elastic properties, are studied. The structure has been calculated using a density functional method. In the case of portlandite the deviation of the c-axis from experimental value is about 1% too small. In the case of foshagite, the positions of hydrogen are calculated as well. The elastic properties of the two crystals have been calculated. In the case of portlandite the calculated values are in good agreement with experimental values. Recent experimental studies by the author of a natural foshagite mineral confirm the order of magnitude for the tensile moduli. These studies are awaiting publication.

References

1   P. Hohenberg and W. Kohn, Phys. Rev., **136**, B864, 1964.
2.   W. Kohn and L. J. Sham, Phys. Rev., **140**, A1133
1   W. Koch and M. C. Holthausen, *A Chemist's Guide to Density Functional Theory*, WILEY-VCH, 2001.
2   DACAPO Version 2.6.1, Density Functional Theory software, CAMP, Technical University of Denmark.
3   J. D. Bernal and H. D. Megaw, Proc. Roy. Soc., **A151**,384, 1935.
4   H. E. Petch and H. D. Megaw, J. Opt. Soc. Amer., **44**,744,1954.
5   W. R. Busing and H. A. Levy, J. Chem. Phys., **26**, 563, 1957.
6   J. A. Gard and H. F. W. Taylor, Amer.Min., **43**, 1, 1958.
7   J. A. Gard and H. F. W. Taylor, Acta. Cryst., **13**, 785, 1960.
8   J. F. Nye, Physical Properties of Crystals, Oxford University Press, 1985.
9   P. J. M. Monteiro and C. T. Chang, Cement and Concrete Research, **25**, 8, 1995.

EXPLORING THE MICRO-MECHANICS OF OPEN-ENDED PILE DRIVING VIA DISCRETE ELEMENT MODELLING.

Catherine O'Sullivan and Kenneth G. Gavin

Department of Civil Engineering, University College Dublin, Earlsfort Terrace, Dublin 2, Ireland

1   INTRODUCTION

Soil is a construction material that exhibits a complex response as a consequence of the micro-scale particle interactions that occur when it is deformed. Recently geotechnical engineers have been applying a relatively new type of analysis called discrete element modelling (DEM) to advance their understanding of the response of soil at the particle scale. In a discrete element simulation analysts can easily model parameters such as individual particle trajectories, the evolution of inter-particle contact forces, and particle rotations. At the current time, measurement of these parameters in a physical model is not tractable, making discrete element modelling an attractive tool for studying the micro-mechanisms of soil response.

Central to the advancement of discrete element modelling technology, is the development of homogenization or smoothing algorithms to relate the micro-scale phenomena to the meso-or macro scale material response. A novel kinematic homogenization method that uses non-linear (cubic) interpolation function and incorporates particle rotation effects was proposed by O'Sullivan[1]. The kinematic homogenization technique has been implemented in both two and three dimensions, and would be useful to develop scientific understanding of how the structure and deformation of any material at the micro-or nano-level influences the meso- or macro-scale properties and response. Homogenization methods are likely to play an important role in numerical analysis to develop construction materials from the bottom up, i.e. specifying the material structure and composition at the nano-level to meet micro- or macro- scale material performance requirements. The details of the non-linear homogenization approach are presented in this paper.

One civil engineering problem where micro scale particle interactions are important is the development of plugs in open-ended piles. Open-ended piles develop their base capacities through a combination of the "internal skin friction" transferred through the internal soil column and resistance beneath the annular area of the pipe. However as noted by Jardine and Chow[2], the "internal skin friction component" is only likely to make a substantial contribution in cases where a practically rigid arch (or plug) develops almost immediately within the base of the internal pile soil column. A DEM analysis would provide detailed information about the micro-mechanics of this plug formation. A simulation of installation of an open ended pile in a two dimensional granular material is presented as a case study in this paper to illustrate the application of the homogenization methods described in this paper to a problem of interest in civil engineering construction.

2   HOMOGENIZATION APPROACHES

2.1 Overview

In a DEM analysis of a particulate material, the individual particles are modelled as rigid disks (in two dimensions) and rigid spheres (in three dimensions). Stiff, linear springs are used to model the interparticle contacts and particle sliding is modelled using Coulomb friction. This method was originally proposed by Cundall and Strack[3] and an overview of recent developments is given by O'Sullivan[1]. A discrete element simulation of a system of particles calculates the system response in terms of particle displacements and inter-particle forces. There has been much research in recent years to develop homogenization techniques to express the analysis results in terms of the more conventional engineering parameters of stress and strain (e.g. Cambou et al[4]).

Kinematic homogenization methods can be applied as a post-processor to the discrete element analysis to obtain an average displacement gradient value to a collection of particles. The strain values are then calculated from the homogenized displacement gradients.   The kinematic homogenization approaches first discretize the domain of particles, then a interpolation method is used to calculate the displacement gradients. As noted by O'Sullivan[1] all of the existing methods use a linear interpolation approach to calculate the displacement gradients. In addition all of these methods consider only the location of the particle centroids, and neglect particle rotation effects.

A linear approach to homogenization is not appropriate for problems involving large, localized deformations, or where the particle response is anyway random in nature. Examples illustrating the inability of linear approaches to effectively capture localization evolution are detailed by Thomas[5] (for two-dimensional problems) and O'Sullivan[1] (for three-dimensional problems). Where there are local erratic fluctuations in the particle displacements, linear interpolation approaches will give strain values with substantial inter-element variation, making quantification of strain within the material tenuous. It may be desirable to average or smooth the displacement gradient, so that one may be able to estimate the overall strain at the region, especially where strong or weak discontinuities develop, for a range of materials and applications.

In response to the limitations of the existing approaches, O'Sullivan[1] proposed a new kinematic homogenization approach, specifically designed to analyze finite deformation problems. This method is also well suited to calculate strain for any materials having local random particle response, at either the nano- or micro-scale. The method uses non-linear, non-local interpolation, and can account for particle rotation effects (important in granular materials). The details of the interpolation approach are given here.

2.2 Non-Linear Interpolation Approach

The framework for the non-linear interpolation approach was found in the literature relating to the "mesh-free" methods.   Mesh-free methods have been used by researchers to simulate strain localization problems (e.g. Li and Liu[6]). The current study considers only the applicability of mesh-free shape functions to homogenize or to smooth the discrete data sets of particle displacements. For a more detailed description of the mesh-free methods refer to Li and Liu[6].

The homogenization method takes as input the particle coordinates and incremental displacement vectors. A rectangular grid is generated to serve as a referential continuum discretization over the volume of particles under consideration.   The interpolated displacements and displacement gradients are then calculated at these grid points using the mesh-free interpolants. As illustrated in Figure 1, a zone of influence is associated with each particle; this is the area over which the particle contributes to the interpolated displacement field, and hence the average strain field. For the analyses discussed here this zone of influence is a circle with a radius of twice the average particle radius.

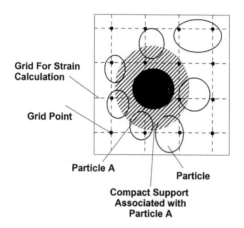

Figure 1   *Schematic diagram of mesh-free interpolation approach.*

At each grid point the interpolated displacement value is calculated by:

$$u(x) \cong \sum_{i=1}^{N_p} K_\rho(x - x_i, x) u(x_i) \Delta V_i \tag{1}$$

where N_p is the number of particles, ΔV_i is the nodal weight (calculated by triangulating the system) , and the term $K_\rho(x-x_i,x)$ is given by

$$K_\rho(x - x_i, x) = C_\rho(x - x_i, x) \Phi_\rho(x - x_i, x) \tag{2}$$

where $C_\rho(x-x_i,x)$ is a correction function to reduce the interpolation error and

$$\Phi_\rho(x) = \frac{1}{\rho^d} \phi\left(\frac{x}{\rho}\right) \quad (d=2, \text{ for 2D and } d=3, \text{ for 3D}) \tag{3}$$

is a compact supported window function, where ρ is the dilation parameter that defines the size of the window function. In this paper, the cubic spline function is chosen as the compact supported window function,

$$\phi(x) = \frac{1}{6}(x+2)^3 \quad -2 \leq x \leq -1 \tag{4}$$

$$\phi(x) = \frac{2}{3} - x^2(1+x/2) \quad -1 \leq x \leq 0 \tag{5}$$

$$\phi(x) = \frac{2}{3} - x^2(1-x/2) \quad 0 \leq x \leq 1 \tag{6}$$

$$\phi(x) = \frac{1}{6}(x-2)^3 \quad 1 \leq x \leq 2 \tag{7}$$

$$\phi(x) = 0 \quad otherwise \tag{8}$$

In higher dimensions, the window function is formed as a Cartesian product, i.e.:

$$\phi(x,y) = \phi(x)\phi(y)$$
$$\phi(x,y,z) = \phi(x)\phi(y)\phi(z) \tag{9}$$

To calculate the mesh-free shape function, $K_\rho(x-x_i,x)$, at each sample point i (tracked point) the size of the compact support must be known. In two dimensions, the compact support is a circle centered at the tracked point with radius $r = 2w_r$, where w_r is the radius of the circular particle associated with the tracked node.

There are several ways to calculate the correction function. The simplest way is to use the condition of partition of unity,

$$\sum_{i=1}^{N_\rho} C_\rho(x-x_i,x)\phi_\rho(x)\Delta V_i = 1 \tag{10}$$

to determine $C_\rho(x-x_i,x)$ at an arbitrary point x. To obtain higher order interpolation accuracy, one may enforce certain "reproducing conditions" and find the correction function by solving a moment equation. For details, readers may consult Liu et al.[7]

Each tracked point is also associated with a weight, ΔV_i, which may be calculated by first triangulating the system in terms of all the tracked points. Then each tracked point is a vertex of a number of triangles. The weight, ΔV_i, can then be determined as

$$\Delta V_i = \frac{1}{N_v}\sum_{k=1}^{N_T} \Delta \Omega_k \tag{11}$$

where $\Delta \Omega_k$ represents the area of a triangle with a vertex at the tracked point i, N_T is the total number of triangles with vertices at point i, and N_v is the number of vertices per triangle (N_v is 3 in two dimensions). Similarly, in three dimensions, the system was divided into tetrahedra, and the incremental volume associated with particle i, was calculated by considering the tetrahedra with vertices at point i (N_v is 4 in three dimensions).

The two-dimensional shape function described above is plotted in Figure 2(a). Visualization of the 3-D shape function is non-trivial. In Figure 2(b), the variation in magnitude of the shape function is illustrated with shading. The region with darker color indicates where the shape function value approaches zero, while white or bright color indicates the region where the shape function is a maximum. As in the finite element method, the displacement gradients are calculated by taking the first derivative of the shape function.

(L_{wp} + L_{up}) can be determined from external measurements, however a determination of L_{wp} is not possible. Measurement and analysis of β and L_{wp} would be relatively trivial in a three-dimensional discrete element analysis of the pile installation. The first stage in development of the three dimensional model is two perform some (less expensive) two dimensional simulations to explore boundary conditions and to examine scale effect issues. In the current study a two-dimensional discrete element model is used to model open-ended pile installation to illustrate the application of the kinematic homogenization approaches to interpret micro scale interactions at the meso and macro scale.

Earlier DEM simulations of relevance to this project include the work of Huang and Ma[12] who performed simulations of cone penetration tests in normally consolidated and overconsolidated disk assemblies. Huang and Ma acknowledged that a 2-D simulation cannot realistically represent a 3-D deposit of granular material. They argued that as an assembly of particles reaches equilibrium by balancing of interparticle forces, whether in 2-D or 3-D, the mechanism of particle movement obtained from a 2-D simulation is expected to be the same as that of a 3-D computation. To take advantage of symmetry only half of the penetrometer and soil mass were used. In a similar study Ke and Bray[13] displaced a rigid footing into a "sand box" of two-dimensional disks to illustrate the capabilities of the disk code DDAD.

3.1 Model Description

The numerical simulations of the performed with the two-dimensional discrete element code PFC-2D developed by Itasca[14]. This program uses the original DEM formulation as developed by Cundall and Strack[3]. The approach in DEM is to model particles as rigid spheres. More complicated particle geometries are accounted for by "gluing" groups of spheres together to form clusters[8]. The computer program can monitor the position of each particle in the system and detect which particles are touching each other. Simple contact models are used to calculate the inter-particle forces. The model comprised 29,458 disks of diameter 5mm, 7.5mm and 10mm. Initially, the assembly of disks was brought into equilibrium under gravity in an open rectangular box. The disk configuration prior to pile installation is given in Figure 4(a). Figure 4(b) is a magnified plot (200 mm x 200 mm) of a representative section of the disk assembly. To simulate pile driving two rigid walls were pushed into the box of disks with a constant velocity. Two simulations were carried out to examine the effects of the coefficient of friction at the interface of the pile with the disks on the response of the disks. For the first ("smooth') simulation the angle of friction at the pile disk interface was assumed to be $0°$, while the interface friction angle was assumed to be $7.5°$ for the second ("rough") simulation.

Parameter	Value
Density	2.6 x 10^8 kg/m^3
Spring Stiffness	1 x 10^{12} N/m
Damping parameter	0.7
Coefficient of friction (disk-disk)	0.487 (26°)

Table 1 *Parameters used in simulations*

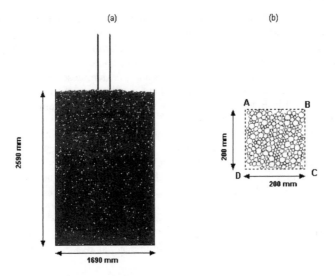

Figure 4 *Initial Configuration of System*

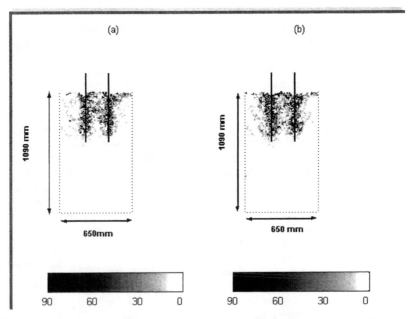

Figure 5   *Deformed particle configuration (a) Pile - -disk friction angle = 0°, (b) Pile - -disk friction angle = 7.5°*

The deformed particle configuration for the zone immediately adjacent to the pile after approximately 450mm of penetration is given in Figure 5. The amount of particle rotation is illustrated by the disk shading, with disks that have experienced more rotation having a darker colour. In both cases significant rotation is observed close to the pile interface. As would be expected, the simulation with perfectly smooth disks yielded disks with higher rotation values. The apparent zone of influence is slightly narrower for the smooth simulation in comparison with the rough simulation. In both cases, it would appear that the zone of influence of the pile over the material is relatively small, suggesting that a relatively small fraction of the problem domain is required to be modelled to gain insight into the micro-mechanics of soil-pile interaction.

The particle displacement vectors are given in Figure 6. The amount of particle displacement is considerably smaller for the simulation with a smooth pile-particle interface. As with the particle rotation plots, these results seem to indicate that the zone of influence of the pile extends further for the rough simulation in comparison with the perfectly smooth case. While particle displacement and rotation plots give some insight into the problem, they are of little use when trying to reconcile the results with our current understanding of the material response which is based upon continuum mechanics

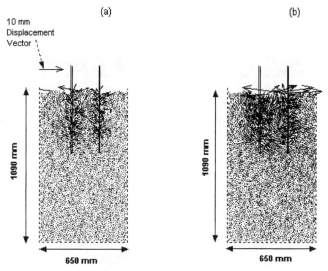

Figure 6 *Particle displacement vectors (a) Pile - -disk friction angle = 0°, (b) Pile - -disk friction angle = 7.5°*

Figure 7 illustrates the volumetric strain values calculated using a simple, linear interpolation approach. The strain contours are plotted at intervals of 2.5 up to a maximum value of 15. The maximum strain is 10 for the smooth simulation and 12.5 for the rough simulation. In Figure 8 the strains calculated using the non-linear interpolation approach, without accounting for rotation effects are illustrated. In Figure 8 the strain contours are plotted at intervals of 0.5 up to a maximum of 5. For both simulations the strain values are significantly lower than for the linear approach due to the smoothing effect of the non-linear interpolation. As before the maximum strain values are slightly higher for the rough simulation (3.5) than for the smooth simulation (3). The non-linear interpolation appears to better capture the extent of the disturbed soil zone in comparison to the linear approach. In fact, the linear interpolation approach appears to indicate that the extent of the disturbed zone is greater for the rough pile than the smooth pile, contrasting with the indications from the particle rotation and displacement plots.

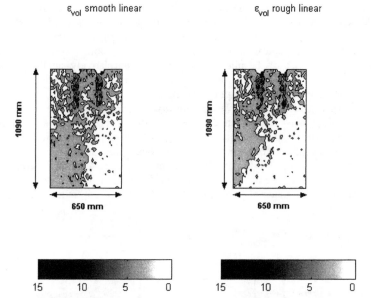

Figure 7   *Volumetric strain contours: Linear interpolation with no rotation effects accounted for (a) Pile - disk friction angle = 0°, (b) Pile - disk friction angle = 7.5° (Contour Interval 2.5)*

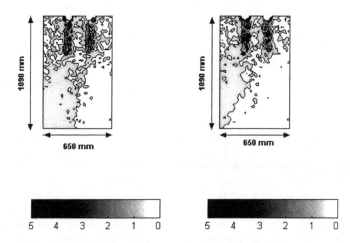

Figure 8   *Volumetric strain contours: Non-Linear interpolation with no rotation effects accounted for (a) Pile - disk friction angle = 0°, (b) Pile - disk friction angle = 7.5° (Contour Interval 1)*

The analysis described here comprises the preliminary stage in a detailed analysis of the micro-mechanics of plug formation in open-ended piles. Many researchers (e.g. Thomas and Bray[15]) have highlighted the fact that using perfectly circular disks is a limited analogue to real non-spherical soil grains, as these particles tend to rotate excessively. Future work will include examining the effects of particle rotation using the overlapping disk clusters proposed by O'Sullivan et al[16]. To limit the computational cost of the simulations it would appear that recognizing the axi-symmetric nature of the problem would be appropriate. Further research must be done to determine the appropriate boundary conditions. While Huang and Ma used a vertical wall in their simulations of CPT tests, for this problem it may be more appropriate to use a periodic boundary as described by Bardet[17]. As demonstrated experimentally by Thomas[5] the response of three-dimensional particles differs from the response of two-dimensional disks, consequently extrapolation of the analysis to three dimensions is important.

4 SUMMARY AND CONCLUSIONS

A new approach to kinematic homogenization for interpreting the results of discrete element simulations was presented in this paper. In contrast to current methods, this new method can capture particle rotations effectively. Importantly, an innovative aspect of this new method is that it uses a mesh-free non-linear interpolation to calculate the displacement gradient values. This approach is particularly applicable to finite deformation problems where strain localizations develop. The new method was applied to interpret the results of a two-dimensional discrete element simulation of open-ended pile installation problem. The non-linear interpolation approach was able to smooth the particle displacement values, resulting in lower volumetric strain values in comparison to the simple, linear interpolation approach.

Acknowledgements

The details of the mesh-free shape functions were provided by Prof. S. Li of the University of California at Berkeley.

References

1   O'Sullivan, C. (2002) "The application of discrete element modeling to finite deformation problems in geomechanics" Ph.D. thesis, Dept. of Civ. Engrg., Univ. of California, Berkeley.

2   Jardine , R.J. and Chow, F. C. (1996) *New Design Methods for Offshore Piles* Marine Technology Directorate, London

3   Cundall, P.A. and Strack, O.D.L. (1979) "A distinct element model for granular assemblies" *Geotechnique*, 29 **47-65**.

4   Cambou, B., Chaze, M. and Dedecker, F. (2000), "Change of scale in granular materials", *European Journal of Mechanics. Vol. A (Solids)* Vol 19, pp. 999-1014.

5   Thomas, P.A. (1997) " Discontinuous deformation analysis of particulate media." Ph.D. thesis, Dept. of Civ. Engrg., Univ. of California, Berkeley.

6   Li, S. and Liu, W.K. (2002) "Meshfree Particle Methods and Their Applications," *Applied Mechanics Review*, Vol. 54, pp **1-34**.

7   Liu, W.K., Li, S., and Belytschko, T. (1997) "Moving Least Square Kernel Method. 1:
     Methodology and Convergence," *Computer Methods in Applied Mechanics and
     Engineering*, Vol. 143, **113-154.**

8   O'Sullivan, C. and J.D. Bray (2002a)   "Methods to Calculate Strain in Discrete
     Element Modeling" *3rd International Symposium on Deformation Characteristics
     of Geomaterials*

9   O'Sullivan, C. and J.D. Bray (2003b)   "Evolution of Localizations in Idealized
     Granular   Materials"   *Workshop   on   Quasi-static   Deformations   of
     Particulate Materials*

10   Randolph, M.F., Leong, E.C. and houlsby, G.T. (1991) "One dimensional analysis of
     soil plug in pipe piles" *Geotechnique* 41(4) **587-598.**

11   Lehane, B.M. and Gavin, K.G. (2001) "Base resistance of jacked pipe piles in sand"
     Journal of Geotechnical and Geoenvironmental Engineering, ASCE 127(6) **473-479**

12   Huang, A. – B. and Ma, M. Y. (1994) "An analytical study of cone penetration tests
     in granular material" *Canadian Geotechnical Journal* Vol. 31 pp 91-103

13   Ke, T.-C., and Bray J.D. (1995). "Modeling of particulate media using discontinuous
     deformation analysis" *J. Engrg. Mechanics,* ASCE, 121(11), **1234-1243.**

14   Itasca Consulting Group (1993-1998). "PFC2D 2.00 Particle Flow Code in Two
     Dimensions". Itasca Consulting Group, Inc., Minneapolis, Minnesota.

15   Thomas, P.A. and Bray, J.D. (1999) "Capturing nonspherical shape of granular media
     with disk clusters" *J. Geotechnical and Geoenvironmental Engineering,* ASCE,
     125(3), **169-178.**

16   O'Sullivan, C., J.D. Bray, and M.F. Riemer, (2002) "The Influence of Particle Shape
     and Surface Friction Variability on Macroscopic Frictional Strength of Rod-Shaped
     Particulate Media" *ASCE Journal of Engineering Mechanics*, Vol 128, No. 11.

17   Bardet, J.-P. (1998) "Introduction to computational granular mechanics", *Behaviour
     of granular materials* . B. Cambou (Ed.) No. 385 in CISM Courses and Lectures,
     Springer-Verlag, Wien New York.

Section 4 : Materials and Products

NANOSTRUCTURE OF SINGLE CARBON FIBRES INVESTIGATED WITH SYNCHROTRON RADIATION

D. Loidl[1], O. Paris[2], M. Müller[3], M. Burghammer[4], C. Riekel[4], K. Kromp[1] and H. Peterlik[1*]

[1]Institute of Materials Physics, University of Vienna, Boltzmanng., 5, A-1090 Vienna, Austria
[2]Erich-Schmid Institute of Materials Science, Austrian Academy of Sciences and Metal Physics Institute, University of Leoben, Jahnstrasse 12, A-8700 Leoben, Austria
[3]Institute of Experimental and Applied Physics, University of Kiel, Leibnizstrasse 19, D-24098 Kiel, Germany
[4]European Synchrotron Radiation Facility, BP 220, F-38043 Grenoble, Cedex, France
* corresponding author

1 INTRODUCTION

In the past years high performance fibres were developed, which significantly improve the mechanical properties of composites and extend their use for constructional purposes. In particular, it was not only the aim to enhance the tenacity of the fibres, but also the modulus: Many light-weight structures are limited not by the strength of the material, but by the resistance to kinking, which is dependent on the Young's modulus and the geometry. The most interesting candidates, i.e. the fibres with the highest moduli and the lowest density are high performance polymeric fibres based on PBO or PIPD[1,2] and carbon fibres[3].

Though carbon fibres were firstly produced in the late 19th century by Edison for the use in incandescent lamps, the first commercial carbon fibres were not produced until the early 1960s[1]. A number of different processing techniques (stabilisation, carbonisation, final high temperature thermal treatment) as well as different precursor routes were developed, the most important among them are polyacrylonitrile (PAN-fibres), mesophase-pitch precursor (MPP-fibres) and rayon.

The Young's modulus of carbon fibres exceeds that of all other types of fibres. It approaches the theoretical value of the planes of single crystal graphite[4] of more than 1000 GPa. The unique mechanical properties are attributed to the highly anisotropic nature of the graphite crystal. Graphite is built up of multiple stacks of sheet-like layers of carbon atoms. In the plane of the sheets, the atoms are linked by strong covalent bonds, whereas perpendicular to this plane, only weak Van der Waals bonds are acting. The carbon fibres consist of small crystallites with a one-dimensional preferred orientation along the fibre axis and a plane spacing significantly larger than that of a perfect graphite single crystals. Different methods were applied to investigate the structure of carbon fibres: Electron microscopy (SEM[5-8], TEM and HTREM[9-11]) as well as scattering methods[12-15].

Ruland and coworkers[12,16] proposed that PAN-based carbon fibres are built up of basic structural units (BSUs), which consist of ribbon shaped layers of sp[2]-type graphene sheets.

The BSUs are forming undulating microfibrils over a range of some hundreds of nanometers and exhibit a preferred orientation along the fibre axis. Another model proposed basket weave structures[17] or crumpled and folded sheets of layer planes, which are entangled, but interlinked only at their boundaries[10].

The elastic properties such as Young's modulus and shear modulus are strongly dependent on the orientation and the arrangement of the BSUs[18]. Different possible arrangements were taken into consideration, such as undulating and zigzag ribbons, series, series rotatable and parallel elements as well as a mosaic model[19]. The situation is still more complicated for MPP-fibres, where additionally a cross-sectional texture such as radial, radial-folded or onion-like, was observed[14]. In the following, we denote the BSUs as crystallites, which we define as the domains of coherent scattering.

In this article, we give an overview about recent results on the investigation of single carbon fibres by using microbeam diffraction with synchrotron radiation, performed at the microfocus beamline ID13 at the European Synchrotron Radiation Source (ESRF) in Grenoble, France.

2 EXPERIMENTAL

2.1 Material

Carbon fibres from PAN- and from MPP-precursor were used for the investigation: The MPP-fibres were chosen so that they cover a wide range in the Young's modulus, the PAN-fibres were heat treated to obtain the same effect (as-received, heat treatment temperature (HTT) of 1800 °C, HTA7-18, of 2100 °C, HTA7-21, and of 2400 °C, HTA7-24). Table 1 shows the tested fibres, the producer and a number of mechanical properties from mechanical tests (single-fibre test)[20]. A remarkable effect is that the Young's modulus increases during loading up to 30 percent. This behaviour was observed in particular for the MPP-based fibres[20].

Fiber name & Type manufacturer		Fibre Diameter [μm]	Density* [g/cm3]	Initial Modulus [GPa]	Initial 002 azimuthal half width [°]
K321 (Mitsubishi)	MPP	10.48	1.9	136	17.6
E35 (DuPont)	MPP	9.7	2.10	197	12.0
E55 (DuPont)	MPP	10.18	2.10	358	7.09
FT500 (Tonen)	MPP	10.0	2.11	380	6.69
K137 (Mitsubishi)	MPP	9.54	2.12	500	3.42
HTA7-AR (Tenax)	PAN	6.84	1.77	198	19.0
HTA7-18 (Tenax)	PAN	7.3	1.77	273	16.3
HTA7-21 (Tenax)	PAN	6.44	1.78	332	11.8
HTA7-24 (Tenax)	PAN	6.2	1.91	349	9.64

* data-sheets

Table 1 *Some properties of the investigated fibres* [20].

2.2 Equipment

The scattering experiments were carried out at the microfocus beamline (ID13) at the European Synchrotron Radiation Facility (ESRF) in Grenoble, France. Three different setups were used in different experiments to focus the monochromatic X-ray beam with a wavelength of 0.975 Å, see Figure 1: 1) A borosilicate glass capillary[21,22], which enables measurements with a full width at half maximum (FWHM) of the X-ray beam of about 3 μm[23] 2) a 10 μm collimator, which was used for investigating the structure in dependence on the applied load using the tensile test equipment developed especially for in-situ experiments[24], and 3) a newly developed X-ray waveguide[25-27], which confines the beam to a vertical beam size of only about 100 nm. In the horizontal direction the beam was compressed by a multilayer mirror to < 5 μm[28]. Due to the vertical beam divergence of about 1 mrad, the beam size is strongly dependent on the distance between the specimen and the respective setup. Therefore, this distance was chosen to be as small as possible (about 200 μm for the glass capillary and 20 μm for the waveguide experiments).

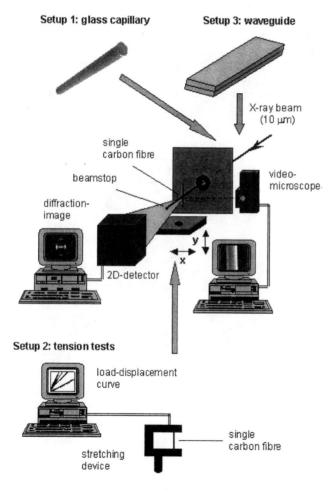

Figure 1 *Three different setups were used to focus the X-ray beam: 1) A glass capillary with a beam FWHM of 3 μm, 2) a 10 μm collimator for in-situ tension tests and 3) a X-ray waveguide with a beam dimension of only 100 nm vertically and 5 μm horizontally.*

An area detector (MAR-CCD, active diameter 130 mm, 2Kx2K pixels, pixel-size 64.5 μm ×64.5 μm) was used for data acquisition. The geometry chosen was the usual fibre geometry with the fibre axis perpendicular to the X-ray beam. All data were normalised with respect to the primary beam monitor and corrected for background. Data evaluation was done by using the ESRF software package FIT2D[29].

3 RESULTS AND DISCUSSION

A typical scattering image of a MPP-carbon fibre is shown in Figure 2. This image shows the following features: The 002-reflection, the 004-reflection, the 10-band and the small-angle X-ray scattering (SAXS). Only the 002-reflection and the 10-band are further investigated in this work. The equatorial position of 002-reflection is inversely proportional to the interlayer spacing of the graphene planes and the angular distribution of the crystallites with respect to the fibre axis is directly related to the azimuthal broadending of the 002-peak. The 10-band, i.e. the ring, is determined by the distances of the atoms within the hexagonal graphene planes. Small layer lines can be seen on the poles of the ring (the 10-band), which are the consequence of the 2-dimensional arrangement of the graphene planes.

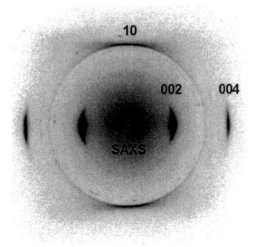

Figure 2 *Typical scattering pattern of a MPP-carbon fibre.*

3.1 Texture

3.1.1 Axial texture: The axial texture is of great interest, because the orientation distribution of the planes with respect to the fibre axis is the main parameter, which determines the Young's modulus of the fibres. Whereas former model calculations were based on results on fibre bundle experiments, where the whole bundle of 6000 and more carbon fibres was measured in the X-ray equipment, the single carbon fibre experiments reveal that the true distribution is considerably smaller (Figure 3). The reason is that within the orientation distribution of the 002-reflection the orientation distribution of the graphene planes cannot be separated from the misalignment of the fibres within the whole bundle, which is inevitable during the production as well as in the course of the preparation for the experiment.

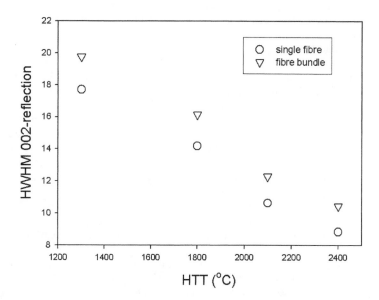

Figure 3 *The orientation distribution of the 002-reflection is smaller for single fibres tests than for bundle tests due to problems in aligning all fibres within the bundle.*

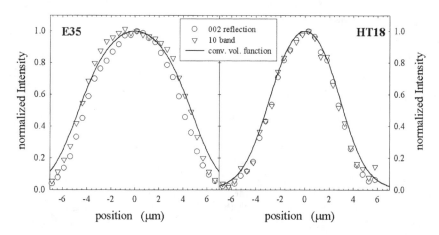

Figure 4 *Integral intensities of the 002-reflection (circles) and the 10-band (triangles) scanned across the fibres. Left figure, MPP-fibre E35, which shows different intensities, right figure, PAN-fibre HTA7 with a HTT of 1800 °C, where all intensities coincide with the convolution of the volume function with the beam profile (solid line).*

3.1.2 Cross-sectional texture: In addition to an axial texture some type of carbon fibres exhibit a distinct cross-sectional texture. Radial, radial folded, onion-like, irregular and amorphous morphologies were reported[14].

It was demonstrated previously that wide-angle X-ray diffraction (WAXD), by scanning across single fibres, can be used to acquire information on the respective type of cross-sectional texture[23,30]. If the size of the beam is not well below the dimension of the sample an accurate knowledge of the beam profile is mandatory. For the two types of investigated fibres a different characteristic of the integral intensity of the 002-reflection in dependence of the position was observed. Whereas for the 'amorphous' PAN-based fibre HTA7 (HTT of 1800°C) the experimental data coincide with the convolution of the volume function with the beam profile (solid line in Figure 4), the deviation of this model, in the case of the MPP-based fibre E35, indicates a more complex cross-sectional morphology. By means of appropriate model calculations, not only qualitative statements on the radial-folded structure for this fibre were obtained, but also quantitative values could be given for the fraction of the random to the radial folded phase and for the ratio of the mean amplitude to the wavelength of the folded phase[23].

3.2 Elastic properties

Single carbon fibres were measured in the X-ray beam during in-situ tension tests. The beam diameter was defined by a 10 μm pinhole. The fibres were glued into a stretching cell, which was especially designed for the in-situ tests of single fibres. The load was increased stepwise: For each loading step, the diffraction patterns were recorded at four positions along the fibre. To obtain the Young's modulus of the graphene planes, the densities of the fibres were corrected for porosity using the densities from the data sheets of the manufacturer and the crystallographic parameters (d_{10}, d_{002}) from the experiment[20].

By integrating the 10-band in the meridional direction in a narrow sector, only the signal from those planes is obtained, which are almost perfectly aligned along the load axis. The microscopic Young's modulus e_{cr} of the crystallites is then obtained from the slope of the stress-strain curves by[20]

$$e_{cr} = d\sigma / d\varepsilon_{cr} \tag{1}$$

with the strain of the crystallite being

$$\varepsilon_{cr} = \left(d_{10}(\sigma) - d_{10}(0)\right) / d_{10}(0) \tag{2}$$

The shear modulus g_{cr} of the crystallites can only indirectly be evaluated from the decrease of the misalignment angle Π of the graphene planes with increasing load:

$$\frac{d\Pi}{d\sigma} = -\frac{1}{2 g_{cr}} \sin\Pi \, \cos\Pi \tag{3}$$

Figure 5 shows the result for the Young's modulus in dependence on the initial misalignment angle. The full symbols denote the MPP-fibres, the open symbols the PAN-fibres: With increasing misalignment the Young's modulus of the crystallites decreases, which indicates that the graphene planes themselves are not perfectly straight, but exhibit the so-called buckled graphite structure. For the fibres with the highest orientation with respect to the fibre axis, i.e. the lowest misalignment angle, a constant value of e_{cr} =1140 GPa is reached[20], which is higher than the generally accepted value for single crystal graphite of 1020 GPa, but coincides with measurements on carbon nanotubes of 1100 GPa from bending[31] and 1200 GPa from vibrational[32] experiments. The possible reason is that the lower value for pyrolytic single crystal graphite was measured with macroscopic samples, whereas our experiments as well as the experiments on nanotubes concern mechanical tests on samples with a length of some tenths of nanometers.

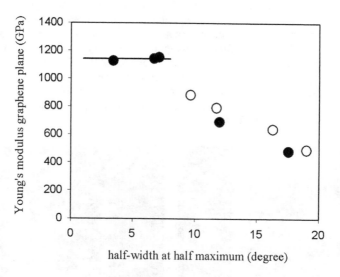

Figure 5 *Young's modulus of the crystallites within the carbon fibres in dependence on their initial orientation. Full symbols, MPP-fibres, open symbols: PAN-fibres.*

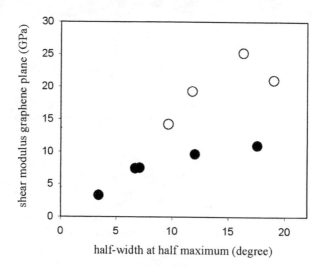

Figure 6 *Shear modulus of the crystallites within the carbon fibres in dependence on their initial orientation. Full symbols, MPP-fibres, open symbols: PAN-fibres.*

3.3 Single fibre bending experiments

The development of the waveguide structure is an example of a nanofabricated optical element. With this equipment, it is possible to measure the atomistic structure (via diffraction) with a spatial resolution of only about 100 nm in one direction depending on the distance of the specimen from the waveguide due to the beam divergence. Fig. 7 shows a comparison of the beam size of the glass capillary with the one of the waveguide used in this study. This figure gives an impression, how small the X-ray beam is, even in comparison to a carbon fibre with a diameter of some microns.

In this work, some preliminary results on the inner structure of carbon fibres loaded in bending are reported. The bending was performed by forming loops and scanning across the fibre with the X-ray beam from the waveguide.

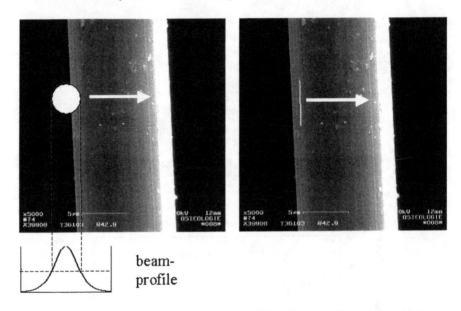

Figure 7 *Comparison of the X-ray beam size of the glass capillary (left) with the one supplied by the waveguide (right): The width of the beam from the waveguide during scanning across the fibre is only 100 nm.*

Figure 8, left picture, shows the azimuthal width of the 002-reflection. This width is defined by the orientation distribution of the graphene planes with respect to the fibre axis. The data shown in this figure were evaluated for the PAN-fibre HTA7 with a HTT of 2100 °C (HTA7-21). Whereas the azimuthal width for the straight fibre (circles) shows no dependence on the scanning position of the fibre, the bended fibre exhibits a strong dependence: In direction towards the tension zone, the distribution of the graphene planes decreases, whereas towards the compression zone a strong increase in the azimuthal width is visible. The orientation with respect to the fibre axis degrades, which is a clear incidence of a buckling of the graphene planes on a nanoscopic level. The neutral axis is not in the centre, indicating an anisotropy of the tension and compression Young's modulus of the carbon fibres.

In the right picture of Figure 8 the integral intensities are depicted, which follow the volume function (line in Figure 8). As the beam is only 100 nm wide, a convolution correction as performed for the glass capillary is not required any more, as in the works with the glass capillary[23,30].

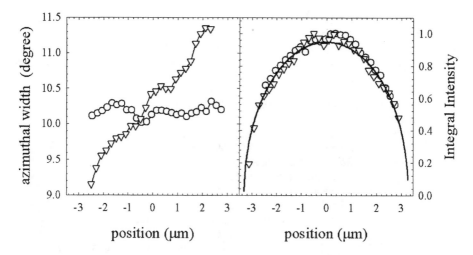

Figure 8 *Left picture: Azimuthal width of 002-reflection with respect to the fibre axis, circles, straight fibre, triangles, bent fibre. Right picture: Both the integral intensities follow the volume function.*

4 CONCLUSION AND OUTLOOK

Due to their high intensity and the possibility to use appropriate focussing devices, X-rays from synchrotron radiation sources allow the investigation of single carbon fibres in a position resolved way. The possibility to test single carbon fibres instead of fibre bundles enables the determination of the axial texture without additionally effects caused by the misalignment of fibres in the bundle. Furthermore, the cross-sectional texture can be measured not only qualitatively, but also quantitatively. In-situ tests with simultaneously performing mechanical tests and structural investigation are also possible as well as a spatial resolution of the measurements down to 100 nm, which is a real step towards the development of characterisation methods on a nanoscopic level.

Acknowledgements

The support from the Austrian Science Funds, proj. Nr. P16315, and from the EU, proj. Nr. G6RD-CT-2001-00523, VaFTeM, is gratefully acknowledged. The waveguide has been developed at the Budker Institute of Nuclear Physics (Novosibirks, Russia) in the context of a collaboration with Sincrotrone Trieste (contact: S. DiFonzo-Jark). We also acknowledge the allocation of beamtime from the ESRF (proposals HS827, ME187 and ME368).

References

1   R.J. Young, R.J. Day and M. Zakikhani, *J. Mater. Sci.*, 1990, **25**, 127.
2   S. Bourbigot and X. Flambard, *Fire and Mater.*, 2001, **26**, 155.
3   E. Fitzer, *High Temp. High Press.*, 1986, **18**, 479.
4   O.L. Blakslee, D.G.Proctor, E.J. Seldin, G.B. Spence and T. Weng, *J. Appl. Phys.*, 1970, **41**, 3373.
5   M. Endo, *J. Mater. Sci.*, 1988, **23**, 598.
6   Y. Huang and R.J. Young, *J. Mater. Sci.*, 1994, **29**, 4027.
7   L.P. Kobets and I.S. Deev, *Comp. Sci. Techn.*, 1997, **67**, 1571.
8   A.B. Barnes, F.M. Dauch N.C. Gallego, C.C. Fain and M.C. Thies, *Carbon*, 1998, **36**, 855.
9   S.C. Benett and D.J. Johnson, *Carbon*, 1979, **17**, 25.
10 M. Guigon, A. Oberlin and G. Desarmot, *Fibre Sci. Techn.*, 1984, **20**, 55.
11 A.Oberlin, *Carbon*, 1984, **22**, 521.
12 R. Perret and W. Ruland, *J. Appl. Cryst.*, 1970, **3**, 525.
13 A. Takaku and M. Shioya, *J. Mater. Sci.*, 1990, **25**, 4387.
14 W. Ruland, *Adv. Mater.*, 1990, **2**, 528.
15 N.C. Gallego and D.D. Edie, *Comp. Part A*, 2001, **32**, 1031.
16 L. Fischer and W. Ruland, *Coll. & Polymer Sci.*, 1980, **258**, 917.
17 R.J. Diefendorf and E. Tokarsky, *Polym. Eng. Sci.*, 1975, **15**, 150.
18 M.G. Northolt, L.H. Veldhuizen and H. Jansen, *Carbon*, 1991, **29**, 1267.
19 M. Shioya, E. Hayakawa and A. Takaku, *J. Mater. Sci.*, 1996, **22**, 521.
20 D. Loidl, H. Peterlik, M. Müller, C. Riekel and O. Paris, *Carbon*, 2003, **41**, 563.
21 P. Engstroem and C. Riekel, *Rev. Sci. Instrum.*, 1996, **67**, 4061.
22 C. Riekel, A. Cedola, F. Heidelbach and K. Wagner, *Macromol.*, 1997, **30**, 1033.
23 O. Paris, D. Loidl, M. Müller, H. Lichtenegger and H. Peterlik,, *J. Appl. Cryst.*, 2001, **34**, 473.
24 C. Riekel, T. Dieing, P. Engström, L. Vincze, C. Martin, and A. Mahendrasingham, *Macromol.*, 1999, **32**, 7859.
25 S. Lagomarsino, W. Jark, S. Di-Fonzo, A. Cedola, B. Müller, P. Engström and C. Riekel, *J. Appl. Phys.*, 1996, **79**, 4471.
26 S. Lagormasino, A. Cedola, S. Di-Fonzo, W. Jark, V. Mocella, J.B. Pelka and C. Riekel, *Cryst. Res. and Techn.*, 2002, **37**, 758.
27 S. Roth, M. Burghammer, A. Janotta, C. Riekel, Macromolecules (2003) 36, 1585-1593
28 M. Mueller, M. Burghammer, D. Flot, C. Riekel, C. Morawe, B. Murphy, A. Cedola, *J.Appl. Cryst.*, 2000, **33**, 1231.
29 http://www.esrf.fr/computing/scientific/FIT2D/
30 O. Paris, D. Loidl and H. Peterlik, *Carbon*, 2002, **40**, 551.
31 S. Akita, H. Nishijima, Y. Nakayama, F. Tokumasu and K. Takeyasu, *J. Phys. D (Appl. Phys.)*, 1999, **32**, 1044.
32 M.M.J. Treacy, A. Krishnan and P.N. Yianilos, *Microsc. Microanal.*, 2000, **6**, 317.

HIGH-PERFORMANCE NANOESTRUCTUERD MATERIALS FOR CONSTRUCTION

I. Campillo*, J. S. Dolado, A. Porro

Centre for Nanomaterials Applications in Construction (NANOC), Construction and Environment Division, Fundación Labein, Cuesta de Olaveaga 16, 48013 – Bilbao, SPAIN
* *Corresponding author. Tel: +34 94 489 2400, Fax: +34 94 441 1749, e-mail: icampillo@labein.es*

1 INTRODUCTION

Nanomaterials are defined as those materials which have at least one dimension (length, width or thickness) below 100 nanometer (nm), 1 nm being a thousandth of a micron, or about 100000 times smaller than a human hair. The Physics and Chemistry of nanomaterials (in their great variety of forms, such as grains, discs, fibres, onions, etc.) are profoundly altered with respect to their macroscopic counterparts. Dramatic changes are brought about not only by size reduction but also by new properties predominant at the nanoscale. This new phenomenology includes surface and interface phenomena, electronic confinement and quantum mechanics in general. One of the most active fields of Nanotechnology is that of nanocomposites. Nanocomposites are composites that incorporate some nanomaterial in a variable percentage not higher than a 10 % by weight.

Polymer based nanocomposites (polymeric matrix) were first introduced back in the seventies, in which sol-gel technology was used to disperse inorganic nanoparticles through a polymeric matrix[1]. Although researched for a couple of decades and first commercially developed by Toyota CRDL in Japan in the late eighties, the field of polymer nanocomposites is still at an embryonic stage of development today. The same can be said about ceramic nanocomposites. In this case, nanoalumina[2] is the most promising nanostructure to give rise to a new generation of ceramics, since its well-balanced mechanical, electrical and thermal properties provide a wide range of applications. As an example, ARGONIDE Nanometal Technologies has recently commercialised an alumina nanofibre (NanoCeram[TM])[3] that can revolutionise the field of ceramics. Other materials such as nano-silica, nano-zirconia etc. are also being researched to improve ceramics based materials.

One of the challenges in the field of construction materials has been (and still is) the development of high performance concrete (HPC), i.e. high resistance and durability. Concrete is a complex multiphase composite material. The properties, behaviour and performance of concrete are dependent of the nanostructure of the cementitious matrix that glues it together and provides integrity. Therefore the study of the structure of cement pastes and phenomena in the nanoscale is crucial for the development of new construction materials and applications[4]. However, the common approach for the development of HPC has mostly consisted of varying the macroscopic parameters that are used to produce concrete, i.e., basically working on the concrete mix design and the reinforcement with different types of fibres[5-15]. To a great extent, this approach, motivated mostly by the inertia of the construction industry itself, has slowed down the advance in the deep understanding of the construction materials. Within the new paradigm of Nanosciences and Nanotechnology this can no longer continue. Construction materials must be investigated within a scientific approach if a new generation of materials that are both of higher performance and more economically viable is to be created. In our particular example of concrete, this can only be accomplished by a proper understanding of the cement nanostructure and by the inclusion of different nanostructures. New tools, such as atomic force microscopy (AFM) and environmental scanning electron microscopy (ESEM) (just to mention two of them), can shed new light on these problems. They not only allow the characterisation of cement paste in the nanoscale, but they also allow the manipulation of the nanostructures (AFM) and the hydration process itself to be followed (AFM and ESEM)[16,17].

The final aim is to create a new class of high performance construction materials: what we will call multifunctional high performance materials. By multifunctionality we understand the emergence of new properties different from the usual properties that define the material. In this way, the material can extend its range of applications. In the particular case of concrete, apart from a better mechanical and durability performance, multifunctional high performance concrete (MHPC) will show additional properties. These can include electromagnetic properties for shielding applications or thermal adaptability to increase energetic efficiency of buildings, just to mention some of them. Furthermore, the proper addition of nanomaterials can lead to a dematerialization of the structures, what can have an impact in the natural resource consumption of the construction industry. This can be achieved either by modifying the nanostructure itself or by incorporation of different nanostructures and exploiting their enhanced reactivity due to their high specific surface or/and their intrinsic properties (such as high magnetic permeability, electrical and thermal conductivity, radiation absorption properties and so on). It is clear that, far from being a science & technology fashion, nanotechnology is a necessary path to achieve real competitive and sustainable growth and innovation within the construction industry.

In this short paper we present some preliminary results on the inclusion of different nanostructures in cement pastes. These are just the first steps on the way to fulfil the above mentioned objectives. Two types of nanomaterials have been considered: amorphous nanosilica and carbon nanotubes

2 NANOSILICA

In the concrete industry silica is a well-known agent that plays an important role as both binding and filling agent in the development of HPC. The usual product is the so-called silica fume or microsilica[18] with diameters ranging from 0,1 μm to 1 μm and a content of silica usually above 90 % by weight. In this way, microsilica is in the limit of nanometer sized materials that are added to cement to increase performance of cementitious composites. Nanosilica is just nanometric size silica (SiO_2). Nanosilica is in the form of spheres of diameter less than 100 nm that are either agglomerated in dry grains or in colloidal suspension stabilised by some dispersive agent. This last procedure is the most usual presentation of nanosilica. Colloidal silica is produced by several chemical companies throughout the world, such as Nyacol Nanotechnologies[49] and Bayer[50] whose products have been used for this particular work. Colloidal silica shows multiple applications: anti-sliding agent, refractory, anti-reflective surface treatments, component in opto-electronic devices, among others. The interaction of colloidal silica with portland cement has been studied by S. Chandra and H. Bergqvist[19] by using XRD and thermogravimetric analysis. They have shown that the reactivity of colloidal silica with calcium hydroxide is much faster in comparison with the condensed silica fume. They have also shown that a small amount of colloidal silica is sufficient to produce the same pozzolanic effect as with the addition of higher doses of condensed silica fume at early ages. They attributed these remarkable effects to the finer particle size of colloidal silica. This is not surprising, since condensed silica fume has typically a N_2 specific surface area of 15-25 m^2g^{-1}, while they use a colloidal silica of 80 m^2g^{-1}. J. Bastien et al.[20] investigated the applicability of nanosilica in its agglomerated form in a grout formulation. They studied the rheological properties of the proposed formulation and compare it to conventional silica fume based grouts and found that the new formulation exhibited no bleeding or segregation and that reached 100 MPa compressive strength after 28 days. Prior to these studies, A. Bjordal et al.[21] used a new cement slurry formulation containing colloidal silica in North Sea cementing operations to help control annular gas migration. They found that the colloidal silica enhanced the rheological properties and slurry stability. Finally, P. Greenwood et al.[22] have recently patented a high-fluidity concrete composition containing colloidal silica and polycarboxylate superplasticiser that could be used as a self-compacting concrete. All the previous works concentrate mainly on the rheological and workability properties of concrete containing colloidal silica. To the best of our knowledge, there is not a systematic work on the mechanical properties of modified cement pastes by nanosilica addition. In this paper we present preliminary results on these topics by using different nanosilica products with diameters ranging from 5 to 100 nm.

3   NANOTUBES

As mentioned in the introduction, concrete is usually reinforced with fibres in order to improve its mechanical performance. Steel fibres are mainly used, although other types of fibres such as glass, polypropylene and carbon fibres are also employed as reinforcement agents[15]. As far as we know, there are no published results on the reinforcement of cement pastes by carbon nanotubes. In this paper we present a preliminary study on the addition of carbon nanotubes in a cement matrix.

Since carbon nanotubes were discovered by Iijima in 1991[23] a lot of work has been done on this particular new nanostructure in such a way that it has become one of the most active field of research of nanosciences and nanotechnology. This is motivated by the outstanding properties of carbon nanotubes[24]. Carbon nanotubes are stronger than steel, but lightweight and can bear torsion and bending without breaking[25,26]. They are believed to be the future substitutes of carbon fibres. Theoretical calculations predict a Young modulus between 1 to 5 TPa for single walled nanotubes[27-29], which means that carbon nanotubes can have a tensile strength higher than any carbon fibre known (high modulus carbon fibre have a Young modulus of about 400 GPa). Furthermore, carbon nanotubes also present extraordinary thermal and electrical properties: they are thermally stable until 2800 °C in vacuum, their thermal conductivity is about two times that of diamond and their electrical conductivity is about 1000 times the conductivity of copper[30]. Besides, depending upon diameter and helicity, they can be conductors or semiconductors[24].

Taking into account these properties it can be stated that carbon nanotubes are a new class of advanced materials that can revolutionise the field of nanocomposites. A new generation of multifunctional nanocomposites can be born by proper incorporation of carbon nanotubes as functional reinforcement. Therefore, carbon nanotubes are key elements to achieve the above mentioned general goal of multifunctional high performance construction materials. This fact has pushed researchers towards the development of such composites. Almost all of the investigation has been directed to the reinforcement of polymers[31-40], although nanotube reinforcement of ceramic[41-46] and metal[47,48] matrix composites is also being studied. Results obtained so far do not show an spectacular efficiency as expected by the so-claimed properties. Problems arise in two stages:

1. It is difficult to obtain a uniform dispersion and disentanglement of carbon nanotubes.
2. Due to their graphitic nature, there is no proper adhesion between nanotube and matrix, causing what is called "sliding". Furthermore, since nanotubes are usually assembled in bundles there is additional sliding inside the bundle that prevents nanotubes from exhibiting their outstanding mechanical properties.

4   EXPERIMENTAL

The pastes used for the study were made from cem I 52.5 R. This type of cement has been selected because it presents the finest granulometry of all the commercial cements, which is best suited for the inclusion of nanomaterials. Reference samples were prepared in order to study the impact of the addition of the nanomaterials.

Different types of nanosilica have been used: in colloidal preparations and in agglomerated precipitated form. In table 1 the main features of the employed nanosilicas are summarised. All of them are products that can be obtained commercially.

In the case of carbon nanotubes, we have used both multi-wall carbon nanotubes (MWNT) and single-wall carbon nanotubes (SWNT). MWNT were purchased from Sun Nanotech Co.[51], which produce the tubes by the CCVD method, giving high yields of nanotubes (>80%) and at reasonable prices (2 \$/gr at the time of writing, and 13 \$/gr at the time we purchased them). These nanotubes present several defects as can be seen in the TEM image of Figure 1, which can be beneficial, as explained in the next section. SWNT were generously provided by the Group of Carbon Nanostructures and Nanotechnology[52] at the Carbonchemistry Institute in Zaragoza (Spain). These nanotubes were produced by the arc discharge method. In Figure 2 an AFM image show the crossing of two bundles of SWNTs over a silica substrate. In table 2 the main features of the carbon nanotubes employed for the study are summarised.

After mixing, different amounts were moulded into prism-shaped specimens (1 × 1 × 6 cm) and compacted by vibration. The specimens were demoulded after 1 day at > 90 % relative humidity. Subsequently groups of six specimens for each mix were stored at the temperature of 21 ± 2 °C under water for 14 and 28 days. The cement quantity was not varied from each mix to another and it was fixed to 100 gr, while water to cement ratios were chosen to be 0,33 and 0,36 for nanosilica and nanotube mixes, respectively. In the case of colloidal silica, the suspension was directly added to the mixing water, while the precipitated silica was added to the cement and homogenised. In the case of nanotubes, they were dispersed in the mixing water, stirred for ten minutes and then placed in a 400W ultrasonic bath for 30 minutes. After that they were mixed with the cement.

No superplasticiser was used for the mixings with nanosilica because we wanted to see the bare effect of the nanosilica on the paste. In the case of nanotubes, we did not use any surfactant for the same reason and because proposed surfactants in the literature for nanotube dispersion are not compatible with cements.

Table 1   *Different types of nanosilica employed in this work*

	Supplier	Average Size (particle)	Average Size (grain)	Presentation
Silica Fume	SIKA	0,1 - 1 μm	0,1 - 1 μm	Black powder
Nyacol® Nyasil 20	NYACOL Nano Technologies Inc.	20 nm	1,44 μm	White powder
Nyacol® Nyasil 5	NYACOL Nano Technologies Inc.	5 nm	2,50 μm	White powder
Nyacol® DP5820	NYACOL Nano Technologies Inc.	20 nm	20 nm	Colloidal silica stabilized in Ethylene Glycol % SiO2 = 30 %
Levasil® Grade 100	BAYER	30 nm	30 nm	Colloidal silica stabilized in aqueous solution by Na_2O. % SiO_2 = 30 % % Na_2O = 0,15 %
Levasil® Grade VP AC 40 38	BAYER	15 nm	15 nm	Colloidal silica stabilized in ammonia % SiO_2 = 15 %

Table 2   *Carbon nanotubes employed in this work*

Supplier	Type	Diameter	Length	Purity	Method of production
Sun Nanotech Co.,Ltd	MWNT	10-30 nm	1-10 μm	80 %	CCVD
ICB	SWNT	1-3 nm	1-10 μm	60 %	Arc Discharge

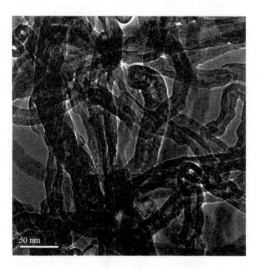

Figure 1   *TEM image of MWNT employed in this work*

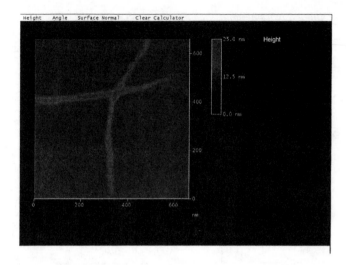

Figure 2   *AFM image of two crossing bundles of SWNT employed in this work*

5   RESULTS

5.1 Nanosilica

Results for the compressive strength at 28 days of cement pastes with the different types of nanosilica presented in Table 1 are plotted in Fig. 3 as a function of the grams of admixture. This coincides with the % by weight of cement, since we have always used 100 gr. of cement as stated in the previous section. Therefore, 0 grams corresponds to plain cement paste. Each value represents the average over twelve measurements, and the bars represent the standard deviation on the mean values. For those cases in which it seemed that a linear behaviour could fit the data, a curve has been plotted in order to help visualisation.

As seen in Fig. 3, the addition of silica fume (white diamonds) tends to reduce the strength of the paste, while nanosilica clearly increases the compressive strength at different rates depending on the nanosilica type. This behaviour of the silica fume is mainly due to the absence of any superplasticizer in the mix, which prevents microsilica from developing its potentiality[18]. It is also due to the low level of addition, which typically is over 8 % by weight of cement, while in this case ranges from 2 to 12 %. In fact for the highest percentages an increase can be observed. The other two types of dry silica are Nyacol® Nyasil 20 (solid circles) and Nyasil 5 (white circles). These products consist of grains of agglomerated nanosilica with primary particle sizes of 20 and 5 nm, respectively. Although the grains of Nyacol® Nyasil 5 (2,5 μm) are coarser than the grains of Nyacol® Nyasil 20 (1,44 μm), the Nyacol® Nyasil 5 ones are less dense and more porous than the Nyacol® Nyasil 20 ones as indicate the data set provided by the producer. Nyacol® Nyasil 5 presents a bulk density of 0,15 gr/cc and an oil absorption of 140 gr. of oil/100 gr. of powder, whereas Nyacol® Nyasil 20 has a bulk density of 0,29 gr/cc and an oil absorption of 80 gr. of oil/100 gr. of powder. This explains the results in Fig. 3, in which Nyacol® Nyasil 5 shows a slope four times larger than that of Nyacol® Nyasil 20.

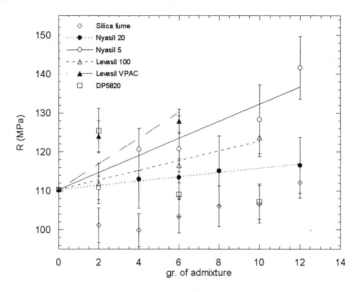

Figure 3   *Compressive strength at 28 days of nanosilica modified cement pastes*

In the case of colloidal silica the results are even more pronounced. This can be understood taking into account that in this case the nanosilica is purely nano, i. e., it is not agglomerated due to the stabilisation of the medium in which is dispersed. Levasil® Grade 100 (white triangles), which is stabilised by a solution of 0,23 % Na₂O in water, shows a remarkable increase with respect to Nyacol® Nyasil 20 although it does not reach Nyacol® Nyasil 5. This can be viewed as a competing effect between particle size and concentration of silica. The particle size in Levasil® Grade 100 is 30 nm and the concentration is 45 %, which means that we have a lower quantity of more reactive silica: in 1 gr of Levasil® Grade 100 we only have 0,3 gr. of silica. This is enough to overcome the agglomerated Nyacol® Nyasil 20 but not the Nyacol® Nyasil 5, which still retains the great reactivity of its 5 nm primary particles. The most dramatic change is shown by Levasil® VPAC 4038 (solid triangles). In this case the nanosilica is ammonia stabilised, the particle diameter is 15 nm and the concentration is 30 %. The reduction in size determines the strength shown in Fig. 3, which is still higher than that of Nyacol® Nyasil 5, although for the same admixture quantity we have less weight of silica. However Nyacol® DP5820 (white squares) does not show the same tendency of the previous products. Its particle size is 20 nm and the concentration is 30 %. Therefore it should behave roughly as Levasil® VPAC 4038. This is the case for low level of admixture, but for higher levels there is a decrease in the strength. This abnormal behaviour is brought about by the stabilisation agent utilised, ethylene glycol. We have found that small amounts of ethylene glycol reduce drastically the strength of cement pastes by avoiding proper hydration of the cement grains. The more Nyacol® DP5820 that is added to the cement the more ethylene glycol that is present in the process until the effect of ethylene glycol is stronger than that of the nanosilica.

5.2 Nanotubes

Results for the compressive strength at 14 days of cement pastes with the MWNT and SWNT presented in Table 2 are plotted in Fig. 4. The quantity of carbon nanotubes is 1 % by weight of cement in both cases. As can be seen in Fig. 4, cement paste containing MWNT shows a higher compressive strength than the one containing SWNT. Cement paste with MWNT has an increase of 30 % with respect to the plain cement paste, whereas cement paste SWNT only has a 6 % increase, i.e., 5 times lower increase than with MWNT. These results can be understood as follows. First, as can be seen in Fig. 1, our MWNT produced by CCVD show a lot of defects along their length. It is well known that SWNT produced by the arc discharge method do not have many defects and are quite straight. The presence of defects in carbon nanotubes reduces their intrinsic mechanical properties. However defects can be beneficial when nanotubes are incorporated in a matrix, since those defects increase the interaction points with the matrix and a better anchorage is obtained. This is important to avoid sliding. Furthermore, in the case of SWNT they are arranged in bundles in such a way that internal sliding between the nanotubes adds to the sliding between the matrix and the bundle. Therefore cement paste with SWNT should exhibit less strength, as our results show. These results are in accordance with data reported about carbon nanotube reinforcement of polymer matrices. In any case, it is obvious that, due to the graphitic nature of the nanotube, it is not trivial to get a high degree of interaction between matrices and nanotubes. A high level of disentanglement, dispersion and functionalisation is needed for the nanotubes to exhibit their outstanding properties.

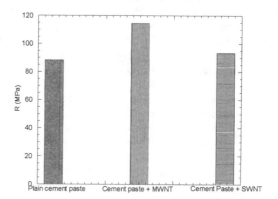

Figure 4   *Compressive strength at 14 days of cement pastes containing carbon nanotubes*

6   CONCLUSIONS

We have presented preliminary results on the modification of cement pastes by two types of nanostructures: nanosilica and carbon nanotubes. The impact of these nanostructures in the compressive strength is analysed.

Two types of nanosilica have been used: colloidal suspension and agglomerated dry silica. We have observed that colloidal silica behaves much better than the agglomerated silica, since with lower weight of silica better results are obtained. This is due to the fact that colloidal silica can exhibit its nano-character, while agglomerated dry silica cannot exhibit the whole specific surface area of its primary particles.

Finally, both MWNT and SWNT have been added to cement paste. A reasonable increase is obtained for MWNT, whereas only a low increase is observed for SWNT. This can be understood in terms of defects present in MWNT and double sliding of SWNT, i.e., of the bundle with the matrix and inside the bundle.

Acknowledgements

The authors wish to acknowledge the Group of Carbon Nanostructures and Nanotechnology at the Carbonchemistry Institute in Zaragoza (Spain). They provided the SWNT and the TEM image of the MWNT. We also wish to acknowledge Bayer Corporation by providing free samples of Levasil®. Finally, we wish to acknowledge Cementos Lemona and Cementos Rezola for providing the cement cem I 52.5 R.

References

1   C. J. Brinker and G.W. Scherer, *The Physics and Chemistry of Sol-Gel Proccesing*, Academic Press Inc., San Diego 1990.

2   Hellming R.J. and Ferkel H., *Using Alumina nanopowder as Cement in Bonding of Alumina Ceramics*, Phys. Stat. Sol. (a) **175**, 549 (1999).

3   Complete information of the Argonide products can be found in http://www.argonide.com

4   Beaudoin J.J., *Why Engineers Need Materials Science*, Concrete International, August 1999, p. 86.

5   Bache H. H., *Densified Cement Ultrafine Cased Materials*. Second International Conference on Superplasticizers in Concrete, Otawa, June 10-12.

6   Alford N. M., Groves G. W., Double D. D. *Physical Properties of High Strength Cement Pastes*. Cement and Concrete Research 12, 349 (1982)

7   Bache H. H. *Densified cement ultrafine cased materials*. Second International Conference on Superplasticizers in Concrete, Otawa, June 10-12 (1981).

8   Alford N.M., Groves G.W., Double D.D. *Physical properties of high strength cement pastes*. Cement and Concrete Research **12**, 349 (1982).

9   Bache H.H. *Compact reinforced composite. Basic principles*. Aalborg Portalnad, Cement-og Betonlaboratoriet, CBL Report No. 41 (1987).

10   Richard P. Cheyrezy M. *Les Bétons de Poudres Réactives*. Annales de líTBTP, No 532, pp85-102 (1995).

11   Rossi P. *High performance multimodal fiber reinforced cement composites (HPMFRCC): the LCPC experience*. ACI Materials Journal **94**, 478 (1997).

12   Lankard D.R., Newell J.K. *Preparation of highly reinforced steel fiber reinforced concrete composites. Fiber Reinforced Concrete*, SP-81, G.C. Hoff, ed., American Concrete Institute, Farmington Hills, Mich., 287-306 (1984).

13   Hackman L.E., Farrell M.B., Dunham O.O. *Slurry infiltrated mat concrete (SIMCON)*. Concrete International, Dec., 53 (1992).

14   Li V.I. *Engineered cementitious composites – tailored composites through micromechanical modelling. Fiber Reinforced Concrete: Present and Future*, Montréal, Canadian Society for Civil Engineering, pp. 64-97 (1998).

15   See, for example, chapter 11: Gosh S.N (Editor), *Cement and Concrete Science & Technology*, Vol. 1, Part II, Abi books (1992).

16   Lesko S. et al., *Investigation by atomic force microscopy of forces at the origin of cement cohesion*, Ultramicroscopy **86** (2001), 11.

17   Gauffinet S. et al., *AFM and SEM studies of CSH growth on C3S surface during its early hydration*, Proceedings of the 20th International Conference on Cement Microscopy, Mexico, 1998.

18   Khayat K.H. and Aitcin P.C., *Silica fume in concrete: an overview*, Fourth CANMET/ACI International Conference on Fly Ash, Silica Fume, Slag and Natural Pozzolans in Concrete, SP-132, V. 2 (1992), 835. *Guide for Use of Silica Fume*, 234R-96 ACI Publications (1996).

19   Chandra S. and Bergqvist H., *Interaction of silica colloid with portland cement*, Proc. Int. Congr. Chem. Cem. Vol. 3 (1997), 3ii106, 6pp.

20   Bastiwn J. et al, *Cement grout containing precipitated silica and superplasticizers for post-tensioning*, ACI Materials Journal **94** (1997), 291.

21   Bjordal A., et al., *Colloidal silica cement: description and use inNorth Sea operations*, Proceedings of Offshore Europe 93 published by the Society of Petroleum Engineers (1993), 431.

22   Greenwood P., et al, Patent Number WO 200190024 A1.

23   Ijima S. *Helical microtubules of graphitic carbon*. Nature **354**, 56 (1991).

24   Ebbesen TW. (Ed.). *Carbon Nanotubes: Preparation and Properties*. Springer-Verlag (1997), and references therein.

25   Endo M, Takeuchi K, Igarashi S, Kobori K, Shiraishi M, Kroto H.W., *The production and structure of pyrolytic carbon nanotubes (PCNTs)* J. Phys. Chem. Solids **54**, 1835, (1993).

26   Falvo M.R., Clary G.J., Taylor R.M., Chi V., Brooks Jr F.P., Washburn S., Superfine R., *Bending and Buckling of Carbon Nanotubes Under Large-Strain* Nature **389**, 582 (1997).

27   Robertson D.H., Brenner D.W., MintmireJ.W., Phys. Rev. B **45**, 12592 (1992).

28   Yacobson B.I., Brabec C.J., Bernholc J. Phys. Rev. Lett. **76**, 2511 (1996).

29   Hernández E, Goze C., Bernier P., Rubio A. in *Electronic properties of novel materials-science and technology of molecular nanostructures*, H. Kuzmany, J: Fink, M. Mehring, S. Roth (Eds.), AIP, Melville, Nueva York 156 (1998).

30   Collins PG, Avouris P. *Nanotubes for electronics*. Scientific American **283**, 62 (2000).

31   Quian D, Dickey EC, Andrews R, Rantell T. *Load transfer and deformation mechanisms in carbon nanotube-polystyrene composites*. Applied Physics Letters **76**, 2868 (2000).

32   Gong X, Liu J, Baskaran S, Voise RD, Young JS. *Surfactant assisted processing of carbon nanotube/polumer composites*. Chemistry of Materials **12**, 1049 (2000).

33   Wagner HD, Lourie O, Feldman Y, Tenne R. *Stress-induced fragmentation of multiwall carbon nanotubes in a polymer matrix*. Applied Physics Letters **72**, 188 (1998).

34   Lourie O, Wagner HD. *Transmission electron microscopy observations of fracture of single-wall carbon nanotubes under axial tension*. Applied Physics Letters **73**, 3527 (1998).

35   Lourie O, Wagner HD. *Buckling and collapse of embedded carbon nanotubes*. Physical Review Letters **81**, 1638 (1998).

36   Lourie O, Wagner HD. *Evidence of stress transfer and formation of clusters in carbon nanotube-based composites*. Composites Science and Technology **59**, 975 (1999).

37   Cooper CA, Young RJ, HalsallM. *Investigation into the deformation of carbon nanotubes and their composites through the use of raman spectroscopy*. Composites Part A: Applied Science and Manufacturing **32**, 401 (2001).

38 Ajayan PM, Schadler LS, Giannaris C, Rubio A. *Single-walled nanotube-polymer composites: strength and weaknesses.* Advanced Materials **12**, 750 (2000).
39 Schadler LS, Giannaris C, Ajayan PM. *Load transfer in carbon nanotube epoxy composites.* Applied Physics Letters **73**, 3842 (1998).
40 Cochet, M., Maser, W.K., Benito, A.M., Callejas, M.A., Martínez, M.T., Benoit, J.M., Schreiber, J., Chauvet, O., *synthesis of a new polyaniline/nanotube composite: "in-situ" polymerisation and charge transfer through site-selective interaction.* Chemical Communication, 1450 (2001).
41 Ma RZ, Wu J, Wei BQ, Liang J, Wu DH. *Processing and properties of carbon nanotubes-nano-SiC ceramic.* Journal of Materials Science **33**, 5243 (1998).
42 Flahaut E, Peigney A, Laurent Ch, Marlière Ch, Chastel F, Rousset A. *Carbon nanotube-metal-oxide nanocomposites: microstructure, electrical conductivity and mechanical properties.* Acta Materalia **48**, 3803 (2000).
43 Peigney A, Laurent Ch, Flahaut E, Rousset A. *Carbon nanotubes in novel ceramic matrix nanocomposites.* Ceramics International **26**, 667 (2000).
44 Peigney A, Laurent Ch, Dumortier O, Rousset A. *Carbon nanotubes-Fe-alumina nanocomposites. Part I: influence of the Fe content on the synthesis of powders.* Journal of the European Ceramic Society **18**, 1995 (1998).
45 Peigney A, Laurent Ch, Dumortier O, Rousset A. *Carbon nanotubes-Fe-alumina nanocomposites. Part II: microstructure and mechanical properties of the hot-pressed composites.* Journal of the European Ceramic Society **18**, 2005 (1998).
46 Peigney A. Laurent Ch, Rousset A. *Synthesis and characterization of alumina matrix nanocomposites containing carbon nanotubes.* Key Engineering Materials **132**, 743 (1997).
47 Chen X, Xia J, Peng J, Li W, Xie S. *Carbon-nanotube metal-matrix composites prepared by electroless plating.* Composites Science and Technology **60**, 301 (2000).
48 Xu CL, We BQ, Ma RZ, Liang J, Ma XK, Wu DH. *Fabrication of aluminum-carbon nanotube composites and their electrical properties.* Carbon **37**, 855 (1999).
49 htpp://www.nyacol.com
50 htpp://www.bayer.com
51 htpp://www.sunnano.com
52 htpp://www.icb.csic.es/nanotubos/primera.html

SYNTHESIS AND CHARACTERIZATION OF NANOPARTICULATE CALCIUM ALUMINATES

L.D. Mitchell, J. Margeson and J.J. Beaudoin

National Research Council Canada, Institute for Research in Construction, 1200 Montreal Road, Ottawa, ON. Canada

1 INTRODUCTION

Nanoscale science and technology has emerged as a very active research area in recent years. Its scope encompasses a wide range of disciplines. The use of nanoparticulate solids is a natural step in the potential application of nanoscale science to the construction sector. Nanoparticles offer the potential for stronger and more flexible cement-based materials. They will have an inevitable influence on the hydration characteristics of the hydraulic phases in cements.

A number of techniques exist for the production of nanocrystalline and nanoparticulate oxide materials. Some of these are relatively complex and/or produce very small quantities of material, e.g. laser techniques. Additional techniques include high energy ball milling, cap precipitation, and other solution based routes. Many of the latter routes are relatively straightforward gel and combustion techniques that rely on complexation and dispersion with rapid heating respectively to achieve small crystallite sizes.

A technique was recently described in the literature that combines both dispersion and combustion in a single synthesis route. This technique relies on the use of a highly acidic sucrose solution to complex cations.[1] The solution is concentrated, foamed and charred to form a combustible carbonaceous foam material that yields an oxide on combustion. The authors repeated the work of Das in producing nanoparticulate alumina,[2] and found the technique to be practical in the laboratory.

To retain the simplicity of this synthesis technique, nitrate salts of the various cations are required. This requirement means that its application to silicate materials would be complex in comparison to aluminate materials. Consequently, this study examined the simplest hydraulic aluminate material, $CaAl_2O_4$; a logical extension of the previous alumina study.[2]

All calcium aluminate cements (CAC's) contain monocalcium aluminate as their principle hydraulic phase in amounts from 40% upwards. Monocalcium aluminate ($CaAl_2O_4$, or CA) is usually monoclinic.

It has been reported that a meta-stable orthorhombic form of CA can be formed under certain conditions.[3] This requires the use of 'chimie douce' type techniques such as the Pechini method.[4] This orthorhombic form has very similar cell parameters to the monoclinic structure described by Hörkner[5] apart from the small monoclinic distortion exhibited by the latter.

2 EXPERIMENTAL

The raw material was synthesised using stoichiometric quantities of $Ca(NO_3)_2 \cdot 6H_2O$ and $Al(NO_3)_3 \cdot 9H_2O$ dissolved in an aqueous solution of sucrose, with a sucrose:cation molar ratio of 4:1. The solution was heated to 65°C with stirring to ensure full dissolution, before dehydration on a hot plate at 250°C for 8 hours to form a brown viscous liquid. The caramelized mass was transferred to an oven at 200°C for 18 hours to foam, dehydrate and char the sample. The carbonaceous precursor was then fired in air inside a muffle furnace at 600°C for 24 hours to oxidise the carbon. The resulting white fluffy powder was removed, purged with dry nitrogen and stored in an air-tight container.

The materials produced were examined using a number of techniques, including the Brunauer-Emmett-Teller method (BET) for surface area determination, powder X-Ray Diffraction (XRD), Scanning Electron Microscopy (SEM), conduction calorimetry, Simultaneous Differential Scanning Calorimetry (DSC) and Thermal Gravimetric Analysis (TGA), and Vickers hardness measurement.

The surface area values were obtained using a Quantachrome Quantasorb analyser, with nitrogen gas as the absorbate.

Powder X-ray diffraction data were obtained using both a Scintag XDS 2000 diffractometer with $Cu_{k\alpha}$ radiation in conjunction with a graphite monochromator, and a Bruker D8 diffractometer equipped with dual mirror parallel beam optics, using $Cu_{k\alpha}$ radiation

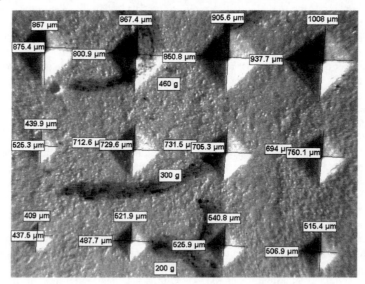

Figure 1   *Hydrated monoclinic CA with pyramid-shaped Vickers Hardness indentations.*

SEM micrographs were taken using a Cambridge Stereoscan 250. Samples were prepared with carbon adhesives on aluminium stubs, and then sputter-coated with gold.

The conduction calorimetry was performed with a laboratory made calorimeter. Readings were taken every ten minutes in °C, with a precision of 0.01 °C.

Simultaneous Differential Scanning Calorimetry and Thermal Gravimetric Analysis (DSC/TGA) measurements were taken with a TA Instruments Q600 SDT. The measurements were carried out under flowing nitrogen, using a heating rate of 10°C per minute.

Microhardness measurements were made with a Leitz Miniload Microhardness tester set up with a Vickers hardness diamond stylus. At least 3 load weights were used to give a range of diamond sizes.

Vickers hardness is determined by producing an indentation with a pyramid shaped diamond, and optically measuring the indentation produced, see Figure 1. Vickers hardness was calculated for each imprint and the average and standard deviation for each load series was calculated.

3 RESULTS

3.1 Characterisation of Anhydrous Materials

The gel-structured-CA fired at 600°C for 24 hours is a very fine, fluffy, white amorphous powder with a surface area of 75 to 150 m^2/g. No detectable changes in the material could be observed with XRD until the material was heated to 850°C, as seen in Figure 2. CA_2 and an orthorhombic monocalcium aluminate were identified at this temperature. The CA_2 persists as a minor phase throughout the whole experiment. The orthorhombic monocalcium aluminate phase was stable up to 1000°C after which a phase transition was observed; the orthorhombic monocalcium aluminate converted to the monoclinic. The observation of monoclinic CA at 1050°C from an amorphous gel broadly agrees with the results of Williamson & Glasser (1962)[6] who made monoclinic CA from an amorphous gel at 1045°C. However these authors did not report the existence of an orthorhombic phase.

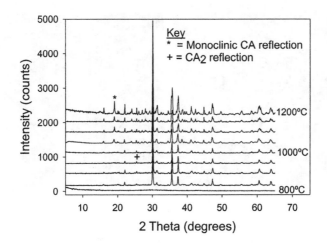

Figure 2   *X-ray diffraction spectra of the amorphous nano-CA starting material heated between 800-1200°C*

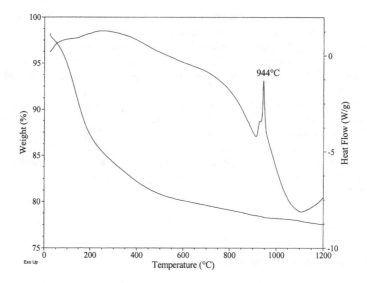

Figure 3   *DSC/TGA curves of the anhydrous amorphous CA calcined at 600°C.*

Figure 3 shows an initial water loss at about 150°C, even though the sample has been previously heated to 600°C, indicating it very quickly reacts with moisture from the atmosphere.   The heat flow curve also exhibits an exotherm at about 945°C.   This is interpreted as a phase change and corresponds well with the XRD data show in Figure 1. The XRD data shows that it is at this point that the sample transforms from a gel structured material to a crystalline one.

20,000X   1,000X

Figure 4   *Scanning Electron Micrograph of gel-structured CA calcined at 600°C*

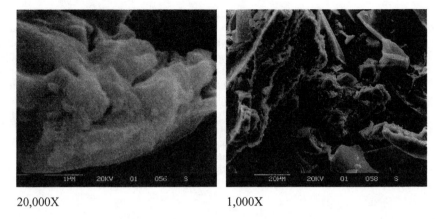

20,000X   1,000X

Figure 5   *Scanning Electron Micrograph of gel-structured CA calcined at 1000°C*

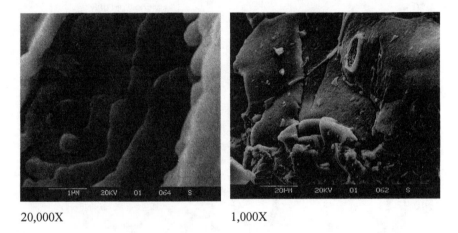

20,000X   1,000X

Figure 6   *Scanning Electron Micrograph of gel-structured CA calcined at 1200°C*

Figures 4, 5 and 6 show scanning electron micrographs of the gel-structured CA calcined at three different temperatures, 600, 1000, and 1200°C. Clear evidence of physical change can be seen in Figure 5, and Figure 6. This is attributed to grain growth akin to sintering. This crystal growth is probably responsible for the observed surface area decreases, and the particle size increase of the material with increasing temperature.

3.2 Characterisation of Hydrated Materials

Amorphous gel-structured-CA material calcined at 600°C was mixed with distilled water at a w/c ratio of 4.0 and left to hydrate in a calorimetry cell. This extremely high water to cement ratio was required due to the very high surface area of the starting material (~100 m^2/g). At this water to cement ratio the sample produced a workable paste. The conduction calorimetry results can be seen in Figure 7 along with those for blends (see table 1) of gel-structured and monoclinic CA, hydrated at a W/C 0.5. The hydrated material was subsequently characterised using both XRD and TGA (Figure 8 & Figure 9).

The results in Figure 7 show that the amorphous gel-structured-CA is hydraulic. The sample exhibits the classical calorimetry curve associated with calcium aluminate cements. This appears to be the first time that CA with an amorphous gel structure has exhibited classical cement characteristics. Nakagawa *et al*[7] heated bauxite and quicklime to 1700°C and quenched the subsequent melt, producing amorphous calcium aluminate. They found their material flash set in a matter of seconds and minutes and did not exhibit the behaviour observed here. Nakagawa *et al* also had to mix their material at a high w/c ratio (4.0 w/c).

Figure 7 shows that the initial peak observed for the gel-structured material is significantly greater than that observed for the blended (see table 1) and monoclinic samples. In addition, the end of the acceleration period occurs approximately 6 hours before the corresponding peak of the monoclinic only sample. The reaction rates of the blends also increase with the fraction of gel-structured material: all this indicates that the gel-structured material is very reactive.

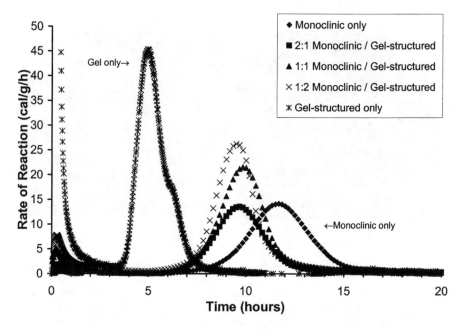

Figure 7   *Calorimetry curves displaying exothermic hydration reactions of CA systems*

The XRD data in Figure 8 show that the main hydration phases are CAH_{10}, C_3AH_6 and gibbsite. There is little evidence of the presence of C_2AH_8. Evidence for formation of CAH_{10} was found in both the XRD and the DSC/TGA curves. The conversion of CAH_{10} to C_3AH_6 has probably occurred, with the increased presence of gibbsite supporting this argument.[8] It is possible that the high surface area and the small particle size of the hydrates accelerate the conversion process; more work on this system is needed to confirm this.

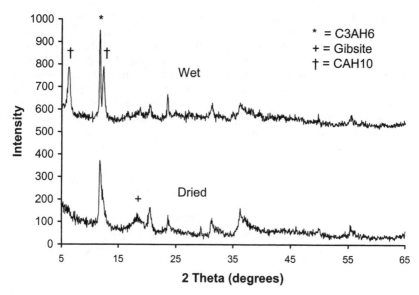

Figure 8   *X-ray diffraction pattern of amorphous CA hydrated for 24 hours*

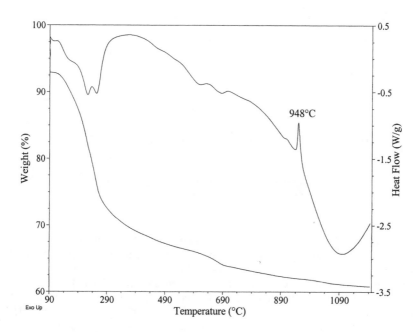

Figure 9   *DSC/TGA curves of the dried 24 hour hydrated amorphous CA material*

Figure 9 shows the TGA curve and the corresponding heat flow curve for the oven dried material (80°C) shown in Figure 8. There is a small peak at 150°C probably due to CAH_{10}. The dual peaks occurring in the temperature range 200-270°C represent dehydration reactions involving gibbsite and C_3AH_6.[9] The small peak at ~675°C has been assigned to calcium carbonate. The phase transition at about 950°C was discussed earlier, and represents the unhydrated material left in the sample (see Figure 10). The heating regime transforms the material from a gel-structured to a crystalline material at this temperature. This spectrum provides further evidence for the hydraulic nature of the anhydrous gel-structured CA created.

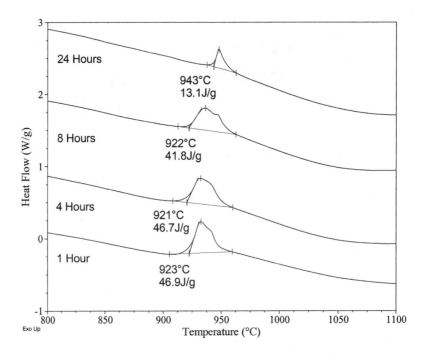

Figure 10   *Heat flow curves (800-1100°C) of gel-structured CA material hydrated for 1, 4, 8, & 24 hours*

In Figure 3 the apparent phase transition in the region of 940°C was clearly demonstrated. Following the decay of this exotherm in the hydrating material facilitates study of hydration kinetics. The results from such measurements are shown in Figure 10. The results show that the dormant period ended sometime after 4 hours, which is in agreement with the calorimetry data. The apparent phase transition for the unhydrated material in Figure 3 had a heat flow (indicated by the exotherm) of 349J/g. Using this figure as a reference for 100% unhydrated material, it would appear that approximately 87% of the original anhydrous material reacted within the first hour of hydration. After 24 hours, only 4% of the original material remained.

3.3 Microhardness - Evaluation of Physical Properties

The high surface area of the gel-structured CA created an enormous water demand. The microhardness measurements were therefore carried out on blends of commercial monoclinic CA (manufactured by CTL) and gel-structured CA.

The approximate blended volume ratios were 1:2, 1:1 and 2:1 monoclinic-CA : gel-structured-CA. This was achieved by using the two materials in the mass ratios shown in Table 1.

Table 1 *Material compositions used for the Vickers Hardness testing*

Volume Ratio	Weight Ratio	
1 : 2 Monoclinic / Gel-structured	1.000g CA	0.100g Gel-CA
1 : 1 Monoclinic / Gel-structured	1.000g CA	0.050g Gel-CA
2 : 1 Monoclinic / Gel-structured	1.000g CA	0.025g Gel-CA

The paste made was thick but workable (w/c 0.5) and the mix was placed into small microforms. The dimension of the microforms was 15 mm diameter and 5 mm height. The form was sealed with parafilm on the bottom of the form and the paste was worked into the forms to remove any trapped air. Then a glass cover was placed over the individual sample. Both sets of samples were marked and placed in a 100% RH environment at room temperature to hydrate.

The results can be seen in Figure 11. There are no statistically significant differences in the properties of the nano-modified cements. This is surprising, given the expected improvements in particle-packing densities. The probable explanation for this, is the tendency of the nanomaterial to agglomerate in water. These agglomerates will act as 'micro-particles.' Thus little change in the Vickers hardness results was observed with changing composition. Further work on the physical properties of nano-modified cement systems is required. Studies to evaluate the efficiency of using various dispersing agents are in progress.

4   DISCUSSION

Major investments both in Canada and around the world will establish nano-science and nano-technology as a growth area of science and engineering. With the construction industry being identified as a potential growth market for nano-materials, new and novel techniques will emerge in the coming years as this sector expands. The successful creation of monocalcium aluminate using a novel sucrose-based polymer dispersion technique is perhaps one of the first examples of innovation to emerge from the application of nano-science to cement research. It has proven to be a cheap and reliable synthesis route.

Before nano-materials can become commercially successful however, the products have to be cheaper, thinner, lighter, stronger or tougher. This has been the goal of many materials scientists throughout the last century. A good or clever technology isn't necessarily sufficient to have an impact in the marketplace. The product has to have significant improvement over competing products.

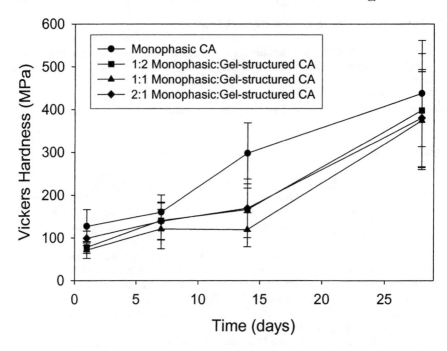

Figure 11   *Graph showing the development of microhardness of several different ratios*
of monoclinic CA to gel-structured CA

5 CONCLUSIONS

A unique synthesis route for producing monocalcium aluminates has been described. It has been shown that the resultant material is hydraulic, has a gel-structure, and is very reactive.

The use of nanoscale hydraulic materials has advantages and disadvantages. The very high surface area of the material affects both the hydration kinetics and water requirements. The latter makes the use of any high surface area material problematic, in terms of adequate mixing at desirable water/cement ratios. Conventional mixing techniques require excessive water to provide an acceptable rheology. The use of admixtures or alternative mixing methodologies, for example high-shear mixing, may reduce the problem. It is unclear whether this problem can be completely overcome.

Rheology and surface science are going to be significant study areas for the future of nano-materials in cement science. Water to cement ratio is an extremely important parameter for cement-based materials used in the construction industry. The addition of high surface area materials, that will raise this ratio significantly, will meet resistance in non-specialist applications. Should the problem with water-demand be solved, then the possibility of high strength, rapid-setting materials could become a reality.

Acknowledgements

The authors would like to acknowledge Mr Gary Polomark for his useful discussions regarding the thermal analysis. Dr Pamela Whitfield provided valuable assistance in the X-ray diffraction analysis and the preparation of this manuscript.

References

1   R.S. Das, A. Bandyopadhyay and S. Bose, *Journal of the American Ceramic Society,* 2001, **84**, 2421-2423
2   L.D. Mitchell, P.S. Whitfield, J. Margeson and J.J. Beaudoin, *J. Mat. Sci. Lett.,* 2002, **21**, 1773-1775
3   S. Ito, K. Ikai, M. Suzuki and M. Inagaki, *Journal of the American Ceramic Society,* 1975, **58**, 79-80
4   P.A. Lessing, *American Ceramic Society Bulletin,* 1989, **68**, 1002-1007
5   W. Hörkner and H.K. Müller-Bushbaum, *J. Inorg. Nucl. Chem.,* 1976, **38**, 983-984
6   J. Williamson and F.P. Glasser, *J. Appl. Chem.,* 1962, **12**, 535-538
7   K. Nakagawa, I. Terashima, K. Asaga and M. Daimon, *Cement and Concrete Research,* 1990, **20**, 655-661
8   J. Bensted and P. Barnes, *Structure and Performance of Cements* (Spon Press, London, 2002).
9   V.S. Ramachandran R.M. Paroli and J.J. Beaudoin, *Handbook of Thermal Analysis of Construction Materials* (Noyes Publications/Williams Andrew, New York, 2003).

EFFECTS OF WATER-CEMENT RATIO AND CURING AGE ON THE THRESHOLD PORE WIDTH OF HARDENED CEMENT PASTE

H.N. Atahan, O.N. Oktar and M.A. Tasdemir

Department of Civil Engineering, Istanbul Technical University, 34469, Maslak, Istanbul, TURKEY. E-mail: hnatahan@ins.itu.edu.tr

1 INTRODUCTION

Normal strength concrete can be considered to be a three phase composite material consisting of a continuous mortar matrix and the interfaces between the cement mortar and aggregate. It is a well-known fact that the most important parameter which directly affects the strength and durability of concrete is the water-cement ratio. The decrease in the water-cement ratio in concretes with sufficient cement content and enough workability leads us to high strength / high performance concretes. As a result of this, there is a growing interest in the use of high strength and high performance concretes in the construction of infrastructure. Their low porosity gives them important durability and low transport properties which makes potentially suitable material for concrete structures. In recent years, two developments have led to great improvements in modern concrete concerning its resistance against damaging mechanisms such as freeze-thaw cycling, embedded steel corrosion and alkali silica reaction. Using superplastizers which enables a rheological control and has an effect of lowering the water-cement ratio of the mixture, and the inclusion of silica fume which enhances the overall durability and strength are those developments.[1] Thus, when the durability is of concern, the pore structure of cement paste has great importance, and the studies have especially been focused on the durability of concrete which is directly related to the pore structure of hardened cement paste (hcp).

The main objective of the presented work is to investigate the combined effects of water-cement (w/c) ratios and curing ages on the "threshold" or "critical" pore width of (hcp). Mercury intrusion porosimetry (MIP) test is one of the techniques that have been used for analyzing the microstructure of cement paste. This study presents the results of the MIP experiments conducted to three different hcps with the water-cement ratios of 0.26, 0.34, and 0.42 which had been cured for 7, 28, and 365 days. The degrees of hydration and the heats of hydration of the pastes were also reported. Effects of water-cement ratio and curing time on the critical pore width in nano-scale were discussed.

2 FROM NANO-SCALE TO MACRO-SCALE

Cement based materials, such as hydrated cement pastes, mortars, plain and fibre reinforced concretes are widely used in the construction of modern structures. During their service life, depending on the environmental conditions they are in, many structures are exposed to various types of aggressive agents. Intrusion of the aggressive agents into the cement based materials, which directly affects durability, is generally controlled by the diffusivity and permeability of the material. In addition, there is a significant decrease in the strength of concrete with increasing porosity of hardened cement paste. Hydration of cement paste is a physico-chemical reaction which develops in time. Gel formation occurs starting from the surfaces of the cement particles. According to Powers[2], each 1 cm^3 of unhydrated cement particles forms into 2.06 cm^3 of cement gel. Capillary pores between the cement particles, which are originally filled with water, are reduced due to gel formation. If the absolute volume of unhydrated cement in 1 m^3 of hardened concrete is denoted by "c", the volume of the gel formed is 2.06αc, where α is the degree of hydration which varies between 0 and 1. Thus, the decrease in capillary pore volume is 2.06αc-αc=1.06αc, and from which the volume of capillary pores can be derived to be w-1.06αc. With respect to these explanations, as the hydration process proceeds, the pores of the paste are gradually filled with hydration products.[3,4] Water-cement ratio, curing conditions, type of binder, and use of mineral and chemical admixtures have significant effects on the value of the total porosity, pore size distribution and the critical pore width.[5,6]

Rössler and Odler[7] prepared cement pastes with different water-cement ratios which were hydrated at different temperatures for different times. They have shown that the main factor influencing the strength properties of the samples is their porosity. In their work, they also conducted MIP tests. According to the tests, they have found that the maximum points on the differential curves were independent from the initial water-cement ratio, but yet depend on the hydration time and shifted to lower values as the hydration proceeded.

It was indicated that the cement paste in concrete and mortar have a pore size distribution different from that of a plain paste hydrated without aggregate.[8,9] They showed that, in mortar and concrete, there was the additional formation of pores greater in size than the threshold pore diameter of the plain cement paste, and they thought that these larger pores are present only in the interfacial zone between the aggregate and cement paste. These pores are most likely to affect the properties such as permeability and durability in an unfavorable way. Reinhardt and Gaber[5] performed an experimental and theoretical study for determining pore size distribution curves and correlating them with water and oxygen permeability. Their studies have shown that knowing equivalent pore size together with the porosity is sufficient to predict the physical properties with an acceptable accuracy.

There are different theories in the literature about concrete which are used in the permeability calculations. Katz-Tompson permeability theory is one of those that attempt to relate pore structure to transport coefficients. The main parameter of this theory is the critical pore diameter, which can be obtained from the mercury intrusion porosimetry tests. Katz and Tompson defined the threshold pore diameter as the inflection point of the curve drawn at the cumulative intruded pore volume versus intrusion pressure graph. They have done their experiments simultaneously by measuring the electrical conductivity of the samples as intrusion took place, and they showed that the electrical conductivity went through a sharp change from zero to nonzero which means there was a metallic pathway across the sample. According to the test results, they found that this point was always at the inflection point of the cumulative intrusion curve.[10] Threshold pore diameter has been defined by Winslow and Diamond as the minimum diameter of pores which are geometrically continuous through all the regions of the hydrated cement paste. A pore channel enables the flow to occur when it is sufficiently large and connected all the way across the sample.[10] Consequently, the importance of the threshold pore diameter for the transport properties becomes clear. Pores which are larger than the threshold pore diameter and have a connected pathway are the most important ones in determining the flow properties of the sample. Therefore, increasing curing time, decreasing water-cement ratio and adding mineral and chemical additives into the cement paste directly decreases the volume of pores which are larger than the threshold pore diameter, and also the diameter of the critical pore. When the durability is of concern, during the hydration process, not only the diameters of pores decrease, but also the probability that gas, ions, or fluid flow through the porous media decreases.

As seen in Figure 1a, an increase in the threshold width from $0.07\mu m$ to $2\mu m$ has resulted in an increase of nearly 10 times in O_2 diffusion coefficient.[11] This means that an increase in the threshold radius leads to a substantial increase in the O_2 diffusion coefficient. Such a relation from nano-scale to macro-scale is a good explanation of the role of threshold pore width in the diffusion characteristics of cement paste and mortar. As the reinforcement of the structures is of concern the oxygen intrusion causes the reinforcement to rust and corrode. Therefore, it can be said that the threshold pore width is also an indication of the long term performance of concrete structures. Schiessl and Hardtl[11] have also shown that there are no significant differences between mixtures with and without fine aggregate.

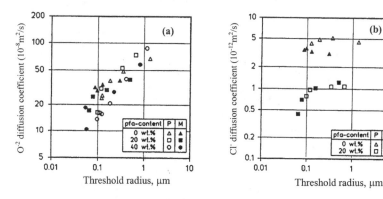

Figure 1 *Relationships between threshold radius and (a) oxygen, (b) chloride diffusion coefficients depending on the fly ash content.[11]*

Figure 1b shows the relationship between the threshold pore width and chloride diffusion coefficient.[11] If the mixtures with and without fly ash are evaluated separately, as in Figure 1b, it can be said that they form two clearly distinct groups. In the mixtures with fly ash, as the threshold pore width increases, chloride diffusion coefficient increases significantly, and then becomes stable. Mixtures without fly ash have diffusion coefficients that are more than four times those of specimens made with fly ash. The results reported by Schiessl and Hardtl[11] showed that reference mixtures have higher capillary pore volumes compared to the mixtures with fly ash. The lower capillary pore volume of mixtures with fly ash indicates that both interfacial zones in the mortar and the pores in the cement paste have been densified. Since the mixtures with fly ash have reduced chloride diffusion coefficients and higher chloride binding capacities, it is expected that the risk of steel corrosion in case of chloride attack may be reduced.

Figure 2 shows that there is a relation between the carbonation depth after three years and the threshold pore width. Thus, it can be concluded that the threshold pore width plays important role in the diffusion of O^{-2} and Cl^- ions, and in the carbonation of concrete cover.

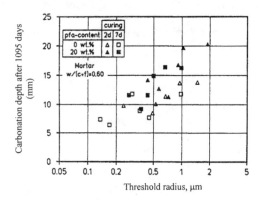

Figure 2 *Relationship between threshold radius at 28 days and carbonation depth after 3 years in climate 20/65.[11]*

3 EXPERIMENTS

In this work, 5 mixes of cement paste were prepared using an ordinary Portland cement with the water-cement ratios of 0.26, 0.30, 0.34, 0.38, and 0.42. All the specimens were cured in water until their test days. The degrees of hydration extend and heats of hydration were determined on specimens cured for 7, 14, 28 and 365 days.

3.1 Hydration Degrees

The degree of hydration of cement paste can be determined by measuring non-evaporable water content. [12] The measurements were made on the specimens at the ages of 7-14-28 and 365 days with the water-cement ratios of 0.26, 0.30, 0.34, 0.38, and 0.42. Non-evaporable water contents (w_n) of hardened cement pastes were determined on the ignited basis. [13] The degree of hydration (α) can be determined based on that the hydration of 1 g of anhydrous cement retains 0.23 g of non-evaporable water. According to this approach α can be calculated as;

$$\alpha = 100 \ [(w_n/C)/0.23] \tag{1}$$

The w_n/C ratios calculated based on the ignited weight are given in Table 1, and the degree of hydration versus water-cement ratio graph can be seen in Figure 3.

Table 1 *w_n/C ratios of hardened cement pastes*

Water/cement ratio	w_n/C			
	7 days	14 days	28 days	365 days
0.26	0.1049	0.1139	0.1106	0.1260
0.30	0.1129	0.1260	0.1228	0.1419
0.34	0.1194	0.1304	0.1300	0.1716
0.38	0.1208	0.1419	0.1444	0.1815
0.42	0.1226	0.1486	0.1491	0.1840

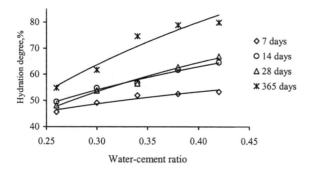

Figure 3 *Water-cement ratio of hcp versus hydration degree*

The non-evaporable water content values have shown that the degrees of hydration were also significantly increased with both increasing the curing time and w/c ratios of hcp's and they varied between 45.8% and 80.0% as shown in Table 2.

Table 2 *Hydration degrees of hardened cement pastes*

Water/cement	Hydration degrees, %			
ratio	7 days	14 days	28 days	365 days
0.26	45.8	49.5	48.1	54.8
0.30	49.1	54.8	53.9	61.7
0.34	51.9	56.7	56.5	74.6
0.38	52.5	61.7	62.8	78.9
0.42	53.2	64.6	64.8	80.0

3.2 Heat of Hydration

The heat of hydration is defined as the quantity of heat, in joules per gram of unhydrated cement, evolved upon hydration at a given temperature.[14] The heats of hydration were measured according to ASTM C186-98[15] on the same samples that had been used for the measurements of the degrees of hydration. Tests were conducted on the specimens at the ages of 7-14-28, and 365 days with the water-cement ratios of 0.26, 0.30, 0.34, 0.38, and 0.42. Test results are shown in Figure 4.

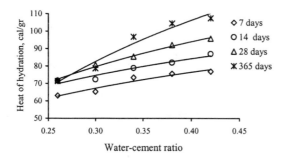

Figure 4 *Relationships between the heat of hydration and the water-cement ratio of hcp*

3.3 Mercury Intrusion Porosimetry Tests

Mercury intrusion porosimetry (MIP) is a widely used method for measuring the pore size distribution of porous media of cement based materials which is an important factor in the durability of concrete. In this test, a porous sample is placed into a chamber, and surrounded by mercury, and then the pressure applied on the mercury is gradually increased. So, as the pressure increases, mercury is forced into the pores of the sample. The fact that the test was based on is that for filling a non-wetting fluid into a pore of the diameter d, a pressure P that is inversely proportional to the diameter of this pore must be applied. This pressure is given by the Washburn equation as seen below[13]:

$$d = -4\gamma(\cos\phi)/P \tag{2}$$

where d is the apparent pore diameter, γ is the surface energy of the mercury, and ϕ is the contact angle between the mercury and the pore wall.

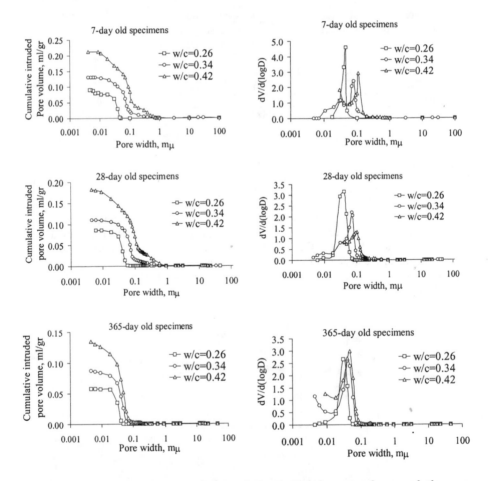

Figure 5 *Relationships between the cumulative intruded pore volume and the pore diameter for 7, 28 and 365 day old specimens, and their differential curves*

In this study, MIP tests were conducted on the specimens at the ages of 7, 28 and 365 days with the water-cement ratios of 0.26, 0.34 and 0.42. The weights of the samples were approximately 2 grams and two samples were used for each porosimetry test. The average of the values for the two samples at the ages of 7, 28 and 365 days were plotted at the pore diameter versus cumulative intruded pore volume diagram (Figure 5).

To obtain the threshold pore widths of the specimens with different water-cement ratios, differential curves at the cumulative intruded pore volume versus pore diameter diagrams were used which were also given in Figure 5. The pore widths corresponding to the peaks of the differential curves were determined to obtain the threshold pore widths. The values that were obtained for threshold pore widths and the relationship between the threshold pore width and the curing time are shown in Table 3 and Figure 6, respectively.

Table 3 *Critical pore widths of the samples determined on 7, 28, and 365 days*

Water -Cement Ratio	Threshold Pore Width, nm		
	7 days	28 days	365 days
0.26	46	40	30
0.34	80	70	40
0.42	110	100	48

As seen in Figures 5 and 6, increasing the curing time and decreasing the water-cement ratio result in lower total porosities and smaller values of the threshold pore width for all pastes. The minimum values of total porosity and threshold pore width were obtained by minimizing water-cement ratio and maximizing curing time.

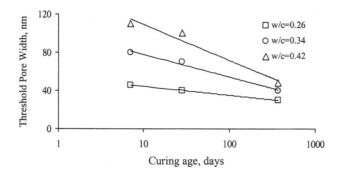

Figure 6 *Relationship between the critical pore width and the curing age*

4 DISCUSSIONS

Test results have shown that, heats of hydration varied between 63.0 cal/gr and 107.4 cal/gr as seen in Table 4. This table shows that, the heat of hydration increases significantly with increasing water-cement ratio and curing time. Results found for hcp with water-cement ratio of 0.42 are also compared with the data obtained by Verbeck[16] for the water-cement ratio of 0.40 which are shown in Table5. As the values at Tables 4 and 5 are compared, it can be said that heats of hydration found for hcp with water-cement ratio of 0.42 are very close to the results obtained by Verbeck.[16]

Table 4 *Heats of hydration of hardened cement pastes found in this study*

Water/cement ratio	Heat of hydration, cal/gr			
	7 days	14 days	28 days	365 days
0.26	63.0	71.6	71.8	71.3
0.30	65.3	72.3	80.6	78.6
0.34	73.4	78.8	85.3	96.6
0.38	75.5	81.9	92.0	104.4
0.42	76.8	86.9	95.7	107.4

Table 5 *Heats of hydration of the Portland type cement paste with water-cement ratio of 0.40 found by Verbeck[16]*

Water/cement ratio	Heat of hydration, cal/gr			
	7 days	14 days	28 days	365 days
0.40	79.2	-	95.6	108.6

The pore width corresponding to the highest rate of mercury intrusion per change in pressure is known as the "threshold", "critical" or "percolation" pore width.[17] The threshold pore widths of the hcp's were determined by using the differential curves at the pore width (μm) versus cumulative intruded mercury (ml/gr) diagrams. Calculations have shown that, the threshold pore width of the hcp's decreased significantly with decreasing water-cement ratio and increasing curing time, and it varied between 30 nm and 110 nm as shown in Table 3. In another words, the minimum value of the threshold pore width is obtained by minimizing water-cement ratio and maximizing curing time. The total intruded pore volume varied between 0.0587 ml/gr and 0.2127 ml/gr as shown in Table 6.

Table 6 *Total intruded pore volumes of the samples found from MIP tests*

Water -Cement Ratio	Total intruded pore volume, ml/gr		
	7 days	28 days	365 days
0.26	0.0901	0.0856	0.0587
0.34	0.1307	0.1098	0.0873
0.42	0.2127	0.1825	0.1348

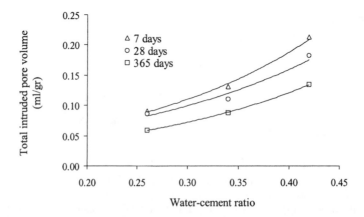

Figure 7 *Relationship between the total intruded pore volume and water-cement ratio for 7, 28 and 365 day old specimens*

In Figure 7, it can be seen that, the total intruded pore volume decreased significantly with decreasing water-cement ratio and increasing curing time which is an indicator of the decrease in total porosity.

In Figure 8, it can be seen that the slopes of the straight lines decrease with increasing curing times. Especially for the 365 day old specimens, the slope of the line decreased significantly. It may be expected that the slope of this line will nearly approach to zero for longer curing times. With the range of the water-cement ratio used in this work, and in case where hydration has been completed, the threshold pore width of the hcp is independent of water-cement ratio, and it is approximately 20 nm. Such a threshold pore width corresponds to the size of the gel pore.[18] In such a case, it is seen that the gel structure is independent of water-cement ratio. At early ages, the reason for the threshold pore width being affected by the water-cement ratio is because of the capillary pores not being segmented enough or not diminishing in these ages.

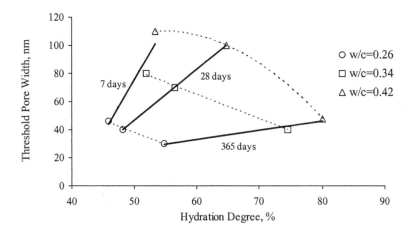

Figure 8 *Relationship between the critical pore width and the hydration degree with different water-cement ratios and curing times*

5 CONCLUSIONS

The results obtained in this work can be summarized as follows:

1. The threshold pore widths varied between 30 nm and 110 nm depending on the curing time and water-cement ratio of the paste. Longer curing times and lower water-cement ratios resulted in lower values of threshold pore width.
2. Degree of hydration increased significantly with increasing curing time and water-cement ratio of the cement paste, and it varied between 46% and 80%. Longer curing times and higher water-cement ratios resulted in greater heat of hydration. The heat of hydration varied between 63 cal/g and 107 cal/g.
3. The threshold pore width determined in nano-scale is of great importance in macro-scale for the durability of cement based composites.
4. The relation between the threshold pore width and hydration degree shows that the longer the curing time the lower the threshold pore width.

5. With the range of the water-cement ratio used in this work, the threshold pore width of the hcp's, in the case where hydration is completed, is independent of the water-cement ratio, and it is approximately 20 nm. Such a threshold pore width corresponds to the size of a gel pore. In which case, it is seen that the gel structure is independent of water-cement ratio.

Acknowledgement

The MIP tests used in this study were carried out at TUBITAK-MAM (The Scientific and Technical Research Council of Turkey - Marmara Research Centre). The authors wish to acknowledge the financial support of THBB (Turkish Ready Mixed Concrete Association). The first author also acknowledges the grant of TCMB (Turkish Cement Manufacturers' Association) for his PhD study.

References

1   J. P. Skalny, (ed.), *Materials Science of Concrete*, The American Ceramic Society, Westerville, OH, 1989.
2   T.C. Powers, Port. *Cem. Assoc. Res. Dept. Bul.*, 1958, **90**.
3   O.N. Oktar, H. Moral and M.A. Tasdemir, *Cem. and Concr. Res.,* 1996, **26**, 1619.
4   H.N. Atahan, O.N. Oktar and M.A. Tasdemir, *6th Int. Symp. on Utililization of High Performance- High Strength Concrete, Leipzig, 2002, 2, 839.*
5   H.W. Reinhard and K. Gaber, *J. Mater. Struct.*, 1990, **23**, 3.
6   J.M. Khatib and P.S. Mangat, *Cem. and Concr. Comp.*, 1999, **21**, 431.
7   I. Odler and M .Rössler, *Cem. and Concr. Res.,* 1985, **15**, 401.
8   D. Winslow and D. Liu, *Cem. and Concr. Res.*, 1990, **20**, 227.
9   D.N. Winslow, M.D. Cohen, D.P. Bentz, G.A. Sydner, and E.J. Garbozci, *Cem. and Concr. Res.*, 1994, **24**, 25.
10 E.J. Garbozci, *Cem. and Concr. Res.*, 1990, **20**, 591.
11 P. Schiessl and R. Hardtl, P.K. Mehta Symposium on Durability of Concrete, Nice, 1994, 99.
12 L.E. Copeland and D.L. Kantro, *Proc. 4th Int. Symp. Chem. Cem.*, 1960, **1**, 443.
13 L. Lam, Y.L. Wong and C.S. Poon, *Cem. and Concr. Res.*, 2000, **30**, 747.
14 A.M. Neville, *Properties of Concrete*, 4th Ed., 2000, 37.
15 ASTM C186-98, Standard test method for heat of hydration of hydraulic cement.
16 G.Verbeck, . *4th Int. Symp. Chem. Cem.*, 1960, **1**, 453.
17 R.A. Cook and K.C.Hover, *Cem. and Concr. Res.*, 1999, **29**, 933.
18 H.M. Jennings and L.J. Parrott, J. Mater. Sci., 1986, **21**, 4053.

EFFECT OF CURING REGIME AND TYPE OF ACTIVATOR ON PROPERTIES OF ALKALI-ACTIVATED FLY ASH

T. Bakharev

Department of Civil Engineering, Monash University, Clayton, Victoria 3800, Australia.
E-mail: bakharev@eng.monash.edu.au

1 INTRODUCTION

This paper presents an investigation of the geopolymer materials prepared using class F fly ash. Geopolymers form a new class of ceramic materials produced by a sol-gel process utilising alumina and silica oxides activated by alkali hydroxides and/or alkali silicates. The starting materials dissolve in high pH alkaline solution and the geopolymers are precipitated; this process is facilitated by heat. In the result of the polymerisation reactions polysialates, polysialate siloxo, polysialate disiloxo are formed[1]. Geopolymer materials have a matrix formed by synthetic minerals belonging to the same aluminosilicates family as zeolites, but unlike zeolites they are essentially amorphous polymers. The mineral polymers have empirical formula: $M_n[-(SiO_2)_z-AlO_2]_n \cdot wH_2O$, where z is 1,2 or 3; M is an alkali cation such as potassium or sodium, and n is a degree of polymerisation[1-2]. Initially, the geopolymer materials were prepared using calcined clay (neokaolin, parakaolin etc.) activated by alkali hydroxides and silicates[2-4]. These materials were reported to have high durability and thermal resistance properties. This paper presents the preparation procedure of the geopolymer materials using class F fly ash, their characterisation using XRD, SEM, FTIR and BET (nitrogen adsorption analysis), and study of durability in acid and sulphate environment. The studied geopolymer materials can be used in structural applications including operating in harsh, chemically aggressive environment.

2 EXPERIMENTAL STUDIES

2.1 Materials

The materials were prepared using class F fly ash from Gladstone, Queensland, Australia supplied by Pozzolanic Enterprises. Table 1 shows its chemical composition. Fly ash is mainly glassy with some crystalline inclusions of mullite, hematite and quartz. Laboratory grade sodium silicate type D with Ms (ratio of silica oxide to sodium oxide) equal to 2, 14.7 % Na_2O and 29.4 % SiO_2 (PQ Australia), 60% w/v sodium hydroxide solution (Sigma) and potassium hydroxide (85+%) (Aldrich) were used as activators. Table 2 shows the details of the geopolymer samples used in the durability studies. Fly ash was mixed with sodium hydroxide, sodium silicate and a mixture of sodium and potassium hydroxide solutions, providing 8-9% Na in mixtures and water/binder ratio of 0.3. The pastes were cast in plastic cylinders and sealed with the lid. The pastes were cured for 24 hours at room temperature, after that they were ramped to 95°C and cured at this temperature for 24 hours, then the materials cooled down with the oven and were cured at room temperature in tap water. The initial strength of geopolymer samples prior to durability tests were: 58 MPa 8FAK, 66 MPa 8FASS, and 59 MPa 8FA.

Table 1   *Composition of fly ash (mass %) by XRF*

Oxide	SiO$_2$	Al$_2$O$_3$	Fe$_2$O$_3$	CaO	MgO	K$_2$O	Na$_2$O	TiO$_2$	P$_2$O$_5$	Loss on ignition
Fly Ash	50.0	28.0	12.0	3.5	1.3	0.7	0.2	-	-	-

Table 2   *Alkali activated fly ash samples*

Sample ID	Type of activator and w/b ratio	Concentration	*Curing Conditions*
8FASS	Sodium silicate, w/b=0.3	8% Na	24 at room temp., 24 h at 95°C
8FA	Sodium hydroxide, w/b=0.3	8% Na	24 at room temp., 24 h at 95°C
8FAK	Sodium hydroxide + Potassium hydroxide, w/b=0.3	8% Na, 1% K	24 at room temp., 24 h at 95°C

2.2 Elevated Temperature Curing

Application of heat had a significant effect on the strength development of the studied materials. Fly ash activated by sodium silicate and by sodium hydroxide, with 2-8% Na in mixtures and w/b=0.3, was used to investigate the effect of elevated temperature curing on compressive strength development. In this study three types of curing were used. In the first case the pastes were cured for 2 hours at room temperature and then were ramped to 75°C and exposed to heat curing at 75°C for 1month (Case I 75C). In the second case the pastes were cured for 24 hours at room temperature, then the pastes were ramped to 75°C (Case II 75C) or 95°C (Case II 95C) and cured at these temperatures for 24 hours, after that the materials cooled down with the oven and were cured at room temperature. In the third case pastes were cured for 24 hours at room temperature, then were ramped to 75°C (Case III 75C) or 95°C (Case III 95C) and exposed to that temperature for 6 hours, after that the materials cooled down with the oven and were cured at room temperature. In all cases the samples were sealed in plastic tubes and cured hydrothermally. Two types of curing were utilised after exposure to heat. In one case the cylindrical samples were covered with hydrophobic film and wrapped in thin plastic to prevent moisture evaporation and tested for compressive strength at indicated times, in the second case the cylinders were stored in tap water and when required were taken out of water and tested.

2.3   Testing Procedures

The compressive strength of ø25 x 50 mm cylinders was measured at 1, 2, 7, 14, 28, 60, 120, 150 days. The materials were analysed by XRD, FTIR and SEM. X-Ray diffraction analyses were made with a Rigaku Geigerflex D-max II automated diffractometer with the following conditions: 40 kV, 22.5 mA, Cu-Kα radiation. The XRD patterns were obtained by scanning at 0.1° (2θ) per min and in steps of 0.05° (2θ). SEM (Hitachi S-2300, Japan) was utilised for microstructural observations of the fracture surfaces, which were coated with gold for examination using secondary electron imaging (SEI). Fourier transform infrared spectroscopy (FTIR) was performed for the samples on Perkin Elmer 1600 FTIR spectrometer using the KBr pellet technique (3 mg powder sample mixed with 100 mg KBr).

The resistance of materials to acid attack was studied by immersion of ø25 x 50 mm cylindrical specimens in 5% solutions of acetic and sulphuric acids. The resistance of materials to sulphate attack was studied by immersion in 5% solutions of sodium sulphate and magnesium sulphate, and in a solution of 5% sodium sulphate + 5% magnesium sulphate. The media were replaced with the fresh solutions monthly. The compressive strength of ø25 x 50 mm cylinders was measured before the test and at 30, 60, 90 and 120 days. Neat Portland cement paste (OPC) and Portland cement paste with 20% fly ash replacement (OPC+FA) with water/binder ratio 0.5 were used for comparison in the tests. The compressive strength of these specimens at the age of 2 months, when the test started, was respectively 45 MPa and 42.9 MPa.

3 RESULTS AND DISCUSSION

3.1 Compressive Strength Evolution

Figure 1(a-b) shows the results of the compressive strength measurements for materials prepared with sodium silicate and sodium hydroxide solutions and cured as described in Case I 75C. There was an increase in strength with increase in concentration from 2% Na to 8% Na in the pastes prepared with the sodium silicate and sodium hydroxide solutions. For sodium hydroxide a gradual increase in strength was observed corresponding to the increase in sodium concentration, while for sodium silicate the strength was low at 2%, 4% and 6%Na and rapidly developed at 8%Na concentration.

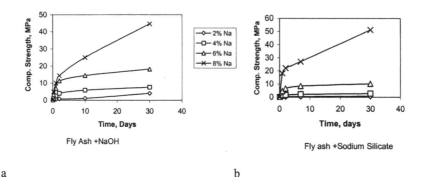

a   b

Figure 1(a-b) *Compressive strength evolution of the samples cured as in Case I 75C: (a) activated by sodium hydroxide and (b) activated by sodium silicate.*

Figure 2(a-b) presents the results of the strength measurements for materials cured according to Case II 75C, Case II 95C, Case III 75C and Case III 95C procedures for fly ash activated by sodium hydroxide solution, and for materials cured according to Case II 75C, Case II 95C, and Case III 95C procedures for fly ash activated by sodium silicate solution. It was found that long precuring at room temperature was beneficial for strength development, and there was about 300% increase in strength after 24 hours of heat curing in Case II 75C and Case II 95C as compared to Case I 75C for materials with both activators. The strength was significantly higher if materials were stored 24 hours at room temperature before application of heat.

Strength of materials was increased in Case II 75C and Case II 95C as compared to Case III 75C due to longer period of elevated temperature treatment, 24 hours versus 6 hours. For both types of activators materials formed at 95°C had higher strength than materials formed at 75°C if 24 hours heat treatment was employed. However, this trend was inversed for materials formed with sodium hydroxide when 6 hours heat treatment was used (Case III 95C and Case III 75C). Materials cured at 75°C (Case III 75C) had higher strength than material cured at 95°C (Case III 95C).

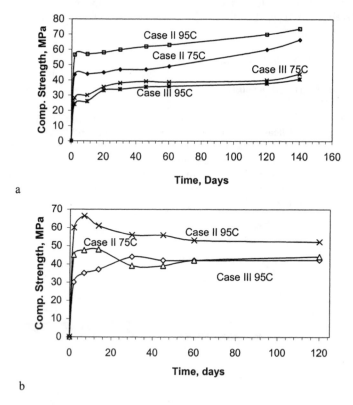

Figure 2(a-b) *Compressive strength evolution of the samples cured as in Case II 75C and Case II 95C, Case III 75C and Case III 95C: (a) activated by sodium hydroxide and (b) activated by sodium silicate. The materials were stored at 23°C in tap water after the heat treatment.*

Figure 2(a) shows that the materials prepared using sodium hydroxide had a steady strength growth after the heat treatment, while the materials prepared with sodium silicate and cured 24 hours at 95°C had some decline of strength. Therefore, strength development after the heat treatment depended on the type of activator and storage conditions. Materials with both activators showed deterioration of strength if moisture evaporation is allowed during or after heat curing. The material's surface has to be covered by thin plastic or hydrophobic film to prevent surface drying.

3.1.1   Fly ash activated by sodium silicate. This investigation revealed that strength of the materials prepared with sodium silicate depends on heat curing procedure and storage conditions after exposure to heat. When stored at room temperature in air, Case II 95C samples had up to 20% loss of strength (Figure 3). These samples were stored covered by hydrophobic film and by thin plastic to prevent water evaporation. When stored in air, materials prepared with sodium silicate and cured as in Case II 75C or Case III 95C had no strength loss and some strength gain was observed for Case III 95C materials.

When stored in tap water after heat curing, materials prepared with the sodium silicate activator and cured as in Case II 75C or Case II 95C had up to 25% loss of strength (Figure 2(b)). However, in the same conditions specimens cured as in Case III 95C did not have strength reduction and had 25% strength gain. Thus, materials activated by sodium silicate and heat cured for 6 hours at 95°C had a steady strength growth compared to samples cured for 24 hours at 95°C, which had up to 25% decline of strength after the elevated temperature curing.

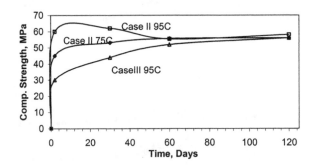

Figure 3   *Compressive strength development in fly ash activated by sodium silicate and cured as in Case II 95C, Case II 75C and Case III 95C. The samples were held at room temperature RH=70% after the heat treatment.*

3.2 XRD

Figure 4 presents XRD traces of the samples prepared with sodium silicate and sodium hydroxide solutions. It was found that the main phases formed in the samples are amorphous, and only in case of materials prepared with sodium hydroxide solution semi-crystalline zeolitic phases were present after 1 month of storage. The zeolitic products formed depended on the curing regime. In materials with 2 hours precuring at room temperature and exposed to elevated temperature curing at 75°C as in Case I 75C chabazite, Na-P1 (gismondine) and hydroxysodalite were found, while materials cured as

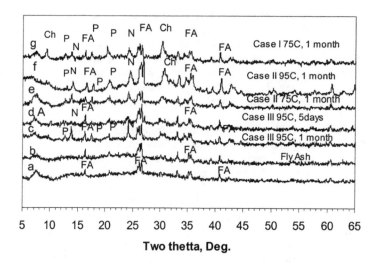

Figure 4 *XRD traces of (a) class F fly ash mixed with sodium silicate and cured as in Case II 95C, (b) class F fly ash, and class F fly ash mixed with sodium hydroxide activator and cured (c) as in Case III 95C , 1 month, (d) as in Case III 95C, 5 days, (e) as in Case II 75C, 1 month, (f) as in Case II 75C, 1 month, (g) as in Case I 75C, 1 month. Ch=chabazite, P=Na-P1 (gismondine), N=hydroxysodalite, A=Linde Type A, FA=unreacted fly ash phases.*

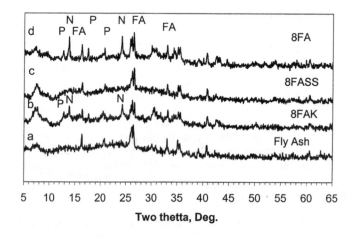

Figure 5 *XRD traces of the geopolymer materials used in durability tests: (a) class F fly ash, (b) 8FAK, (c) 8FASS, (d) 8FA. P=Na-P1 (gismondine), N=hydroxysodalite, FA=unreacted fly ash phases.*

in Case I 75C, Case II 95C and Case III 95 C contained traces of Linde-Type A, Na-P1 (gismondine) and hydroxysodalite. Regardless of the regime of elevated temperature curing all materials prepared with sodium silicate were essentially amorphous for X-Ray.

Figure 5 present XRD traces of the samples used in durability studies. The 8FASS and 8FAK materials were mainly amorphous with some residue of crystalline phases of quartz, mullite and hematite left from the fly ash. In the 8FAK samples some semicrystalline hydroxysodalite and Na-P1 (gismondine) zeolites were present. The 8FA specimen contained more ordered hydroxysodalite and Na-P1 (gismondine) zeolites in addition to amorphous aluminosilicate gel. XRD was found to be a useful technique that indicated that different degrees of intrinsic ordering exist within polymer gel in the 8FA, 8FASS and 8FAK materials.

3.4 SEM

Figure 6(a-c) presents SEI of the 8FA, 8FASS and 8FAK samples used in durability studies. The 8FA sample activated by sodium silicate has more crystalline appearance than

Figure 6(a-c) *SEI of geopolymer samples before the test: (a) 8FA, (b) 8FASS, (c) 8FAK.*
the 8FASS sample activated by sodium silicate and 8FAK sample activated by sodium and potassium hydroxide solutions.

3.5   FTIR

Figure 7 presents IR spectra of the geopolymer materials. The attribution of the IR spectra was performed using zeolite IR assignments given in Breck[5]. The strongest vibration at 960 cm^{-1} is assigned to asymmetrical T-O stretch, here T is a silicon atom. The next strongest band at 426 cm^{-1} is assigned to a T-O bending mode. The T-O-T symmetrical stretching vibrations are assigned in 688 cm^{-1} region. The T-OH stretching modes are assigned in the region 850 cm^{-1}. Vibrations assigned to double ring are in 520-532 cm^{-1} region.

The stretching modes are sensitive to the Si-Al composition of the framework and may shift to a lower frequency with increasing number of tetrahedral aluminium atoms. Thus, asymmetric stretch Al-O-Si was assigned in 960 cm^{-1} , while Si-O-Si stretch was assigned in 1000 cm^{-1} region. The bending mode, that is, the 420-500 cm^{-1} band is not as sensitive to aluminium substitution. The IR spectra are consistent with the formation of the aluminosilicate network. In all geopolymeric materials the bands appeared in the regions of 1600 and 3450 cm^{-1}, which were attributed to bending vibrations (H-O-H) and stretching vibration (-OH), respectively.

Comparison of IR spectra of 8FASS, 8FA and 8FAK samples shows that the intensity of the bands at 960-1000 cm^{-1}, 750cm^{-1}, 600 cm^{-1} and 500cm^{-1} frequencies are significantly enhanced in case of the 8FA (fly ash activated by sodium hydroxide) and 8FAK (fly ash activated by a mixture of sodium and potassium hydroxides) samples, both cured as in Case II 95C compared to the spectra of the 8FASS material formed with sodium silicate activator. The increase in intensity indicates an increase in the polymer chain length in the 8FA and 8FAK materials.

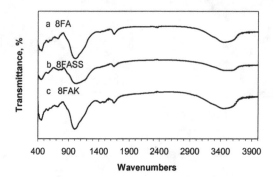

Figure 7   *IR spectra of the geopolymer samples: (a) 8FA, (b) 8FASS and (c) 8FAK.*

3.6 Durability Studies

3.6.1 Acid Resistance. Geopolymer materials 8FA, 8FASS and 8FAK had only small changes in their appearance after months of immersion in the sulphuric acid solution, such as some softening and insignificant lightening of the surface cover. In the acetic acid solution there was no change in appearance of the 8FAK specimens, and very small change in appearance of the 8FASS and 8FA specimens. Visual examination of specimens exposed to the sulphuric acid solution showed severe deterioration of the OPC and OPC +FA specimens. Within days OPC samples had a thick layer of white precipitate formed on the surface. After two weeks the surface layer of the samples was converted to the reaction products to a depth of 8 mm. After one month OPC samples were severely deteriorated and converted to reaction products to a significant extent. The surface layer of OPC +FA samples deteriorated to a depth of 4 mm after 1 month, and 7 mm after 3 months. In the acetic solution there was softening of the surface layer and significant loss of the material in the OPC specimens, and softening and much less deterioration in the OPC+FA specimens.

Table 3 shows the observations of the weight changes for the specimens exposed to acidic tests. The most significant weight change in both acetic and sulphuric acids was in the OPC and OPC+FA samples. Both OPC and OPC+FA samples had weight loss in acetic acid solution, 10% in case of OPC samples and 5.47% in case of OPC+FA samples. In sulphuric acid both, OPC and OPC+FA samples had weight gain, 40% and 19.15% respectively, and severe deterioration.

Table 3   *Weight changes of the samples exposed to tests in the 5% solutions of sulphuric and acetic acids.*

Sample ID	Acetic acid	Sulphuric acid
OPC	-10%	>40%
OPC+FA	-5.5%	19.2%
8FA	-0.45%	-1.96%
8FASS	3.83%	-2.56%
8FAK	-1.15%	-12.43%

Among geopolymer samples the best performance in both tests had the samples activated by sodium hydroxide with the weight loss in acetic and sulphuric acid solutions respectively 0.45% and 1.96%. Next in performance were samples activated by sodium silicate solution, which had 3.83% weight gain in acetic acid solution, and 2.56% weight loss in sulphuric acid solution. The performance of the samples prepared with sodium hydroxide and potassium hydroxide solutions was good in acetic acid solution with 1.15% weight loss, and rather low in sulphuric acid solution with 12.43% weight loss.

Figure 8(a-b) shows the compressive strength evolution of the samples subjected to the durability tests in 5% solutions of sulphuric and acetic acids, respectively. The studied materials had very different resistance to acidic solutions. Figure 8(a) shows performance of the samples in the sulphuric acid solution. In this test the 8FA samples were the most durable with 18% strength loss after 3 months of the test. The 8FAK and 8FASS samples had very little change in appearance after 2 months of the test, but the samples had 89% and 82% strength loss, respectively. The OPC samples completely deteriorated in the first month of the test, while OPC+FA samples had 77% strength loss and severe deterioration. All the samples except the 8FA rapidly lost their strength on immersion into sulphuric acid solution.

Figure 8(b) shows the evolution of the compressive strength of the samples exposed to the acetic acid solution. The 8FAK samples activated by sodium and potassium hydroxide solutions, performed well with 2.8% strength decline in the first month of the test and about 38% strength decline after 5 months of exposure. Next in performance were the 8FA samples activated by sodium hydroxide solution, which had no change of strength in the first month of the test and about 42% strength loss was observed after 2 months of immersion. The 8FASS samples activated by sodium silicate had 46% strength decline in the first month of the test, and 61% strength loss after 5 months of immersion. The OPC and OPC+FA paste samples had 91% and 69% strength loss respectively after 2 months of exposure, and 92% loss after 5 months.

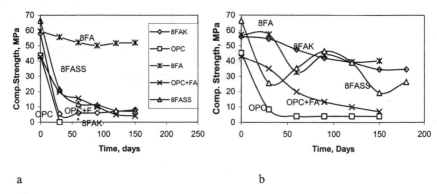

a   b

Figure 8(a-b) *Compressive strength evolution of specimens immersed into (a) 5% solution of sulphuric acid, (b) 5% solution of acetic acid.*

3.6.2   *Sulphate Resistance.* Geopolymer samples exposed to sulphate solutions had no signs of deterioration. After months of exposure to sulphate solutions the surface of the samples had no deposits and was as smooth as immediately before the test. However, there were some weight changes in the samples after 5 months of immersion (Table 4).

Table 4 *Weight changes of the samples exposed to tests in the 5% solutions of sodium and magnesium sulphates, and solution of 5% sodium sulphate+5% magnesium sulphate.*

Sample ID	5% Na_2SO_4	5% $MgSO_4$	5% Na_2SO_4+5% $MgSO_4$
OPC	3.2%	6.16%	9.1%
OPC+FA	2.35%	3.17%	7.3%
8FA	3.1%	1.4%	2.1%
8FASS	4.7%	5.3%	0.4%
8FAK	1.3%	-1.02%	1.5%

In sulphate tests OPC and OPC+FA samples had some changes in appearance: in the solution of magnesium sulphate the samples became covered by 1 mm thick white cover, while in the solution of sodium sulphate there was also cracking along the corners of the specimens. The OPC+FA samples had less deterioration than OPC samples. The OPC and OPC+FA samples had the most significant changes in weight after sulphate tests. In solution of sodium + magnesium sulphate OPC and OPC+FA samples experienced the most significant deterioration and gained weight, 9.1% and 7.3% respectively. Geopolymer samples had no visual signs of deterioration, but all of them had some changes in weight. The 8FA samples gained 3.1% in sodium sulphate, and 1.5% in magnesium sulphate solution, while the 8FASS samples gained 4.7% in sodium sulphate and 5.3% in magnesium sulphate solution. The 8FAK samples gained 1.3% in sodium sulphate and lost 1.5% in magnesium sulphate solution. In the solution of sodium sulphate + magnesium sulphate geopolymer samples had the least weight changes: 2.1% weight gain in 8FA, 0.4% in 8FASS and 1.5% in 8FAK samples. On contrary, weight changes of the OPC and OPC+FA samples were the most significant in this solution.

Figure 9(a-c) shows the compressive strength evolution of the specimens in the sulphate attack tests. In the solution of sodium sulphate the 8FA samples performed better than the 8FAK and 8FASS samples. Strength of the 8FA samples was gradually growing over the time of experiment. The 8FASS and 8FAK samples had fluctuating strength, which significantly declined from the initial value, particularly in case of the 8FAK specimens (83% reduction). In the solution of magnesium sulphate all geopolymer samples

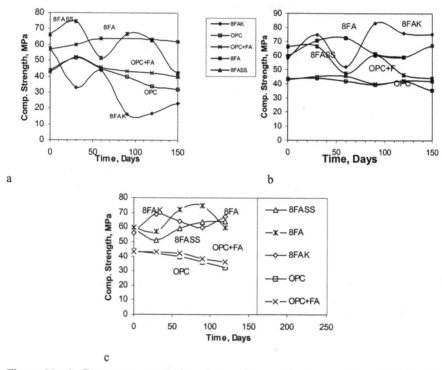

a   b

c

Figure 9(a-c) *Compressive strength evolution of specimens immersed into (a) 5% solution of sodium sulphate, (b) 5% solution of magnesium sulphate, (c) 5% sodium sulphate+5% magnesium sulphate solution.*

had fluctuating strength, which in case of the 8FAK specimens gradually improved and had 33% increase compared to the initial value, in case of 8FA there was no significant loss, and in case of the 8FASS samples there was 33% decline of the compressive strength. The best overall performance of the geopolymer samples was in the solution of sodium sulphate +magnesium sulphate. The fluctuations of strength did not exceed 17% in all specimens. After three months of experiment, strength of the 8FAK samples did not decline below the initial value.

4 DISCUSSION

The results of the experiments show a remarkable difference between the studied geopolymer materials in strength and durability. SEM and XRD results also indicated that the materials have different degrees of intrinsic ordering within the polymer gel. From the XRD and FTIR data it is evident, that geopolymer structures of the 8FA material prepared using sodium hydroxide had longer chains, and it was also more crystalline than 8FASS sample prepared using sodium silicate. Materials prepared using sodium hydroxide had a steady strength growth after the heat treatment, while the materials prepared with sodium silicate and cured 24 hours at 95°C had some decline of strength. The 8FA material also proved to be more durable in the solution of sulphuric acid and sulphate solutions than the 8FASS material. Introduction of potassium ion lowed crystallinity of gel in the 8FAK material and significantly reduced its resistance to the sodium sulphate and sulphuric acid solutions. In order to assess changes that occurred due to use of different activators, nitrogen adsorption experiments were performed. Table 5 shows BET surface area and average pore size diameter for the 8FA, 8FAK and 8FASS materials.

Table 5 *Results of BET nitrogen adsorption analysis.*

Specimen ID	BET Surface Area, m^2/g	Average Pore Diameter (4V/A by BET)
8FA	42.9	45.2 Å
8FAK	14.2	116 Å
8FASS	13.8	62.8 Å

Adsorption experiments indicated that the studied materials belong to nanoporous materials: the 8FA specimens had the smallest average pore dimeter of 45.2 Å, while 8FAK and 8FASS specimens had the average pore diameter respectively 116 Å and 62.8 Å. The average pore diameter is connected to permeability of the materials and thus, their durability in aggressive media. The results of durability tests in the sulphuric acid solution correlate with the measured average pore diameters. The 8FA samples have the smallest pore diameter and also have the best durability in the sulphuric acid solution. However, besides permeability, the reaction of the geopolymer surface with aggressive solution will determine resistance of geopolymers in aggressive environment. Therefore, in spite of fairly large pores, the 8FAK sample may perform well in the acetic acid solution. The difference in the performance of the geopolymer materials must be attributed to the distribution of the active centres on the surface of the aluminosilicate gel and to the morphology of the aluminosilicate polymer matrix. The active centres on the surface of the aluminosilicate gel influence its reactions with the aggressive media. Evidently, introduction of potassium ions in the 8FAK sample created active sites, which reacted with the sulphuric acid solution and caused rapid deterioration of strength. However, the surface reaction with the acetic acid solution caused decline of the rate of reaction and ensured good durability. In solutions of both acids 8FASS samples initially rapidly lost strength, then in case of acetic acid the strength loss stabilised, while in sulphuric acid the loss of strength was continuous. However, chemistry is not the only factor that affects properties. The morphology is also important for the durability performance of the materials. XRD and SEM suggested different degrees of intrinsic ordering in the geopolymer samples. The presence of traces of poorly crystalline zeolites in 8FA showed that there were regions of the ordered structures in this material. High durability of the 8FA samples was attributed to the presence of cross-linked polymer structures that successfully resisted the acid attack. In 8FASS sample only amorphous phases were present. Chemical stability of the polymers is improved when intrinsic order is present in one or all phases.

However, differences between the geopolymer samples appeared at nanometer-scale, which was difficult to access using equipment used in this study. Attempts to use other characterisation techniques were not successful. NMR may provide some information about structure of the aluminosilicate gel, but because of 12% Fe_2O_3 present in the fly ash this method could not be used in the investigation. An attempt to use atomic force microscopy for materials characterisation was not successful due to significant heterogeneity of the studied materials. Small Angle X-Ray Scattering (SAXS) provides opportunity to study the ordering in the material with the scattering sites of the size of 50-3000 nm. However, SAXS experiment may require a very strong source of radiation, and an attempt to do SAXS experiment using available equipment was not successful at this stage.

Alkali aluminosilicate gel is a metastable phase, and its performance depends on the number of cross-links in the structure. Geopolymers in the fly ash activated by sodium hydroxide have more cross-links and longer chains than geopolymers in the fly ash activated by sodium silicate and in the fly ash activated by a mixture of sodium and potassium hydroxides. Therefore, the 8FA samples had more stable strength development and were more resistant to acid and sulphate solutions than 8FASS and 8FAK samples.

5 CONCLUSION

The investigation shows that the strength and durability in an aggressive environment of geopolymer materials depend on the degree of intrinsic ordering within polymer gel. A regime of elevated temperature curing and type of activator can be used to affect the degree of intrinsic ordering within aluminosilicate gel. Geopolymers prepared using sodium hydroxide activator were more durable in acidic and sulphate solutions than geopolymer materials prepared with sodium silicate activator. The introduction of potassium hydroxide had an adverse effect on durability of geopolymer materials in sulphuric acid and sodium sulphate solutions.

References

1   J. Davidovits, Properties of geopolymer cements, Proceedings First International Conference on Alkaline Cements and Concretes, 1994, 131-149.
2   J. Davidovits, Mineral polymers and methods of making them. Patent US4349386, 1982.
3   J. Davidovits, Synthetic mineral polymer compound of the silicoaluminates family and preparation process. Patent US4472199, 1984.
4   J. Davidovits, M. Davidovits, N. Davidovits, Process of obtaining a geopolymeric alumino-silicate and products thus obtained. Patent US5342595, 1994.
5   Donald W. Breck, Zeolite molecular sieves: Structure, Chemistry and Use, Wiley-Interscience, New York, 1974, 415-418.

TAKE A CLOSER LOOK: CALCIUM SULPHATE BASED BUILDING MATERIALS IN INTERACTION WITH CHEMICAL ADDITIVES

B. Middendorf, C. Vellmer and M. Schmidt

Department of Structural Materials, Faculty of Civil Engineering, University Kassel, Mönchebergstrasse 7, D-34125 Kassel, Germany. E-mail: midden@uni-kassel.de

1 ABSTRACT

Calcium sulphate based materials are traditional and well accepted building materials. However, we are still far from an understanding of what happens in detail during conversion of hemihydrate to dihydrate. In gypsum technology for practical application it is necessary to use different types of additives as accelerators or retarders. Until now the usage of chemical additives is based on empirical studies. It is well known that additives not only affect the setting of the gypsum paste but also influence the morphology and habit of growing gypsum crystals resulting in microstructures with different physico-mechanical characteristics. This occasionally leads to unforeseen or unexpected behaviour in the field of application.

By using atomic force microscopy (AFM) the growth of crystal faces can be visualized in-situ. AFM studies have shown that additives block the initial growing crystal faces. As a consequence crystal morphology - shape and size - changes and causes a modified microstructure of the gypsum based building material. Scanning electron microscopy (SEM) was used to assess the relationship between crystal habit and type and concentration of used additives. On the basis of mercury intrusion porosity measurements (MIP) 3D-porosity models were calculated. The study is a first attempt to incorporate nano-, micro- and macro technologies to study calcium sulphate building materials.

2 INTRODUCTION

Gypsum plaster is prepared by autoclaving or calcining calcium sulphate dihydrate (gypsum). Due to partial dehydration calcium sulphate hemihydrate (bassanite) is produced. The re-hydration of calcium sulphate hemihydrate can be described by the following reaction:

$$CaSO_4 * \frac{1}{2}H_2O \; + \; \frac{3}{2}H_2O \; \Rightarrow \; CaSO_4 * 2H_2O$$

The reaction occurs via a solution mechanism since hemihydrate is more soluble in water than gypsum. The re-hydration of calcium sulphate hemihydrate results in a mass of entangled, needle-like gypsum crystals.

The use of calcium sulphate based materials can be traced back for over 4500 years. In Europe in the vicinity of natural gypsum outcrops, calcium sulphate based materials were widely used for joints in exterior walls of sacred buildings. The high water solubility of gypsum ($CaSO_4*2H_2O$) and its low wet compressive strength are disadvantages that make it difficult to use calcium sulphate based building materials in areas exposed to weathering. The durability of historic gypsum building materials is due to its composition[1] – binder and aggregate made out of the same kind of substance – as well as to its preparation with extremely small quantities of water. A water/binder value of less than 0.4 and a bulk density of 2.0 g/cm^2 is mentioned for historic gypsum mortars. If so in the past the use of additives was absolutely necessary otherwise workability could not be achieved [2].

It will be shown that there is a close correlation between microstructure and physico-mechanical properties of set pastes. Because of empirical studies it is well known that chemical additives alter the crystal morphology. In a reconnaissance study [3] the influence of 108 different chemicals on the crystal morphology was evaluated. To exploit the possibilities of the alteration of dihydrate crystals it is necessary to establish a sound knowledge of the underlying principles. Among them are reaction kinetics, nucleation, crystal growth. etc.

3   MATERIALS AND METHODS

Industrial produced α- and β-hemihydrate produced from natural granular gypsum were used as starting materials. In both cases no crystallizers or inhibitors - agents which modify the crystal shape - were used in the production of the hemihydrates. But it must be considered that these agents may be contained in the original gypsum. For the formation of α-hemihydrates with compact morphologies usually various crystallizers or inhibitors are used resulting in different crystal habits (Figure 1). Since chemical additives from the manufacturing process or impurities from the original gypsum can have a great effect on nucleation and growth rate, α-hemihydrate was produced in cooperation with Grenzebach BSH in a 200 l research autoclave without using crystallizers. This approach produced α-hemihydrates with needle-shaped crystals (Figure 1). Experiments using this α-hemihydrate are still in progress. Therefore the results of this study are based on experiments using β-hemihydrate produced from natural gypsum in a rotary kiln.

Various pastes were produced following the guidelines given in the German standard DIN 1168. Mechanical properties were determined after 28 days storage at 20° C and 65 % relative humidity. The microstructure were studied using scanning electron microscopy (SEM) and mercury intrusion porosimetry (MIP). The void geometry of the hardened pastes were evaluated using a pore size simulation programme. Furthermore atomic force microscopy was used to image dihydrate surfaces of model substances.

Figure 1   *SEM-pictures of different α-hemihydrate crystals,*
             upper left: produced without chemical additives,
             others: produced with chemical additives

4   EXPERIMENTAL APPROACH AND RESULTS

4.1   Variable water/binder ratio; no chemical additives

The workability of gypsum pastes can be adjusted by the quantity of water. With increasing water/binder ratio of the pastes the total porosity increases and strengths and dynamic modulus of elasticity decrease [4]. An increasing water/binder ratio not only affects the total porosity but also shifts the maximum of the pore size distribution to higher values and increases the amount of capillary pores (Table 1, Figure 2).

SEM micrographs (Figure 3) show that prismatic gypsum crystals elongated along the crystallographic c-axis prevail in pure hemihydrate-water systems. The crystals have a typical length of 10 μm, an aspect ratio of about 13 and show an intensive intergrowth and entanglement. Since the SEM-micrographs do not exhibit significant differences it can be stated that the water/binder ratio does not affect the morphology of gypsum crystals.

Table 1   MIP-data of water/hemihydrate mixtures with different water/binder ratios

Parameter	*w/b = 0.44*	*w/b = 0.54*	*w/b = 0.64*
maximum of pore size distribution [μm]	1.10	1.19	1.79
total porosity [%]	42.4	48.6	52.7
air voids* [rel.-%]	11.0	9.2	8.87
capillary pores* [rel.-%]	86.8	90.5	91.1
gel pores* [rel.-%]	2.2	0.3	0.0
median pore radius [μm]	0.98	1.17	1.63
bulk density [g/ml]	1.335	1.207	1.086
skeletal density [g/ml]	2.310	2.348	2.296

*) air voids ($10^4 - 10^6$ nm); capillary pores ($10 - 10^4$ nm); gel pores (< 10 nm)

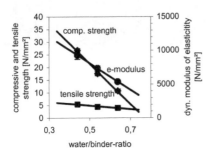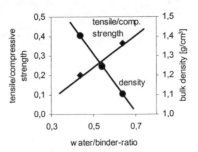

Figure 2   *Left: correlation between strengths and water/binder ratio (w/b),*
Right: correlation between tensile/compressive-strength ratio respectively bulk density and water/binder ratio

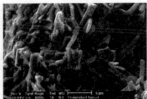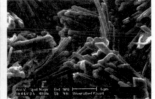

Figure 3   *SEM-micrographs of fracture surface of gypsum pastes prepared with different water/binder-ratios (w/b),*
left: w/b = 0.44, centre = 0.54, right = 0.64   Scale bar 5 microns

4.2   Constant water/binder ratio; variable chemical additive content

Chemical additives are widely used in gypsum technology regulating the setting and hardening, workability, water demand, water retention, adhesion, volumetric constancy, water resistance and porosity. For example the retarding effect of organic acids especially hydroxy carboxylic acids and their water-soluble alkali salts is well known [3, 5-8]. One common retarder of this group is citric acid [9]. Since this a well known molecule and its geometric properties are available we choose it as a model substance.

To study the influence of citric acid on the microstructure and the physico-mechanical properties of hardened gypsum pastes, samples with a constant water/binder ratio of 0.44 but various citric acid concentrations were prepared (Table 2). Beside the retarding effect the addition of citric acid also affected the fresh mortar properties. For example the retarding effect of 0.2 wt.-% citric acid relative to the β-hemihydrate content was about 3 hours.

Table 2   *MIP-data of water /hemihydrate mixtures with constant water / binder ratio (w/b=0.44) and different concentrations of citric acid as retarder*

Parameter	\-	\t citric acid concentration added to the binder			
		0.02 wt.-%	*0.05 wt.-%*	*0.10 wt.-%*	*0.20 wt.-%*
maximum of pore size distribution [μm]	1.10	1.11	1.62	1.77	2.51
total porosity [%]	42.4	40.7	41.4	40.4	40.2
air voids* [rel.-%]	11.0	8.5	9.1	9.1	9.6
capillary pores* [rel.-%]	86.8	91.0	88.5	89.9	90.2
gel pores* [rel.-%]	2.2	0.5	2.4	1.0	0.2
median pore radius [μm]	0.98	1.00	1.58	1.71	2.41
bulk density [g/ml]	1.330	1.380	1.369	1.393	1.369
skeletal density [g/ml]	2.310	2.328	2.334	2.338	2.288

*) see table 1

In figure 4 SEM micrographs of set gypsum pastes prepared with constant water/binder ratio but different citric acid concentrations are shown. It is quite obvious that with increasing citric acid concentration the morphologies of dihydrate crystals formed on setting of the gypsum had changed. Gypsum crystals grew larger but having a lower aspect ratio. With an increasing content of citric acid the median pore size also increased. Both changes modified the microstructure and altered the physico-mechanical properties of the hardened pastes(Figure 5).

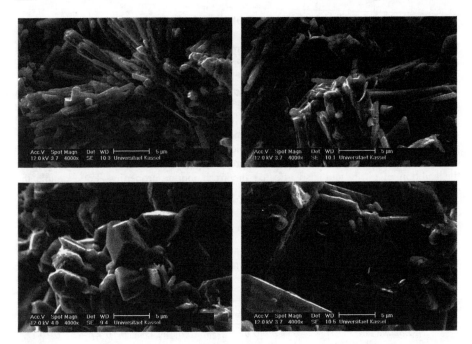

Figure 4   *SEM-micrographs of fracture surfaces of gypsum pastes prepared with constant   water / binder-ratio of 0.44 and different citric acid concentrations*
upper right: citric acid concentration 0.01 % to hemi hydrate,
upper left: 0.02 %, lower right: 0.05 %, lower left: 0.1 %

While in pure hemihydrate-water systems strength properties are a function of the total porosity and hence on the amount of excess water used for mixing, in the presence of chemical additives the relation is more ambiguous. In view of the fact that both the size and shape of contained gypsum crystals as well as the pore-size distribution changed due to the effect of chemical additives. However, good correlations between strength properties and median pore radius provide evidence that at constant total porosity higher strength properties are related to smaller median pore radii (Figure 5). With increasing median pore radius both strength properties and dynamic modulus of elasticity decrease. There are linear relations among these parameters (Figure 5).

Pore-Cor, a commercial computer code was used to evaluate and visualize the pore structure of the hardened pastes based on MIP data. Figure 6 shows the modelled pore structures of two hardened gypsum pastes with the same total porosity but different median pore radii. The read out of this program can be used to model and predict physico-mechanical properties of porous solid materials like water suction, water retention, etc. [10].

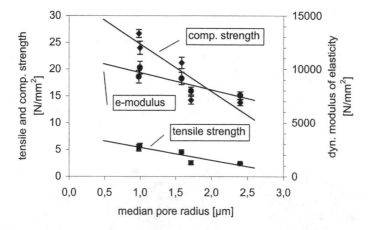

Correlations between strengths and median pore radius on the basis of linear equation systems

tensile strength = -2.365 * pore radius + 7.762	$R^2 = 0.8043$
comp. strength = -8.858 * pore radius + 33.557	$R^2 = 0.8069$
dyn. modulus of elasticity = -1603 * pore radius + 11271	$R^2 = 0.8029$

Figure 5   *Correlation between strengths and median pore radius*

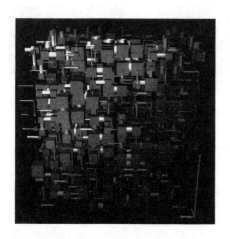

Figure 6   *Modelled pore structures of set gypsum pastes, w/b = 0.44,*
left: prepared of β-HH without additives
right: prepared of β-HH with 0.2 % citric acid as additive

4.3   Constant water/binder ratio; variable chemical additives

As mentioned above there is a variety of chemicals which have a retarding effect on the
setting of gypsum. To ensure a similar retarding effect, depending on the used chemical
additive different concentrations were required (Figure 7). In case of citric acid also a ten
times higher concentration was used to envisage the consequence of high additive
concentrations. Compared to the reference sample the samples with additives have much
lower compressive strength properties. Samples with large stubby crystals have the lowest
strength.

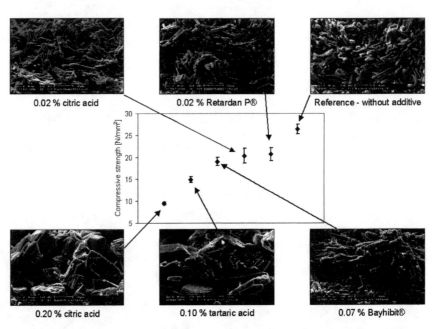

Figure 7   *Correlation between microstructure and strength using different additives*

4.4 AFM studies - assessment of crystal growth

The above results show that even small quantities ($< 10^{-3}$ mol) of chemical additives can
influence crystallisation processes, pore structures and crystal morphologies. For an
assessment of gypsum crystal growth and the interaction of chemical additives and crystal
faces AFM is the method choice. It allows in-situ imaging of crystal faces and growth
layers as shown in figure 8. The image shows a fresh cleavage plane of gypsum (001)
faces. Steps of 1.5 nm represent half of the unit cell perpendicular to the layered lattice
arrangement of gypsum consisting of double layers of $CaSO_4$ alternating with double
layers of H_2O molecules.

Changes in crystal morphology usually are related to a decrease of the growth rate due
to the slow down of ion migration to the growth site, a decrease of the lateral advancement
velocity of growth layers or a reduction of kinks available for the growth[11]. The influence
of a given chemical additive on the growth rate of individual crystal faces varies due to

their different structures and energetically situations. Depending on the site on crystal face (kink, step or terrace), chemical additives affect the relative interfacial energy of individual faces or block active growth centres [12].

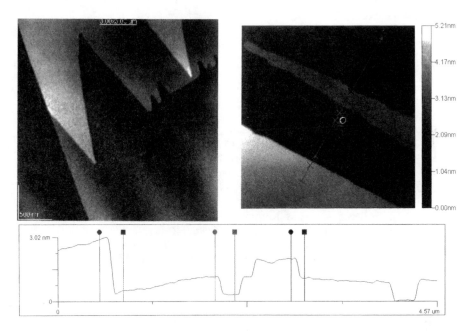

Figure 8 *AFM measurement on gypsum (001) face*

5 SUMMARY

In the form of paste-like suspensions calcium sulphate hemihydrate reacts to calcium sulphate dihydrate and set to hard masses. The hardened paste is not a compact solid, but is a highly porous material consisting of numerous entangled needle-like gypsum crystals. The total porosity and hence the internal surface of the hardened paste is determined by the amount of excess water used for mixing. In case of pure hemihydrate-water mixtures strength properties (E modulus, compressive and tensile strength) decrease with an increasing water/binder ratio.

The influence of chemical additives on the properties of hardened pastes was examined preparing pastes at a constant water/binder ratio of 0.44. Even small amounts of chemical additives affect the setting time, the morphology of dihydrate crystals formed on setting and the pore-size distribution of the hardened paste. While the total porosity is determined by the water/binder ratio an increasing citric acid content shifts the median pore radius of the hardened paste to higher values.

The strength properties of hardened gypsum pastes depend on the pore-size distribution, crystal morphology and interlocking of contained crystals. Hardened gypsum pastes with stubby dehydrate crystals have lower strength properties than pastes with needle-shaped crystals and a high degree of entanglement. The latter also determines the tensile strength of hardened pastes. A sound knowledge of the effect of chemical additives on crystal morphology will lead calcium sulphate based binders which show greatly improved properties.

Acknowledgement

The authors gratefully acknowledge the German Research Foundation (DFG)and the German Environmental Foundation (DBU).

References

1   M. Steinbrecher, *Bausubstanz*, 1992, 59.
2   B. Middendorf, *Geological Society, London*, 2003, Special Publication **205**, 165.
3   Th. Mallon, *Zement- Kalk- Gips*, 6/1988, **41**, 309.
4   M. Rößler and I. Odler, *Zement-Kalk-Gips*, 1989, **42**, 96.
5   J. P. Boisvert, et al., *Journal of Crystal Growth,* 2000, **220**, 579.
6   F. Brandt and D. Bosbach, *Journal of Crystal Growth*, 2001, **91**, 1.
7   L. Amathieu and R. Boistelle, *Journal of Crystal Growth*, 1998, **88**, 183.
8   M.C. Van der Leeden and G.M. Rosmalen, 1987, *Desalinaion*, **66**, 185
9   Th. Koslowski and U. Ludwig, *ZKG International*, 5/1999, **52**, 274.
10   G.P. Matthews, A.K. Moss and C.J. Ridgway, *Powder Technology*, 1995, **83**, 61.
11   W. Kossel, *Nachr. Ges. Wiss. Göttingen, Math. Phys*. Kl. K1,1927, 135.
12   M. Broul and J. Nyvlt, *Chem. Listy*, 1980, **74**, 362

INVESTIGATION OF THE MICRO-MECHANICAL PROPERTIES OF UNDERWATER CONCRETE

Mohammed Sonebi and Wenzhong Zhu

Advanced Concrete and Masonry Centre, University of Paisley, Paisley PA1 2BE, UK

ABSTRACT

This paper aims to investigate the micro-mechanical properties of underwater concrete (UWC) by means of the microindentation method. A novel nanotechnology based, depth-sensing microindentation apparatus was used to evaluate the elastic modulus and micro-hardness of the interfacial transition zone (ITZ) and to estimate its extent around the aggregate-matrix interface. Underwater concrete was made without any anti-washout admixtures. The water-binder ratio of the UWC was 0.41 and the dosage of the superplasticizer was adjusted to have a slump value of 220 mm. The washout mass loss was measured according to CRD C 61 standard. The underwater compressive strength was determined by casting concrete into 100-mm cubes immersed in water without any compaction. These results were compared to compressive strengths of cubes normally cast and compacted in air. Similarly, the results of the micro-mechanical properties of the ITZ around aggregate-matrix of the concrete cast in air and in water were compared.

1 INTRODUCTION

The stability of fresh concrete is characterized by its resistance to washout loss, segregation, and bleeding and is affected by the mix proportioning, aggregate shape and gradation, the extent of vibration compaction and placement conditions.[1,2] Whenever concrete is cast through water, the differential velocity at the interface between the fresh concrete and surrounding water can erode cementitious materials (CM) and other fines from the concrete. Such erosion can dramatically impair the strength and abrasion erosion of the concrete.[3] The magnitude of the differential velocity depends on the method of placement and water movements. The susceptibility of concrete to water erosion depends on the ability of the fresh concrete mix to retain its water and fines. Antiwashout admixtures (AWA) can be used to enhance the water retentivity of the concrete.[1-5] The washout resistance can also be improved by using a rich and cohesive concrete that incorporates a high concentration of silica fume.[5] When a concrete does not possess an adequate level of stability, the cement paste may not be cohesive enough to retain individual aggregate particles in a homogeneous suspension. This causes the different concrete constituents to separate, thus resulting in a significant reduction in mechanical properties and durability.[3-6] Some of the measures used to enhance stability involve the reduction in water/cementitious materials ratio (W/CM), the increase of CM content, the addition of silica fume and fly ash, the incorporation of an AWA, as well as the reduction in fluidity.[1-3] A large decrease in aggregate volume or increase in water content can reduce the cohesiveness and lead to wet segregation. A relatively high sand-to-total-aggregate content of 42 to 50% is often used to enhance cohesiveness and reduce the risk of segregation and water dilution.[3-5]

The improvement of washout resistance of cement-based material is advantageous in underwater placements where high strength, good durability, and sound bonding to reinforcing steel and adjacent surfaces are often required. The casting of fluid, yet washout-resistant concrete is especially advantageous in the repair and rehabilitation of existing structures, and can be necessary to improve the constructibility, performance, and cost effectiveness of the repair. The resistance of a cement-based material to water dilution can be enhanced by reducing the fluidity of mix. However, this can limit the ability of the cast material to spread readily into place and around obstacles.

Few studies have investigated the ITZ in underwater concrete and particularly the effect of washout resistance on the micromechanical properties. A new nanotechnology based test method was previously used to study bond in glass fibre reinforced cement[7] and also the interfacial zone around aggregate and reinforcement bars in conventional and self-compacting concrete.[8-9] The depth-sensing nano/microindentation test method has shown significant advantages over other existing methods in studying micromechanical properties and particularly ITZ characteristics in cementitious composites and practical concrete.[7-11]

The objective of this study was to investigate the micromechanical properties of the interface zone between aggregate and the paste of underwater concrete using the depth sensing nano/micro-indentation method. The elastic modulus and microhardness of the ITZ and the extent of the ITZ around the aggregate-matrix interface were evaluated. Results for concrete cast above water (i.e. in air) and in water were compared and discussed.

2 EXPERIMENTAL PROGRAMME

A concrete mix was proportioned with W/C of 0.41 as conventional underwater concrete. The dosage of SP (superplasticiser) was adjusted to secure an initial slump of 220 ± 10 mm. The washout mass loss, the density and the compressive strength of cubes of 100 mm cast above water with compaction and into water without any compaction were measured.

2.1 Materials Properties

The concrete mix investigated in this study was prepared using a 42.5 N standard Portland cement. A coarse aggregate consisting of crushed basalt with a nominal aggregate size of 20 mm was used. A well-graded quartzite sand with a fineness modulus of 2.74 was also employed. The relative density values of the coarse aggregate and sand were 2.90 and 2.56, respectively, and their absorption rates were 1.7 and 1%, respectively. A polycarboxylic-based superplasticiser was used which has solid content and specific gravity of 30% and 1.05, respectively.

2.2 Mixing and Test Methods

The mix was prepared in 25 litre batches and mixed in a drum mixer. The mixing sequence consisted of homogenising the sand and coarse aggregate for 30 s, then adding 50% of the mixing water in 15 s. After mixing for 2 - 3 min, the mixer was stopped for 5 min while the contents were covered. The cement was then added along with the remaining solution of water and SP. The concrete was then mixed for a further 3 min.

The workability of concrete was evaluated using the slump test. The washout test consists of determining the mass loss of a fresh concrete sample weighing 2.0 ± 0.2 kg, which is placed in a perforated basket and allowed to freely fall three times through a 1.7 m-high column of water (CRD C 61).[12]

The underwater strength was determined by casting concrete into 100-mm cubes immersed in water without any compaction. These results were compared to strengths determined on cubes normally cast and compacted in air. For the underwater casting, the moulds were positioned in a box filled with water to a depth of 300 mm. The cubes were then slowly retrieved from the water, and their surfaces were struck flat and covered with wet hessian and plastic sheeting. Cubes cast in air were cast in two lifts and compacted by rodding before covering them with wet hessian and plastic sheeting. All specimens were demoulded after 24 ± 1 hours and cured in lime-saturated water at 20 ± 3°C. They were tested for compressive strength according to BS 4550-3 at 28 days.

Specimens for microindentation test were extracted from the 100-mm cubes after 28 days' curing. Small samples with dimensions about 20 mm x 20 mm were first cut using a diamond saw. The samples were then subjected to treatments including resin embedding (but not impregnated), precision sectioning, grinding, polishing and ultrasonic cleaning to obtain the final disk specimens (Φ30 x 3 mm) for the microindentation test.

2.3 Nanoindentation Testing of Micromechanical Properties of ITZ in Concrete

Evaluation of micromechanical properties of materials on the micron to submicron scale, particularly within the ITZ of cementitious composites and concrete, poses special difficulties. One major problem is that it is not possible to isolate the material in the ITZ for mechanical testing. As a result, for a long time the evaluation of properties of ITZ has relied on either indirect methods (e.g. SEM image analysis and theoretical modelling, etc.) or test of a model composite specimen that does not represent the actual material used in practice.

To overcome the difficulties and limitations of existing techniques, significant progress has recently been made in the development of nanotechnology based, depth-sensing micro/nano-indentation apparatus and the associated methodology.[10,13] The apparatus monitors load and displacement (or depth) continuously during indentation: this enables the mechanical properties to be determined even when the indentations are too small to be imaged conveniently. Such an apparatus was used in this study to examine the properties of ITZ around coarse aggregates.

The operating principle and special features of the apparatus are described in detail elsewhere.[10,13] A typical outcome of the nanoindentation testing is an indentation load-depth hysteresis curve as shown in Fig. 1. As a load is applied to an indenter in contact with a specimen surface, an indent/impression is produced which consists of permanent/plastic deformation and temporary/elastic deformation. Recovery of the elastic deformation occurs at the start of the unloading. Determination of the elastic recovery by analysing the unloading data according to a model for the elastic contact problem leads to a solution for calculation of elastic modulus E and also microhardness H of the test area.

Details of the theoretical background and methodology for the elastic modulus determination have been reviewed and presented elsewhere.[10,13]

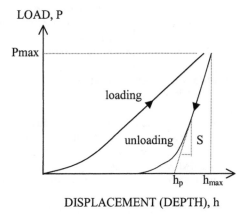

Figure 1 *A schematic diagram of an indentation load vs. displacement curve*

Briefly, the specimen elastic modulus is determined using equations (1) and (2):

$$S = \frac{dP}{dh} = \frac{2}{\sqrt{\pi}} E_r \sqrt{A} \tag{1}$$

$$\frac{1}{E_r} = \frac{\left(1 - v^2\right)}{E} + \frac{\left(1 - v_i^2\right)}{E_i} \tag{2}$$

where, $S = dP/dh$ is the experimentally measured stiffness of the upper portion of the unloading data; E_r is a reduced elastic modulus defined in equation (2); A is the projected area of the elastic contact; E and v are Young's modulus and Poisson's ratio for the specimen; and E_i and v_i are the same parameters for the indenter. For the diamond indenter used in this study, $E_i = 1141$ GPa and $v_i = 0.07$. The projected area A can be derived from the plastic depth h_p obtained using the unloading data and the indenter shape function, which is dependent on its geometry. For an ideally perfect 90° (corner of a cube) indenter, $A = 2.6\ h_p^2$. In the case of imperfect indenter tip geometry, the shape function can be determined by using electron microscopy techniques or through calibration of hardness (or elastic modulus) - plastic depth curve using a homogeneous material of constant hardness (or elastic modulus). The microhardness, $H = P_{max}/A$, can also be calculated, where P_{max} is the maximum load applied.

2.4 Mix Proportions

The study seeks to determine the micromechanical properties of the ITZ around coarse aggregate in underwater concrete in comparison with normal concrete. The investigated mix was prepared with fixed W/C of 0.41 corresponding to high-quality underwater concrete. The mix was made with 410 kg/m^3 cement, 1050 kg/m^3 coarse aggregate, 730 kg/m^3 sand and 3.3 L/m^3 superplasticizer.

3 RESULTS AND DISCUSSION

3.1 Fresh and Hardened Properties

Results of slump, washout mass loss, compressive strength and density of concrete cast in air and in water are given in Table 1. The properties presented in this paper outline the difference of the performance of concrete in air and in water.

The compressive strength of cubes cast in air was higher than that cast in water. The reduction of in-situ compressive strength compared to reference cubes cast above water (in air) is due to the lack of full compaction coupled with water dilution of the cast concrete. The residual compressive strength at 28 days $f_{c\,uw}$ /$f_{c\,air}$ was 61%. This value was inferior to the specific value for underwater-cast concrete containing an AWA; standards of the Japan Society of Civil Engineers recommend up to 80%.[14] The low ratio achieved in this study is likely due to the high value of washout mass loss of the concrete (23%).

The density of concrete cast in water was also lower than that cast above water owing to the lack of compaction and dilution of fines materials in water. The ratio of the density of concrete cast in air and in water was 97.3%.

Table 1 *Fresh and hardened properties of underwater concrete*

Slump (mm)	Washout mass loss (%)	Density		Compressive strength at 28 days	
		air	water	air	water
220	23	2566	2549	82.3	48.6
		2608	2507	67.6	46.2
		2590	2503	73.2	40.7
	Average	2588	2520	74.4	45.2

3.2 Micromechanical Properties of the ITZ

More than 80 valid indentations were identified and selected from 210 indentation test points made at areas close to the coarse aggregates (varying from 0 to 90 μm from the actual interface), for concretes cast in air and in water. Since some sand particles and large voids also existed in the tested area, the test points which lie closer to the sand particles or fall into the voids were considered invalid and discarded. The identification and screening of all the indented points were carried out under the scanning electron microscope (SEM).

Upon analysing the valid indentation results and their profile of distribution in the interfacial area, the width of ITZ for both concretes was found to be 40~45 μm from the actual coarse aggregate interface. Therefore, the indentation results varying between 5 to 45 μm were used to calculate the properties of the ITZ while the indentation points which fell between 45 and 90 μm were considered to represent the properties of the bulk matrix in this study.

The SEM photographs for concrete cast in air and in water, which show indents left on the specimen surface following the indentation testing are given in Figures 2 and 3, respectively.

Figure 2   *SEM image of a tested area for concrete cast in air (scale bar = 20 μm)*

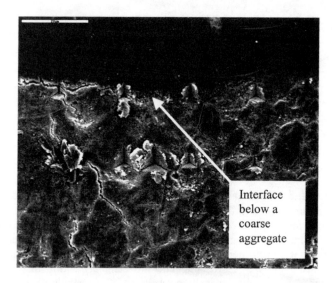

Figure 3   *SEM image of a tested area for concrete cast in water (scale bar = 20μm)*

Figures 4 and 5 show the elastic modulus, and the microhardness of the ITZ (i.e. < 45 μm) and the bulk matrix (i.e. > 45 μm) for the concrete cast in air and in water, respectively. The results revealed that the E modulus and the microhardness (H) of both the ITZ and the bulk matrix were higher for concrete cast in air than for concrete cast in water. This could be due to the lack of compaction and the loss of part of the cement and fines particles due to dilution for concrete cast in water, which lead to an increase in porosity and a reduction of strength.

The results also indicated that in the ITZ the E and H values were 10~25% higher for concrete cast in air than for concrete cast in water. In the bulk matrix the difference seemed to be greater, with the E and H values being 40~80% higher for the concrete cast in air. This seemed to suggest that the ITZ around coarse aggregate was less affected than the bulk paste matrix for concrete cast in water.

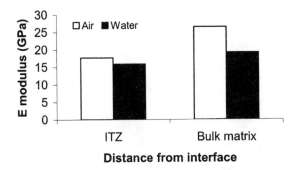

Figure 4   *E Modulus of ITZ and bulk matrix for concrete cast in air and in water*

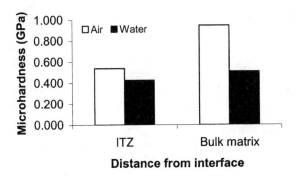

Figure 5   *Microhardness of ITZ and bulk matrix for concrete cast in air and in water*

3 CONCLUSIONS

Based on the above results, the following conclusions can be drawn:

- Concrete cast in water had lower compressive strength at 28 days and density than that cast in air and compacted. This is due to the lack of full compaction coupled with water dilution of the cast concrete (washout).

- The widths of the ITZ were found to be around 40~45 µm for both concrete mixes. The E and H values were considerably lower in the ITZ than in the bulk matrix for both concretes. The difference was more significantly for concrete cast in air than for concrete cast in water.
- The micromechanical properties (E and H) of ITZ around coarse aggregates were lower for concrete cast in water than for concrete cast in air. This is attributed mainly to the dilution of paste cement and fines particles in water causing reduction of strength and increasing the porosity of concrete.
- The E and H values in the ITZ were 10~25% higher for concrete cast in air than for concrete cast in water. In the bulk matrix the difference seemed to be greater, with the E and H values being 40~80% higher for the concrete cast in air. This suggested that the ITZ around coarse aggregate was less affected by the washout than the bulk paste matrix for concrete cast in water.

References

1   K.H. Khayat, M. Sonebi, 'Effect of Mixture Composition on Washout Resistance of Highly Flowable Underwater Concrete', *ACI Materials Journal*, 2001, Vol. 98, No. 4, 2001, pp. 289-295.

2   M. Sonebi, 'Testing Washout Resistance of High Performance Underwater Concrete', *Concrete*, 2001, Vol. 35, No. 1, pp. 32-35.

3   M. Sonebi, K.H. Khayat, 'Effect of Mixture Composition on Relative Strength of Highly Flowable Underwater Concrete', *ACI Materials Journal*, 2001, Vol. 98, No. 3, pp. 233-239.

4   M. Sonebi, K.H. Khayat, 'Effect of Free-Fall Height in Water on Performance of Flowable Concrete', *ACI Materials Journal*, 2001, Vol. 98, No. 1, pp. 72-78.

5   M. Sonebi, 'Effect of Silica Fume, Fly Ash and Water-to-Binder Ratio on Bond Strength of Underwater, Self-Consolidating Concrete', in *Proceedings of the 7th CANMET/ACI International Conference on Fly Ash, Silica Fume, Slag and Natural Pozzolans in Concrete*, SP 199-33V. M. Malhotra, Madras, India, 2001, pp. 595-610.

6   N. Hasan, E. Faerman, D. Berner, 'Advances in Underwater Concreting: St Lucie Plant Intake Velocity Cap Rehabilitation High Performance Concrete', ACI SP 140. M. Malhotra, 1992, pp. 187-213.

7   Zhu, W and Bartos, PJM, 'Assessment of Interfacial Microstructure and Bond Properties in Aged GRC Using a Novel Microindentation Method', *Cement and Concrete Research*, Vol.27, No.11, 1997, pp.1701-1711.

8   Zhu, W., Bartos, P.JM, 'Application of Depth-Sensing Microindentation Testing to Study of Interfacial Transition Zone in Reinforced Concrete', *Cement & Concrete Research*, 2000, Vol. 30, No., pp. 1299-1304.

9   Sonebi, M., Zhu, W. & Gibbs, J., 'Bond of Reinforcement in Self-Compacting Concrete', *Concrete*, Vol. 35, No. 7, July-August 2001, pp. 26-28.

10 Zhu, W, Trtik, P., Bartos, P.J.M. 'Evaluation of Elastic Modulus at Interfacial Transition Zone in Reinforced Concrete by a Microindentation Technique', Proceedings of the 6th International Symposium on Brittle Matrix Composites, Eds: Brandt, AM, Li, VC and Marshall, IH, Warsaw, 2000, pp.317-325.

11 Trtik, P and Bartos, P.J.M., 'Micromechanical Properties of Cementitious Composites', *Materials & Structures*, Vol.32, 1999, pp.388-393.

12 CRD C61 'Test Method for Determining the Resistance of Freshly-Mixed Concrete to Washing Out in Water', US Army Experiment Station, Handbook for Concrete, Vicksburg, Mississippi, Dec. 1989, 3 p.

13 Oliver, W.C., Pharr, G.M. 'An improved technique for determining hardness and elastic modulus using load and displacement sensing indentation experiments', *Journal of Materials Research* **7** (1992) 1564-1579.

14 Japan Society of Civil Engineers, 'Recommendations for Design and Construction of Antiwashout Underwater Concrete', Concrete Library of JSCE, No. 67, 1991, 89 p.

Part 5 : Applications

THIN FILMS AND COATINGS: ATOMIC ENGINEERING

Frank Placido

Thin Film Centre, University of Paisley, Paisley PA1 2BE, Renfrewshire, Scotland

1   EXTENDED ABSTRACT

1.1 What is a thin film?

Thin films here are taken to be thin coatings (5 to 500 nm thick) on some solid substrate deposited in a vacuum process. Multi-layers of different materials are often used.

1.2 Why use thin film coatings?

- Many bulk properties depend on the atomic composition of <u>surface layers</u>
- e.g. roughness, hardness, friction, corrosion, colour, reflectivity, electrical and thermal conductivity
- Thin films deposited on the surface of bulk materials can provide a <u>cost-effective</u> way to modify surface properties of bulk material
- Interference effects between light waves reflected from different layers allow optical filters to be designed and manufactured.

There are a huge range of application areas for thin films (Figure 1).

1.3 Thin Films versus Bulk Materials

- Thin films are <u>*not*</u> simply thin slices of the corresponding bulk materials.
- Composition, density, porosity, crystallinity, conductivity, hardness, optical properties etc. may be different and can be controlled by the deposition method and conditions.

Non-equilibrium compositions are possible.

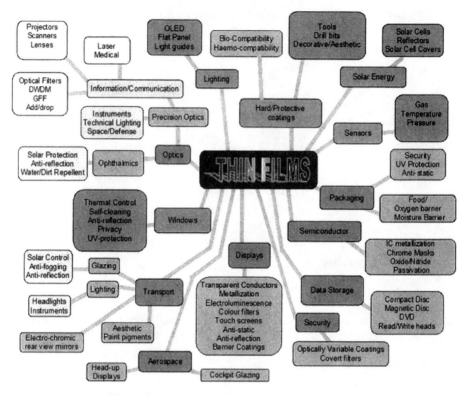

Figure 1   *Potential applications of thin films*

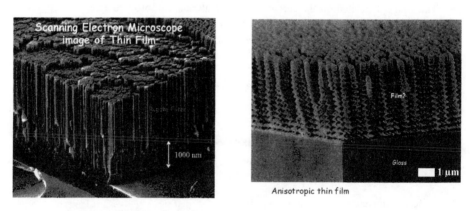

Figure 2   *SEM images of thin film structures.*

2   HOW ARE VACUUM-DEPOSITED THIN FILMS MADE?

- Chemical Vapour Deposition - Gaseous sources

- Physical Vapour Deposition - Solid sources, Sputtering, Electron-beam evaporation.

Coating plants can be small-scale, coating microscope slides, or huge, coating 2m wide rolls of polymer. We now consider a few architectural/building applications in more detail

3   APPLICATIONS

3.1 Films for Decorative Use

Titanium Nitride produces a pleasing range of gold colours. These coatings adhere well to plastics and are very robust. They are widely used on bathroom fittings.

3.2 Hard, Protective Coatings

Hard, wear-resistant coatings include:-
- Chromium, chromium alloys   Cr, CrCo, NiCr,
- Titanium, titanium alloys   Ti, TiAl,
- Carbides, borides, nitrides,   SiC, CrN,TiAlN,TiB, TiBN,TiBC, BN,WC,
- CVD diamond
- Diamond-like carbon

These coatings are most likely to be found on machine tools, drill bits etc.

3.3 Films for Energy Generation

Photo Voltaic slates or roof-tiles now available, harvesting energy from sunlight. Thin film solar cells (Photo-voltaics) are commonly made from *amorphous silicon* on glass substrates. New materials include *Copper Indium Sulphide* on glass. Future possible developments include all polymer PV's based on *semiconducting polymers*, ink-jet printed onto polyester substrates.

3.4 Films for Windows

Electrically-switchable opaque/ transparent films for window glass are available. This privacy glass contains a *Polymer Dispersed Liquid Crystal* layer between two glass or plastic panels coated with ITO – a transparent electrically conducting film.

Also available are electrochromic_ films which can have their transparency changed electrically. These are complex multilayer coatings and are expensive as yet for large areas. Their most common use is in rear-view mirrors in cars.

Multi-layer coatings of metals and dielectrics can be used to selectively reflect and transmit light in different wavebands. In the diagram below, films of differing refractive indices (n,k) reflect the incident light at each interface. The various reflected (and transmitted) beams can interfere, destructively or constructively.

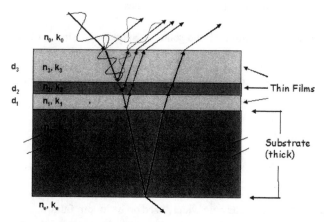

Figure 3 *Example of multilayer thin film with light reflections and transmissions.*

Knowing the optical constants (n and k) and thicknesses of each layer, the Fresnel equations allow us to easily calculate the transmittance, reflectance and absorptance of the assembly as a function of light wavelength. The *inverse problem*, where we need to determine the thicknesses, optical constants and number of layers required to obtain a required transmittance, reflectance or absorptance is a much more difficult situation. However, this is what thin film designers do! Some examples follow.

3.5 Low-e coatings for glass

A common requirement is for a thin-film filter coating for glass that will transmit visible light and reflect infrared light (heat). A three-layer coating of tin oxide/silver/tin oxide on float glass can produce the following performance:-

Design

SnO_2 163.95 nm
Ag   12.41 nm
SnO_2 28.86 nm
Float Glass

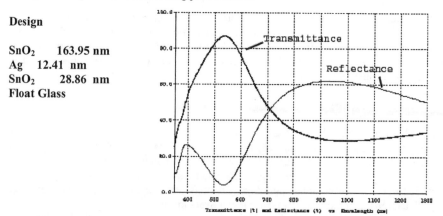

Figure 4 *Example transmittance and reflectance of multiple layer thin film with design.*

This coating has high visible transmittance (81%), but also high visible reflectance (10%) and has a yellowish tinge in reflection and a purplish-pink tinge in transmission.

Much better performances are possible. There is an inevitable trade-off between performance and coating complexity (price).

The six-layer coating below has high visible Transmittance (78.4%), low visible reflectance (3.42%), v. high near IR and thermal IR reflectance and has no colour tint in reflection or transmission (it is a neutral gray).

Design

SnO₂	36.83 nm
Ag	17.28 nm
SnO₂	74.39 nm
Ag	13.34 nm
SnO₂	54.63 nm
Ag	4.79 nm
Float Glass	

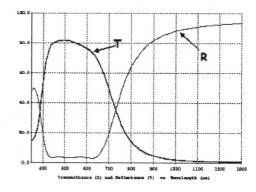

Figure 5 *Further example of thin film design. See text for details.*

More recently, titanium oxide and silver have been used for low-e coatings. Suitable designs can produce very effective performance, as shown below for a three-layer coating. This design has high visible Transmittance (83.1%), lowish visible reflectance (8.11%), high near IR and thermal IR reflectance. It has a reddish purple colour in reflected light but nearly neutral gray in transmission.

Design

TiO₂	26.80 nm
Ag	20.92 nm
TiO₂	136.80 nm
Float Glass	

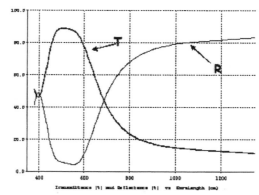

Figure 6 *Further example of thin film design. See text for details.*

TiO₂ top coatings have another interesting and useful property, self-cleaning" of organic contaminants. This is due to the formation of free radicals when TiO₂ is exposed to UV radiation. Free radicals (OH and H) break down organic molecules to remnants that are soluble in water. The *hydrophilic* nature of the coating causes rain to "sheet" on the surface and wash contaminants away.

3.6 Energy saving coatings for Lighting

Video projector lamps have a complex multilayer all-dielectric coating (no metal) that reflects heat back towards the filament and transmits nearly all of the visible light. The complexity of the coating is justified because this allows the lamp to run at much lower power and simultaneously reduces the problem of heat dissipation in the projector.

Organic light-emitting diodes (OLED's can be vacuum-deposited onto ITO coated glass to produce a flat-panel light source (and even a display). Soon, these may be ink-jet printed onto polymer substrates to produce cheap large area flat light sources.

3.7 Light-guiding

Metal-coated mirror surfaces are surprisingly inefficient reflectors for light-guiding applications, where the light has to make multiple reflections of the surfaces. For example, after only 10 reflections from aluminium coated surfaces, 60% of the initial light intensity is lost and the remaining light has a yellow tinge. A recent product from the 3M company is co-extruded polymers – hundreds of layers are rolled and pulled to give very thin layers. These are designed as an optical filter to give very high reflection over the visible range, for all angles of incidence. This has the appearance of a "perfect" mirror, but contains no metal layers.

THE NANOHOUSE™ – AN AUSTRALIAN INITIATIVE TO DEVELOP THE HOME OF THE FUTURE

J. Muir[1], G. Smith[2], C. Masens[2], D. Tomkin[1] and M. Cortie[2]

[1]Faculty of Design, Architecture & Building, University of Technology Sydney, PO Box 123, Broadway NSW 2007, Australia
[2]Institute for Nanoscale Technology, University of Technology Sydney, PO Box 123, Broadway NSW 2007, Australia

1   INTRODUCTION

Nanotechnology encompasses an array of technologies, all sharing the common attribute of arising from the science of the scale of nanometres. At this scale many materials exhibit physical properties different from those observable in larger quantities of the same materials. This presents a vast number of opportunities to develop new materials and systems leading to a corresponding array of new products and processes. There is great interest in exploring how these new materials can be applied in existing and new buildings. A multidisciplinary team led by The Institute for Nanoscale Technology at the University of Technology, Sydney (UTS), that includes people from a number other institutions within Australia such as the Commonwealth Scientific Industrial Research Organisation (CSIRO) are developing the Nanohouse™. The Nanohouse is at the time of writing a concept house, existing in the form of architectural drawings, mathematical models and as a 3D computer simulation. The Nanohouse™ is being designed to illustrate what uses various nanotechnologies (and other recent innovations) have to offer within the context of a domestic dwelling and also to note the wider applications of these technologies in commercial structures. It is naturally a dynamic project, with the design being modified as new technologies and materials become available. In this paper we describe the methodology used to create the Nanohouse™, and evaluate some aspects of its performance. The aspects that touched on include the architectural design and the Nanohouses overall energy efficiency.

2   THE HOUSE OF THE FUTURE

The Nanohouse™ is planned to be as much of a house of the present as of the future. The aim is incorporate products that already exist and commercially available alongside prototypical examples of more recent developments. Before we describe the elements of the Nanohouse™, it is necessary to make some general comments about the nature of future housing.

Some broad trends are already apparent. First among these is an inclination in many developed countries, including Australia, towards larger family homes. Currently, this trend is coupled with a movement towards more open planning admitting more light through extensive use of large window openings and the installation of features (such as swimming pools or home theatres) that were previously only enjoyed at public venues. Paralleling these trends is the desire for homes to be as 'environmentally-friendly' as possible, while allowing for a comfortable and convenient lifestyle. One barrier to the uptake of designs that are energy conscious is the up-front price with consumers often opting for short term savings on energy efficient systems in favour more immediate luxuries, even if a full analysis of the lifetime costs indicates an overall saving.

It is worthwhile to consider developments in two ways:

1. in materials terms, and
2. in terms of systems.

New materials are at the heart of change in the way we construct our built environments. New materials enable new systems to be developed from their combinations. As materials are developed that have, for example, useful properties of optical interaction such as spectral selectivity in the transmission of visible light and the attenuation of infra-red, we find ourselves incorporating these materials in our window systems and hence we can contemplate the implications of this on the thermal characteristics of a building. Materials that have useful electronic and photonic properties can be incorporated into information systems. Materials that have useful mechanical properties can be used either as structural elements in their own right or as part of a composite material. One outcome of research at the nanometre scale is the development of capability in making materials suitable for use in systems of extreme miniaturisation, such as microscopic sensors and actuators.

There are a wide variety of materials being developed currently that have application in nearly every element of the built environment. Photocatalytic self-sterilising and self-cleaning materials such as TiO_2 are already being employed on work surfaces in kitchens [1] and bathrooms as well as being incorporated into windows. There is already a system in Japan based on catalytic nano-particles of gold that destroys toilet odours [2]. Indeed, the engineering of surfaces is about to yield a domestic environment that may not appear to the naked eye to be any different to current surfaces, yet impossible to mark, scratch, wet, grease, foul, photodamage or burn. This may be equally true for textiles, ceramics, polymers, metals and all other materials that one may encounter in the built environment.

In terms of systems, it is evident that new residences need to be wired with considerably more electrical and information infrastructure than ever before. The single electrical outlet per room of our parents has now been replaced with multiple outlets consisting of, co-axial cable, doorbells, intercoms, temperature sensors, multiple telephone lines and security systems. In terms of cost-efficiency this infrastructure needs to be built into future houses at the earliest stage of construction. In addition, the trend toward the home office and working remote from ones employer under flexible time arrangements makes items such as his-and-hers SOHO workstations with access to high-speed Internet connections highly desirable. This convergence of information technologies such as telecommunications, television, internet and wireless technologies will see future houses having fully integrated systems with all modes available in all rooms. Furthermore, the widespread use of sensors of various types, monitoring air and water quality in addition to the monitoring of power supply fluctuations, mechanical and thermal stresses in the building will ensure that the home of the future will be "smart" in comparison with the homes of all previous generations.

Extreme reduction in the sizes of systems means that they can be incorporated into materials such as paints or other more prosaic building materials to provide unobtrusive enhanced functionality. Very small systems also require very low power in order operate, enabling much greater exploitation of photovoltaic technologies such as organic molecular solar cells. This then impacts on the overall energy demands of the building.

There are a multitude of possible developments in this field and in this paper we have confined ourselves to addressing only two of these:

1. the opportunities offered by pre-fabricated modular housing systems, and
2. the improvements in energy efficiency and comfort available through the use of modern glazing systems engineered with nanoscale coatings.

3 DESIGNING THE HOUSE OF THE FUTURE

3.1 Architectural Design Determinates and Modular System

One of the most important functions of any building is to modify the external environmental conditions through its fabric to enable its occupants to perform their tasks comfortably. Over most of human history this was achieved by the manipulation of the elements and components of the building, producing a rich array of appropriate vernacular forms across most of the worlds climatic regions. "Although from time to time these forms have been distorted by architectural metaphors and spiritual requirements the basic need for the physical performance of a building to respond to the climatic region in which it is placed remains a valid form determinate." [3] Throughout the 20th century with cheap energy viewed as an abundant source, scant attention was paid to the latter, which in general has resulted in the construction of an energy inefficient built environment, with recurrent high running costs to the owner/occupant.

To change the existing paradigm, requires the adoption of a holistic approach by designers, component manufacturers and contractors to enable the development of new building prototypes, and for the purpose of this paper new housing prototypes, taking into account the nature, and number of resources required to achieve the desired result, accompanied by a cradle to grave analysis of all material and labour inputs. Adopting this approach requires full utilization of our present knowledge base, from identifying climatic constraints; co-ordinating infrastructure, planning and layout; to the application of appropriate materials and manufacturing technologies, the integration of the landscape with the built environment and a complete energy audit of the resulting built object to achieve the various environmental goals. Concurrent benefits that flow from this approach enhance the marketability and profitability of the end product, by offering to the consumer a package of benefits and in particular energy saving benefits that will save money over the economic and operating life their investment.

From the outset the Nanohouse™ has been conceived as a mass customised factory built item, assembled using close tolerance components and assemblies utilizing the latest manufacturing technologies and numerically controlled machine tools to replace outmoded craft building techniques that have too large a tolerance of fit, are labour intensive uneconomical and wasteful of material. Treating the house of the future as a manufactured consumer item allows for an ecomanufacturing approach - orderly end of life disposal and recycling under a buy back policy in addition to it being able to be constantly upgrade by changing the accommodation and servicing requirements over its lifetime and the ability to disassemble and rearrange its components by virtue of its modularity and dimensional control.

Energy efficient building design has a dramatic impact on both heating and cooling needs. Overseas experience has shown that "super-insulated houses" built with heavily insulated walls and ceilings, tight fitting components, which often contain ventilating systems that recover heat from the exhaust air have reduced primary energy consumption by 70%. [4] For example remarkably low heating bills result because super-insulated houses store the "free" heat from people, lighting appliances, and passive solar heating through windows and other devices designed into the system by insulting the building mass from external environmental influences in a controlled way.

Houses with conventional insulation tend to float about 2-3 degrees C above the outdoor temperature because of the internal free heat. Therefore, when a thermostat is set in a conventional house at 21degrees C the heating source will not operate until the outdoor temperature falls below 18 degrees c., the "balance point" of the house. The three degree difference known as the free temperature rise of the building.

Super insulated houses exploit this property where the free temperature rise can be as much as 12 degrees C, so that if the thermostat is set to activate at 22 degrees C, heating will not commence until the temperature drops below 10 degrees C, making heating of houses in the Australian environment virtually unnecessary except for the alpine regions and a very few days per year in other areas.

On the cooling side the super-insulated house works in a similar manner, but this time using the thermally insulated mass of the house as a heat sink to stabilize the indoor temperature up to 12 degrees below the ambient outdoor temperature, depending on whether it is located in a hot dry, or a hot humid climate zone, and also provided that the free air flow into the house is controlled, and its temperature modified as it passes into the house. These houses also exploit radiant heat loss from the occupants to the surrounding surfaces as an additional aid to the feeling of comfort.

Understanding the above principals was central to setting the base parameters for the design of the Nanohouse™ and for it to accommodate a wide range of climates, ranging from hot humid, hot arid, to temperate cool and cold. The above principles also informed the distribution and amount of thermal mass contained in the external envelope and internal subdivision for each climatic regime.

To achieve the degree of flexibility required in taking this bio-climatic approach to design required the various components making up the external envelope to be carefully selected and dimensionally coordinated in such a way to allow their incorporation into any desired part of the building's envelope, dependent on their function, i.e.; whether they are required to act as a heat sink, a heat store, a radiative panel or a heat filtering and control system using high tech and smart glass as part the internal environmental control mechanism. The housing system has also been designed to allow a flexibility in its internal planning by virtue of the dimensional control contained in the base computer generated model, which allows for various locations of the fully factory assembled rooms containing key service and wet areas forming an integral part of the structure of the building and associated internal subdivision. The issue of acoustic privacy between key areas of the dwelling are an inbuilt attribute of the total system. The patterns of ventilation into the dwelling is controlled by either the external envelope or by passive solar means to ensure correct air temperature, which can be monitored by temperature sensors imperceptibly located throughout the house, and available for real time inspection via computer by the occupant at any time. All components assemblies, sub-assemblies and building elements have also been co-ordinated to fit to standard shipping and trucking container sizes.

Also fully integrated into the buildings fabric are specific solar energy conversion systems which in association with transparent insulting materials (TIM's) enhance the performance of spectrally selective coating systems using nano-particles on various substrates, which are outside the scope of this paper. These conversion systems designed to match the climatic zone in which the building is to be placed, having environmental monitoring processes embedded in their assembly.

The adoption of this approach to the house as a major component of our built environment offers the following benefits to the end user:-

- Low maintenance once only cost incorporating climatic control mechanisms in its external envelope.
- Low recurrent energy cost, power, gas, heating oil.
- Flexible planning able to adapt to later change.
- Ability to acoustically isolate various areas of the house
- Quality control through the offsite manufacture of components
- Shorter time for erection, causing fewer disturbances to the site, by reducing on site labour
- Reduction of labour content, and substituting this with fully assembled factory components allows better quality finishes within any given budget
- Computer design and modelling program allows full 3D visualisation of finished product prior to construction
- Cost control is built into the initial model thus reducing the likelihood of cost blowouts by reducing cost variations to individual site conditions.

3.2 Thermal Performance

To optimise the thermal performance of the Nanohouse™ over a broad range of climatic regimes requires the marriage between advanced materials technology, mass affect and insulation to ensure that the most efficient solar energy conversion was achieved for internal climate control utilizing various materials and surface coating systems radiative and heat storage properties in association with the glazing elements, with the following material properties of particular relevance in its development.

- Porous or cellular materials that prevent air circulation by suppressing convective heat transfer and acting as thermal insulators while allowing high transmittance of solar radiation.
- Materials that allow high transmittance of solar radiation combined with low emittance, while allowing high transmittance of visible light, combined with high reflectance of infrared solar radiation.
- Chromogenic materials with radiative properties that can be changed to match different demands of heat and light transmittance.
- Materials with angularly selective radiative properties.
- Materials that have high absorption and that can be combined with low emittance coatings using spectrally selective surfaces allowing for efficient photothermal conversion of solar energy.
- Materials that exploit the high atmospheric transmittance in the 8-13ηm wavelength range to channel energy from sky facing surfaces into space realizing that the ability to exploit this phenomenon is dependent on the amount of water vapour contained in the atmosphere.

With no one material able to bring about the amount of environmental control necessary for the various climatic regimes, research is ongoing to develop various composites that act in unison with the glazed elements in the buildings envelope allowing it to perform as an effective environmental filter.

The materials able to be applied to internal thermal control, range from transparent through translucent to opaque acting in concert with the mass elements of the building.

A group of materials of particular interest in the development of the thermal control of the Nanohouse[TM] were various transparent and translucent composites used as façade, roof lighting and roof elements falling into two groups.

1. Solar energy conversion systems incorporating transparent insulating materials (TIM's) used to enhance the performance of other spectrally selective coatings on either transparent or solid substrates using nano-particles performing the function of radiating thermal energy to the interior environment and outside the scope of this paper.
2. Solar control glazing employing heat reflecting, heat absorbing, angularly selective coated and electrochromic glasses treated in section 4 below

4   NANOTECH PRODUCTS FOR ARCHITECTURAL ENERGY EFFICIENCY

The drive to make any Australian 'house of the future' as energy-efficient as possible is fuelled by the observation that current domestic and commercial buildings draw 6 and 12% respectively of Australia's total electricity generated [5].   A proportion of this is for heating water and for cooking, two areas which will not be further considered here, except to note that some improvements are possible by making greater use of thermal solar power and natural gas. A further proportion of the power is consumed for climate control and lighting and it is specifically these two items that can benefit from the application of nanotechnologies. The situation in other countries is broadly similar, and, for example, it has been stated that about one third of the energy consumed in the USA is used to heat, cool or light buildings [6]. Since the issue of staying warm in cold climates has already received much attention in the literature, the focus of the present paper will be on nanotechnological solutions for staying cool in hot climates, an issue which is particularly relevant in Australia. It has been claimed that efficient use of existing technologies can reduce the annual electricity bill of an average Australian household from US$800 to US$250 [7]   We will attempt to show below how nanotechnologies can assist in this process.

4.1 Climate control

Once it is recognised that up to 30% of the heat transfer in and out of typical buildings is through their windows, then it is reasonable to accept that about 5% of all electricity produced is 'wasted' by windows [5]. It has been estimated that there are about 27 billion m^2 of architectural window in the world and an worldwide annual production of flat glass approaching two billion m^2 [8], so there is evidently both a large market and large savings to be had by addressing this opportunity.

In cold climates, the main factor determining the rate of energy loss through windows is the convective transfer of heat from interior to exterior through the boundary layers on either side of the glass [9]. This is because solid-state conduction of heat through the glass pane is rapid and does not significantly retard heat transfer. In this situation, constructing double or even treble glazed systems, with these systems increasing the number of boundary layers from the traditional two layers to four and six layers respectively, may aid in reducing heat transfer from inside to outside. However, in tropical climates, the problem is the opposite with heat transfer from outside to inside. In this respect the worst offender is heat input from solar radiation, and convection and conduction are comparatively unimportant, contributing, for example, no more than 10% of the energy transferred into a Singaporean residence [10].

This is because the infra red (IR) radiation from direct sunlight passes through glass windows into the interior of the building with minimal loss. There it heats the exposed surfaces of the room and thence its interior atmosphere. Ordinary double or treble glazing is of very little value in this situation, because, for most of the time, the convective/conductive heat input is much smaller in magnitude than that due to solar radiation. What is needed in this situation is a means to prevent the IR from entering the room in the first place. One (trivial) solution is not to have glass windows, and this is an element common in much indigenous architecture in hot climates. However, in general, modern homeowners and office workers require windows and, it seems, the more the better. In, this situation the combination of double glazing/IR blocking window glazing systems (to be described below) will give the best of both worlds in sub-tropical and temperate regions, that have hot summers and cold winters.

A clear glass window exposed to sunlight on the east or west side of a building at 30° southern latitude can transmit up to 550 W/m^2 in summer [12], and allows through about 90% of the incident radiation [Sydney Latitude: 33d52'00" S Longitude: 151d12'00" E]. A typical home might have a total 12 m^2 of such window on both the east and the west sides, that are illuminated alternatively, which nevertheless yields a peak heating rate per aspect of nearly 7 kW. When it is considered that a typical electrical bar heater runs at about 1 kW, it can be appreciated that a significant amount of energy can enter the building through the windows. In winter the sun is lower in the sky slightly reducing energy input from the east and west windows to 400 W/m^2 during this season while, conversely, that of a north-facing window increases from a summer value of less than 100 W/m^2 to over 500 W/m^2, for the same reason. This increase is naturally very welcome in winter, and explains the desirability of north-facing glazing to buildings at temperate latitudes in the southern hemisphere.

It is in this area where nanotechnology can, and has, played a role. One of the most readily recognisable implementations of this technology is its incorporation in coated glass. A material with a high free electron density (such as gold, silver or copper) has a very high reflectivity with respect to light above some critical wavelength, and is transparent below that wavelength [5] (Figure 1).

The cross-over wavelength in the case of nano-scale gold coatings occurs at about 500 nm, and provides a cool, filtered light which is ideal for many architectural applications. Silver is shown above as reflecting most visible light, but actual performance depends on subtle details such as the presence of anti-reflection coatings and in fact some types of silver coating have a better transmittance in the visible than gold - a 5 nm thick film of Ag will theoretically allow through 85% of visible light while reflecting almost all IR [5]. The radiative energy in the visible part of the solar spectrum is roughly half of the total, so a target to aim for would be systems that have T_{vis}/T_{sol} of ~2, where T_{vis} is the proportion of visible light transmitted (e.g. 100%) and T_{sol} is the portion of the entire solar spectrum transmitted (e.g. 50%) [Bell 1998]. The performance of thin noble metal films may be significantly improved by sandwiching them between two nano-scale layers of dielectrics with high refractive index such as ZnO, Nb_2O_5, Bi_2O_3, TiO_2 and SnO_2 [Bell 1998, Nadel & Hill]. Complex coatings with two or more layers have been demonstrated with close to the theoretical maximum efficiency [Bell 1998, Nadel & Hill]. Alternatively, some other material with appreciable density of free electrons may be used, for example TiN and ZrN, Cr_xN, Si_3N_4, SnO_2 or In_2O_3 [5, 13]. These compounds are harder and more abrasion resistant than the noble metals, which is a useful advantage. The industrial-scale production of large areas of such coated glasses using magnetron sputtering began in the early 1980s and is still growing strongly [14]. It has been estimated that between 30 and 40% of the climate control energy costs of a residence or commercial building can be saved by using these coatings in combination with double glazing systems [15].

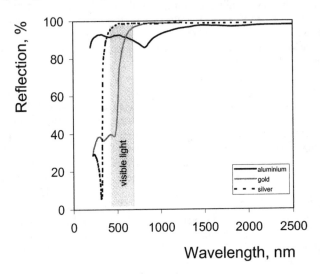

Figure 1   *Reflectivity of simple metallic films on glass as a function of wavelength showing how the use of gold permits simultaneous partial transmission of visible light and full reflection of infra-red (data from reference [11]).*

Unfortunately, various factors conspire against the wider use of such reflective coatings. These factors include aesthetics - the slight colour-cast and mirror-like appearance are not liked by some people and the high cost of applying these nanoscale layers, which of necessity must be achieved by deposition in large, high-vacuum chambers. In addition, the deposition tends to occur as little islands of metal rather than as a continuous film, which considerably reduces the efficiency [5]. The high cost of applying these coatings (e.g. US$10 per m^2) is a dominating aspect of the economics, while the actual cost of the actual raw materials used (e.g. US$1 per m^2) is low.

Therefore, the alternative strategy of absorbing IR (rather than reflecting it) has attracted interest. This is achieved by dispersing either a nano-particulate material with strong IR absorptive characteristics such as FeO_x or indium tin oxide or antimony tin oxide [17] or LaB6 [18] in the glass itself ('supertinted' glass), into a polymer film applied to the glass ('plastic-filmed glass') or in the PVB laminate use din windscreens and safety glass. A variety of materials are capable of absorbing IR while transmitting visible light, but the optimum performance, once again, requires that nano-scale dimensions are achieved. Some of the substances in current or impending use include FeO_x, CoO_x, CrO_x, TiO_x, SeO_x, phthalocyanine dyes, and blends of rare earth hexaborides with indium and tin oxides [19]. The performance of a modern absorptive system, applied onto glass as part of a plastic film, is compared to that of a state-of-the-art reflective coating system in Figure 2.

Since about two thirds of the heating from solar radiation comes from wavelengths above about 1000 nm, it is evident that both systems could be theoretically quite effective in screening out the sun. It is clear that, the energy absorbed by the LaB$_6$ film must go somewhere, and it is released as heat in the film on the glass. From there it is transferred inwards by conduction and subsequent convection at the inner boundary layer (an undesired outcome) and outwards by convection. In practice it appears that, in single glazed systems, only about 40% of the heat is shed into the external environment [19], but this is still a considerable benefit. However, the absorption of heat on the glass leads of course to a rise in temperature and a thermal expansion. Care must therefore be taken in the design of such windows to allow for considerable diurnal stress cycling.

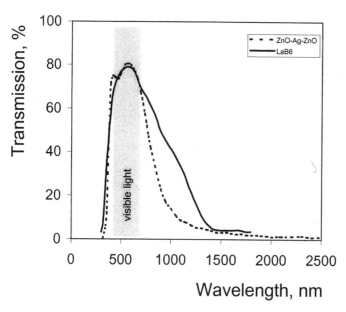

Figure 2 *Comparison of the spectral selectivity of two state-of-the-art coatings that block infra-red radiation. The ZnO-Ag-ZnO reflects the IR, and the LaB$_6$ coating absorbs it. Data from refs [15], and [16] respectively.*

Solar control by a combination of reflection and double glazing can reduce the heat input through a sunlit window to as little as 20 to 30% of its 'normal' value while the combination of absorption and double-glazing can typically reduce energy input by about 50% [20]. It may even be worthwhile to retrofit with a plastic film of absorptive material, with improvements of 20 to 40% being likely over untreated glass, and a saving in air conditioning costs of nearly 30% being possible [10]. However, reflective glasses are generally the superior technology from a performance point-of-view due to the problem of IR-absorbing schemes causing the surface of the glass window to heat up. Obviously, a double glazed system could help to reduce conductive-convective heat transfer into the house, and a combination double-glazed/IR absorptive system, while expensive, can half the input of solar heat, while avoiding the 'mirror'-like appearance of reflective systems.

4.2 Lighting

It has been claimed that a typical home in Australia spends about US$50 per annum on electricity for lighting [21]. This is equivalent to about 800 to 900 kW.h. and is obviously only a small proportion of the average household's annual electricity bill. However, the utilisation of this electrical energy is relatively inefficient, with the major portion of the energy purchased actually being released as heat rather than light. As far as cold conditions are concerned, the excess heat generated usually contributes to the overall heating of the home and is therefore not completely wasted. But, the situation in a hot climate or season is quite different. The excess heat is undesirable, and would probably at the very least need to be vented to the outside world. However, should it be remediated by air conditioning then the householder suffers a double loss; the electrical energy that was not converted to light, and the electrical energy used to remove the extra heat (which has its own inefficiencies). It is better by far in a hot climate to prevent wasteful generation of indoors heat.

Fortunately, the generation of interior light is currently experiencing diverse technological improvements, many based on nanoscale phenomena and technologies. The traditional tungsten filament incandescent light bulbs are notoriously inefficient, producing only around 10 to 15 lumens per watt of energy drawn (about 5 to 10% of the theoretical efficiency possible, depending on wavelength). The wasted energy is naturally rejected to the environment as heat. Two technologies, one old and the other new, are in contention to make inroads into the territory of incandescent bulbs. These are the fluorescent light, actually first conceived in 1855 as the Geissler tube, and the white light emitting diode (LED), a technology which first emerged only in the late 1990s. The fluorescent tube can deliver up to 95 lumens per watt excluding ballast losses (although the colours more popularly used in domestic lighting and lower wattages yield only about 60 lumens/watt. Commercial white LEDs are already at 25 to 30 lumens per watt [22] after only a few years of development, with a probable further improvement to 100 lumens per watt or better envisaged. In both cases the colour of light emitted by the device is obtained by engineering materials at the nanoscale. In the case of fluorescent tubes it is the phosphors applied as a coating to the inside of the tube that largely control the colour, whereas in the case of the LEDs it is a complex epitaxial structure of nanoscale films of III-V semiconductors plus a phosphor. RGB combinations that can yield higher luminous efficacies which are better than incandescent bulbs, but good light mixing is needed to achieve colour uniformity . Special polymer composite materials and special light guide structures can be used for this purpose [23]. Although current white LEDs are not more efficient than incandescent lamps, they are said to offer 100 000 hour lives and user-variable intensities and colour tones. Due to their geometry they also allow for innovative designs of lighting systems.

It is also worth commenting on the issue of lighting the inside of buildings during daytime. High tech sky-lighting is also an option, with the light being obtained from fluorescence and directed down a flexible light guide and thence into a luminaire in the interior of the building. Finally, a very pleasing form of interior illumination that has been developed at UTS is the polymer-in-polymer light pipe [23]. These structures consist of thick, flexible polymer extrusions, containing a nano-scale dispersion of a second polymer. The result is controlled scattering of the light from the sides of the polymer extrusion with an overall effect much like a low intensity neon light tube.

3.1 Related technologies

Although not strictly intended as a means of saving energy, it is worth briefly mentioning the range of other interesting nano-technologies currently available for the windows of a future Nanohouse™. There are various kinds of 'smart' window, for example, in which the ability of the window to transmit visible light can be changed by application of some external stimulus, such as an electric potential, chemical signal or change in temperature. These types of glass have been termed 'electrochromic', 'gaschromic' [24] and 'thermochromic' [25] respectively. In one version of the electrochromic effect, a plastic sheet consisting of magnetic nano-particles sandwiched between two transparent but electrically conducting surfaces is applied to a window. When a potential difference is applied the particles line up causing the sheet to become transparent [26]. This system can apparently be retrofitted to existing glass windows. In another scheme the effect is achieved by an electrochromic substance applied to the surface of the glass. A commonly studied system involves the compound tungsten oxide, WO_3, which can be switched between a transparent state (WO_3) and a coloured 'tungsten bronze' state (e.g. Li_nWO_3) by application of an electrical potential [27]. The change takes a few minutes to occur and the resulting colour change can vary transmission of visible light by a factor of seven (Figure 3, from [27]).

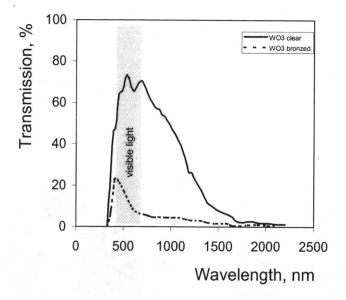

Figure 3   *Spectral selectivity of the two states of a prototype electrochromic window produced by Pilkington Glass using tungsten oxide (redrawn from [27]).*

A gaschromic system is also interesting, and could be implemented by coating one of the internal surfaces in a double glazing unit with WO_3 plus a thin coating of platinum catalyst [24]. A central unit in the building would flood the internal space of the double glazed unit with a (non-combustible) gas containing 1% hydrogen whenever the transmission of light through the window must be restricted. In this state the coating turns dark blue. The effect can be reversed by purging the hydrogen and replacing it with clean air, which causes the coating to become transparent again after a few minutes. While at first sight the logistics of the gas supply system seem daunting, it should be noted that gas streams of appropriate composition may be readily derived from existing urban gas reticulation systems, or perhaps one day, from in-building fuel cell plants.

The thermochromic effect is achieved by laminating a sheet containing temperature-sensitive dyes between two sheets of glass. When the glass becomes too hot, the dyes change colour and darken (presumably making the glass even hotter and darker, so the effect will be autocatalytic). Eventually, once the glass has cooled below some critical temperature the window becomes clear again.

However, all the currently available 'smart' glass systems are very expensive, and apparently suffer from a gradual decline in performance with use, and they do not seem yet to be able to offer the benchmark of twenty years service desired for architectural applications. Nevertheless, they are apparently becoming well established in the niche area of automobile mirrors [26] where smaller surface areas, shorter lives and higher costs can be tolerated.

Another technology that has been recently commercialised is that of super-hydrophobic coatings on windows (self cleaning surfaces). [28, 29]. As a result of the application of such coatings, the contact angle of water on glass is increased from about 15° to over 90° [28]. The idea here is that such films of water will readily run off such surfaces (if inclined of course) taking with them entrained dirt as well as any corrosive alkalis that they may happen to have taken up into solution from the concrete of the surrounding building. Even if the water does not drain completely off the glass, a quick wipe with a soft cloth should be sufficient to clean it.

The hydrophobic films may be combined with a photocatalytic layer of TiO_2 to produce a 'self-cleaning' window. Any organic grease or dirt that did not wash off with the water is decomposed by the combined action of ultra-violet radiation from the sun and the TiO_2 catalyst. The result is said to be a window that requires only intermittent rain, or perhaps a quick hose down, to stay clean and clear [29].

Figure 4   *Two rendered images of the first example Nanohouse designed by James Muir. The left image is the northwest view, and the right image is the southeast view.*

4 CONCLUSIONS

Nanotechnology-enhanced fenestration products are now available for use by the building industry. In particular, these provide the opportunity to design comfortable, energy-efficient buildings with unprecedented large areas of window. This availability allows designers to exploit the aesthetic and psychological properties of windows without having to be overly concerned about unpleasant indoor climatic effects. On the other hand, a saving of around 20% in air conditioning costs is readily achievable in conventional buildings if they are retrofitted with the new products. If the new types of window are combined with a latitude-appropriate building design then the need for air-conditioning or heating will in any case be drastically reduced. In general, it is desirable to design houses with a reasonable heat capacity or thermal inertia, in order that the diurnal fluctuations can be smoothed out as far as possible by passive means. This implies a need for internal masonry, or perhaps a concrete slab or two. Windows should be designed so that they are shielded as far as possible from direct solar radiation during summer, but receive it in winter.

In general, the more important problem in Australian housing is to keep the indoor temperatures low during summer, and in this is a very important difference to the situation in much of the Northern Hemisphere where more emphasis is placed on keeping warm during winter. This difference increases the value of energy efficient lighting in Australia, since any excess heat generated indoors in summer is unwelcome. The ordinary incandescent light bulb is not very efficient in this regard and far better performance may be obtained with a fluorescent light. The new generation of white light emitting diodes currently offer efficiencies intermediate the between incandescent and fluorescent lights, but are undergoing a rapid evolution. Therefore, it appears feasible that they will eventually surpass the performance of fluorescent tubes.

It seems likely that the state-of-the art residence of the future will be a true 'Nanohouse™' with sophisticated fenestration, self-cleaning surfaces and internal lighting of unprecedented versatility.

Acknowledgements

This paper is published by permission of the University of Technology, Sydney and CSIRO, Australia.

References

1   Deutsche Steinzeug Cremer & Breuer AG, *Self-Cleaning Facades, Ceram. Forum Int./Ber.DKG.* **Vol.77**, No.8, 2000, E15.
2   World Gold Council, *Gold Research*, Summer 2002.
3   Greenland J, *Foundations of Architectural Science*, UTS Sydney, 1991
4   Ridley L, in *Solar Building Architecture* ed; Anderson B, MIT Press Mass. 1990, 36.
5   J.M. Bell and J.P. Matthews, *Materials Forum*, 1998, **22**, 1.
6   V. Block, in *Glass Processing Days 2001*, 18th-21st June 2001, Tampere, Finland, pp.826-829.
7   S. Peatling, *The Sydney Morning Herald*, 30th July 2002, 5.
8   C. M. Lampert, *Glass Processing Days*, 13-16 June 1999.
9   D.G. Stephenson, *Canadian Building Digest*, CBD-101, National Research Council Canada, May 1968
10 Y. Dai, in *Glass Processing Days 2001*, 18th-21st June 2001, Tampere, Finland, 156.

11 Oriel Instruments. *Metallic Reflector Coatings*, pamphlet, Connecticut, USA, undated, circa 1999.

12 Air Conditioning Systems Design Manual, in *Handbook of Australian Department of Housing and Construction*, Australian Government Publishing Service, Canberra, 1974.

13 R. Manfred, C. Braatz, *Selective and non-selective solar control coatings*. http://www.appliedfilms.com/precision3/09%5Fsolar%5Fcontrol%5Flayers/solar_control_layers_print.htm, accessed 12th Feb. 2003.

14 F Kühnel and T. Paul, in *Glass Processing Days 2001*, 18th-21st June 2001, Tampere, Finland, pp.760-761.

15 S.J. Nadel and R.J. Hill, *in Glass Processing Days 1997*, 13-15th September 1997, Tampere, Finland, pp.209-212.

16 H. Takeda, K. Yabuki, K. Adachi, United States Patent 6319613, 2001.

17 G.B. Smith, C. A. Deller , .D. Swift, A. Gentle, P.D. Garrett and W.K. Fisher, *J. Nanoparticle Research* 4, 2002, 157.

18 S. Schelm and G. B. Smith, submitted *App. Phys Lett*.

19 G. James Body Tint, pamphlet of G. James Pty. Ltd., www.gjames.com.au/gjames/cdrom/misc/bodytint.html, accessed 16th September 2002.

20 *Energy Saving for Windows*, pamphlet, Sustainable Energy Authority Victoria, Australia, August 2001.

21 *Energy Smart Lighting*, pamphlet, Sustainable Energy Authority Victoria, Australia, August 2001.

22 Lumiled press release : Lumileds announce the Luxeon 5-watt, the world's brightest LED light source, www. panelx.com/news/. posted 16th April 2002, accessed 6th January 2003.

23 C. Deller, G.B. Smith, J.Franklin and E. Joseph, *Proc. Australian Institute of Physics, (AIP) 15th Biennial Congress*, July 2002, 307.

24 V. Wittwer, W.Graf, A. Georg, in *Glass Processing Days 2001*, Tampere, Finland, 725.

25 O. Yanush, article in Library of Glassfiles.com

26 C. M. Lampert, *Glass Processing Days*, 13-16 June 1999.

27 H. Wittkopf, in *Glass Processing Days 1997*, 13-15th September 1997, Tampere, Finland, pp.299-303.

28 M. Hüber, in *Glass Processing Days 2001*, Tampere, Finland, pp.775-779.

29 anon, Pilkington Aktiv™, pamphlet of Pilkington Glass (Australia), 95 Greens Road, Dandenong South, Victoria 3164, Australia, www.pilkington.com.au/Australasia/Australia/English/default.htm, accessed 4th September 2002.

BUILDING FAÇADE INTEGRATED QUANTUM DOT CONCENTRATED SOLAR ELECTRICITY PRODUCTION

S. Gallagher, B. Norton and P.C. Eames

Centre for Sustainable Technologies, School of Built Environment, University of Ulster, Newtownabbey, N. Ireland, BT37 0QB.
email: s.gallagher@ulster.ac.uk

1   INTRODUCTION

Photovoltaics convert solar energy directly into electricity. Building Integrated Photovoltaics (BIPV) is now recognised widely as the most cost-effective form of grid-connected photovoltaic power generation in industrialised countries[1]. While maintaining or exceeding present performance, the cost of wide-scale implementation as a building façade cladding can be reduced by economies of scale associated with increased manufacturing output, improvements in solar conversion efficiency and/or by the substitution of expensive photovoltaic materials by lower cost concentrating systems[2].

A novel, non-tracking concentrator is described, which uses nano-scale Quantum Dot technology rendering the concept of a fluorescent dye solar concentrator (FSC) a practical proposition. The Quantum Dot Solar Concentrator (QDSC) comprises quantum dots (QDs) seeded in materials such as plastics and glasses that are suitable for incorporation into buildings. QDSC systems would be used as a building façade with photovoltaic (PV) cells attached to the edges. The PV converts direct and diffuse collected solar energy into electricity for use in the building.

QDSC[3] is similar in operation to FSC[4]. Insolation incident on the surface is refracted into a transparent carrier material seeded with quantum dots and after absorption by a QD leads to the emission of photons at lower frequencies. The number of photons emitted depends on the absorptivity of the carrier material and on the QD quantum efficiency. Emitted photons may leave the carrier material or be reflected at the surface and remain within the concentrator. If the refractive index of the carrier material is much higher than that of the surrounding medium (in this context, air), a large proportion of the emitted photons will travel by total internal reflection to the edges[4]. Silvering of three of the edges and the back surface ensures light can only emerge at the top surface and at a fourth edge where a photovoltaic (PV) cell may be located, as illustrated in figure 1.

QDSC plates would make it possible to separate the solar spectrum if plates are doped with different sized QDs, such that each plate absorbs only part of the solar spectrum and is transparent to the rest[6]. The lower plates absorb at increasing wavelengths and their QDs emit further down in the infrared. They can be arranged as shown in Figure 2a. The main advantage of using a stack is that the concentrated light from the individual collectors can then be transformed by correspondingly spectrally optimised solar cells, with a higher overall efficiency.

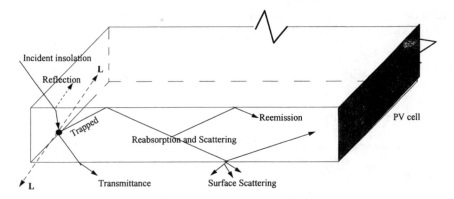

Figure 1   *Principle of the QDSC from [4]. Where:* • *Quantum Dot and L   Loss due to light leaving the collector through two boundary planes*

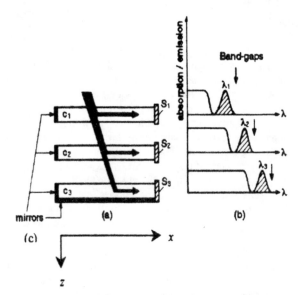

Figure 2   *(a) shows a schematic of a stack of three collectors c1-c3 with three solar cells s1-s3 in series. The absorption and emission spectra of the molecules in the collectors are sketched in (b)[7].*

The development of FSC was limited by the stringent requirements on the dye, namely high quantum efficiency (QE), suitable absorption spectra and red shifts and stability under illumination[4]. A typical measured electrical efficiency with a two-stack concentrator with GaAs solar cells was 4%[4], whereas the original predictions were in the range 13-23%[5]. These problems are addressed by replacing the organic dyes with QDs. The advantages of QDs over dyes are:

- QDs are crystalline semiconductors and therefore should degrade less than organic dyes.

- High luminescence QE has been observed in QDs at room temperature. Colloidal CdSe/CdS heterostructure dots have demonstrated luminescence quantum yields above 80% at room temperature[9].
- The absorption threshold can be tuned simply by choice of dot diameter. Colloidal InP quantum dots have thresholds which span the optical spectrum[10].
- Red shift between absorption and luminescence is determined primarily by the spread of dot sizes, which in turn can be optimised by choice of growth conditions. Re-absorption therefore can be minimised and high efficiencies and high concentration ratios achieved.[3]
- Luminescent peaks and absorption thresholds are well separated and the red shifts are comparable with those assumed even in the more optimistic predictions for fluorescent concentrators[5].

If the QDs with high QE can be incorporated into a suitable transparent media then QDSC, which performs in the upper range of efficiency predicted for dye concentrators, may be realised.

2 BACKGROUND

Artificially made semiconductor structures show a variety of new and interesting properties that are completely different from solid-state bulk materials[9]. With high fluorescence QE, and optical, structural and self-organisational properties all tuneable to a certain extent with the size of the crystal, these materials are promising constituents of nanometer scale devices[12]. It is possible to engineer QDs of different sizes that absorb photons at all solar radiation frequencies and re-emit photons at frequencies at which the solar cell is most efficient. To develop the QDSC, it is necessary to use QDs which efficiently absorb solar energy, are intensely luminescent in the spectral region of the PV cell's maximum sensitivity, mechanically strong, chemically stable and preserve their spectral-luminescent characteristics with sharp variations in temperature in field conditions[15]. However, for QDs to be useful they have to perform well not just under laboratory conditions but also in real environment. QDSC must be packaged to keep them in place and protect them from the atmosphere. The packaging must not interfere with their operation. Working out QDs' interaction with their surroundings means using theoretical models and due to this a QDSC ray trace model has been developed.

3 MODEL AND ASSUMPTIONS

The development of a highly efficient QDSC requires QDSC geometries to be optimised. Performing these investigations experimentally is not only extremely time consuming but also provides an unsatisfactory physical understanding of the system. To address this a theoretical model has been developed. Analytical approaches have been developed to describe Fluorescent Solar Concentrators[11-14] but they use semi-empirical approximations or include assumptions that render them inappropriate for application to a QDSC. A distinctive feature of a QDSC is the interdependence of statistical processes and is illustrated in a flowchart in figure 3.

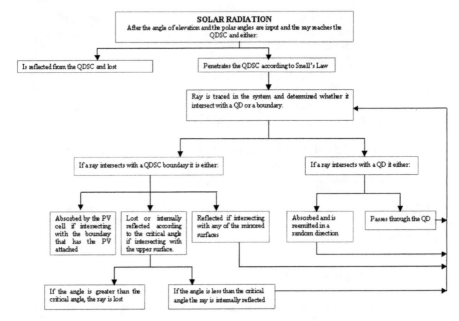

Figure 3 *Processes within a QDSC*

3.1 Model Assumptions

- Incident rays are equally spaced over the material.
- Rays represent discrete photons and are reemitted in a random direction after intersecting with QD
- QDs are spherical in shape and are seeded randomly.
- QDs are the same size and have the same absorption and emission spectra.
- Wavelength of the incident radiation is within the absorption spectrum of the QD.
- Only one boundary face is covered with the PV cell.
- There is no absorption by the carrier material

3.2 Fixed Parameters

- Refractive index of the carrier material is 1.492.
- Reflectivity of all the other surfaces =1
- Quantum efficiency, number of photons emitted is equal to the number of photons absorbed, =0.97.
- Incident angle is 90. Since after a ray enters the system and intersects with a QD, it is reemitted in a random direction there is the equal probability of reaching the PV. It is therefore unnecessary that all ray angles are traced through the system.

Parameters such as size and thickness of the material, radius and concentration of the QDs, QD absorption and emission spectra and the incidence angle are variables, allowing the determination of the optimal set of these parameters for a of the QDSC.

4   RAY TRACE MODEL DESCRIPTION

Rays equally spaced over the material are traced. For example, for 100 rays the material is equally divided into a 10x10 grid and the ray start point is set at each grid centre. The rays are either reflected from the material (and lost) or undergo refraction at the boundary and penetrate the material according to Snell's Law[15]. To incorporate a large number of QDs the material has been subdivided into a set of up to one million zones. For specific sizes and percentage volume fractions of QDs a random distribution of QDs is generated in each zone.

The QDs are shifted by random amounts in x, y and z directions to generate 999,999 other distributions, illustrated in figure 4. This procedure has enabled up to 1.5×10^9 QDs to be modelled. To allow reduced but more efficient ray tracing, the model traces the ray path within each zone and then moves between zones. Calculation speed is thus greatly improved. Figure 5a & b shows a sample path of a ray in the QDSC system.

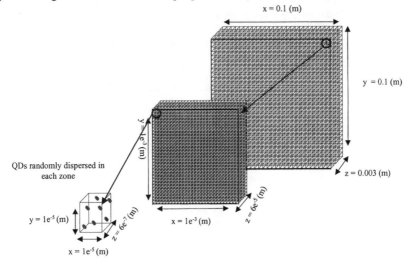

Figure 4   *A three-dimensional illustration of the material after subdividing into zones*

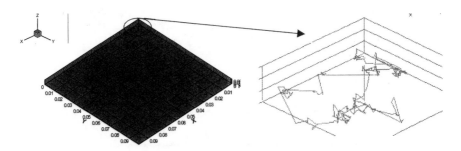

Figure 5 (a)*Left - the ray path of a specific ray. (b)Right- shows it's enlargement.*

5   INCORPORATION OF SPECTRAL CHARACTERISTICS

One of the features of QDs is the almost universal occurrence of a redshift of the emission relative to the absorption; an example is illustrated in Figure 6. One reason for the redshift is the existence of a residual size distribution: the larger dots in a sample have lower band-edge energies. If the sample is excited with photons of sufficiently high energy above the band edge of the smallest dot, the emission will be redshifted[20]. Figure 6 shows the experimental results and predictions for the absorption and emission spectra of CdS/CdSe core-shell QDs in an acrylic slab[21]. Whereas the absorption peak centred around 500nm, the emission spectrum has a peak centred at approximately 645nm.

In the model, each ray is assigned a specific wavelength. The QDs have homogenous wavelength characteristics, as they are assumed to be homogenous in size. If the spectral characteristics of the ray correspond to that of the absorption spectrum of QD, the ray is absorbed and reemitted in a random direction. If the spectral characteristics of the ray do not correspond to that of the absorption spectrum of QD it is assumed that the ray is transmitted through or reflected by the QD. Spectral characteristics are represented in the model by normalised absorption and emission spectral data[17]. The absorption data is the energy absorbed at a particular wavelength and the emission data is the energy emitted (equal to 1). The energy input to the model is assumed to be (i) isotropic; (ii) at a specific wavelength that of the laser used in the experimental work[21]; (iii) discretised solar irradiance spectrum for air mass 1.5[22]. All rays are assumed to be within the absorption range of the QD. The absorbed energy of each ray is emitted over the same probability of emission as illustrated in figure 6.

For each intersection, the amount of energy absorbed and re emitted by the QDs is calculated as follows:

$$I_{new_x} = I_{reabsorb_x} + I_{rem_x}$$

Where $I_{reabsorb_x} = (I_{emis_x} * I_{absorb_x}) * I_{reabsorb_{x-1}}$ $I_{rem_x} = (1 - I_{absorb_x})I_{emis_x}$ (1)

At specific wavelengths x, I_{emis_x} is QD emission energy, I_{absorb_x} is QD absorption energy, $I_{reabsorb_x}$ is energy reabsorbed, I_{rem_x} is energy remaining, and I_{new_x} is new ray input. I_{new} is looped back into the program and is the input energy for the next intersection with a QD.

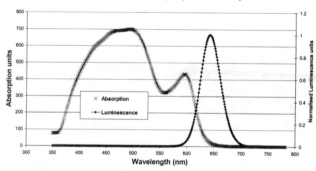

Figure 6   *Normalised absorption and emission for CdSe/CdS dots in acrylic, adapted from [17]*

Figure 7 shows the shift in the emission spectra after 100 intersections with a QD. The decrease in energy, at approximately 450nm to 500nm, corresponds with the absorption peak of the QDs in figure 4. Also the shift in the emission spectrum to a longer wavelength of 650nm corresponds to the emission peak in figure 6. With increased intersections with a QD, more energy is shifted to an amplified peak at 650nm. To mimic experimental work[21], based on the thermodynamic modelling of the QDSC, the energy input is assumed to be at one wavelength, that of the laser used, (530nm). Figure 8 shows the shift in the emission spectra after one, 15 and 100 intersections with a QD. The energy input at 530nm is absorbed by the QDs and the emission is shifted to a longer wavelength corresponding to the emission peak of the QD in figure 6. From the figure it can be seen that there is very little change in the emission spectra after 15 intersections with a QD.

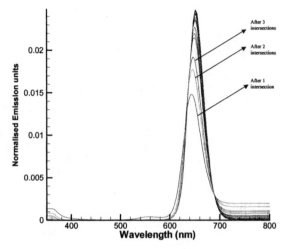

Figure 7   *The shift in the emission spectra after 100 intersections with QD, assuming isotropic radiation.*

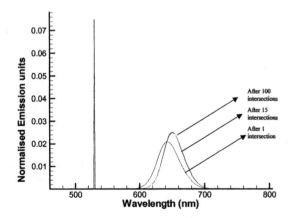

Figure 8   *The shift in the emission spectra after 1, 15 and 100 intersections with a QD*

Using discretised solar irradiance spectrum air mass 1.5[22] (290-1200nm), as the energy input, the emission spectra were calculated for different numbers of intersections with a QD. Figure 9 illustrates how the solar spectrum and absorption and luminescence of QD relate.

In figure 10 a & b, there is a clear shift in the emission spectra to longer wavelength, peaking at approximately 650nm, correlating to the emission peak of the QD in the figure 9. It can be seen from figure 10a & b that there is very little difference between the emission peak after 15 intersections with a QD. We can assume then for our model that after 15 intersections with a QD, the ray is no longer absorbed and reemitted randomly but passes through the QD.

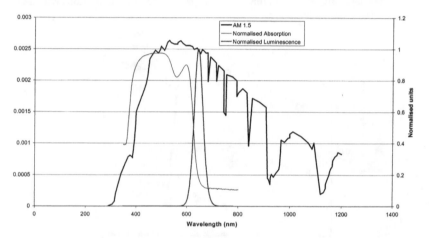

Figure 9   *Solar irradiance spectrum air mass 1.5, with normalised absorption and luminescence of QD.*

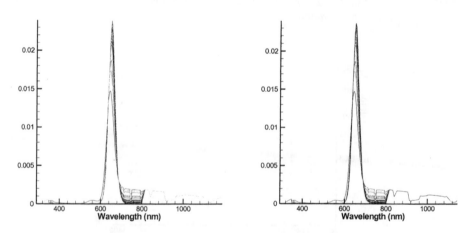

Figure 10   *(a- left)The emission spectra from 1-100 intersections with a QD. (b-right) The emission spectra from 1-15 intersections with a QD.*

The reabsorption of the rays by the QDs is of huge significance to the efficiency of the system. It is of great importance that the absorption and emission spectra are well separated with little overlap. If the absorption and emission spectra have a large overlap, the rays will be absorbed and reemitted at the same wavelength. The ray will be absorbed by a large number of QDs, greatly dissipating its energy, resulting in a low possibility of the ray reaching the PV with substantial energy remaining. However, if the spectra are well separated, the ray will be absorbed at one wavelength and reemitted at another. The ray will only be absorbed once by a QD giving higher system efficiency.

6   RAY PATHLENGTH AND EXTINCTION COEFFICIENT

Ray pathlength, (the distance that the ray travels in the medium) and extinction coefficient, (i.e. the fraction of light lost to scattering and absorption in the medium) are critical in the design and optimisation of a QDSC system. With a long pathlength and a low extinction coefficient, the ray has an increased probability of reaching the PV with a significant energy content. However, if the rays entering the system have a long pathlength it is essential that the extinction coefficient of the material is low. This will decrease absorption by the material and allow the rays to reach the PV cell with minimal energy dissipation.

For ray pathlength calculations, the system considered has dimensions 10cmx10cmx0.3cm, QD radius $5e^{-8}$m and quantum efficiency, (QE) of QD=0.97. 100 rays were traced through the system and the total distance travelled of each ray between every intersection with a QD or a material boundary was determined. In the programmed algorithm calculation is terminated when the ray exits the medium, reaches the PV or energy dissipates. For QD percent volume fraction of $2e^{-4}$ ($1.5e^{+13}$QDs), the total distance travelled is determined for each ray. The pathlength distribution found is presented in figure 11, with an average pathlength of all rays of 7.4cm. Figure 12 shows the pathlengths calculated for the rays reaching the PV after intersecting with a QD, with an average pathlength of 4cm. Under these conditions 8% of rays reach the PV as the energy of the other rays dissipate to less than 5% of the energy entering the system due to QE of QD.

At decreased percent volume fractions the pathlength of the rays increase, as can be seen from figure 13 illustrating the calculated pathlength distribution for QD percent volume fraction of $1.5e^{-4}$ ($1e^{+13}$QDs). The average pathlength of rays is 9cm, with the average pathlength for those reaching the PV, 9.2cm, shown in figure 14. For these parameters there is a calculated 11% increase in the percentage of rays reaching the PV.

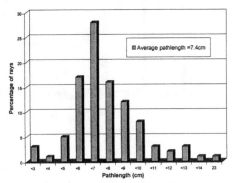

Figure 11   *Percentage of rays of specific pathlengths*

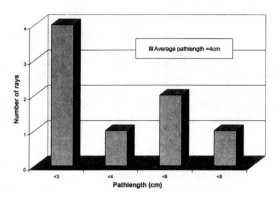

Figure 12   *Pathlengths of rays reaching the PV.*

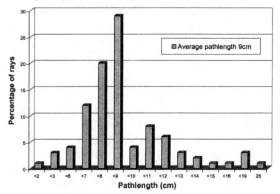

Figure 13 *Percentage of rays of specific pathlengths.*

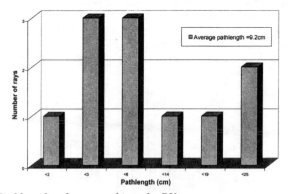

Figure 14 *Pathlengths of rays reaching. the PV.*

6.1 Extinction Coefficient

Diamant glass[20] was considered as a possible carrier or encapsulation material. Using percentage absorption at particular wavelengths of material with 3.19mm thickness and average pathlengths calculated previously, the extinction coefficient was calculated using the following equation.

$$I_{absorbed} = \left(1 - e^{-KL}\right) I_{incident} \qquad (2)$$

where L=pathlength, K = extinction coefficient, assuming $I_{incident} = 1$

Results for extinction coefficient for different wavelengths can be seen in table 2.

Table 2 *Extinction coefficient at specific wavelengths.*

Wavelength (nm)	Extinction coefficient (m⁻¹)
300	306
400	0.22
500	1.382
600	0.848
700	5.216
800	8.580
900	9.451
1000	10
1100	10.357
1200	8.548

As can be seen from figure 15, absorptance at a particular wavelength increases with increasing pathlength, L. At a wavelength of 300nm there is a high extinction coefficient, K, of 306, which would lead to total absorptance by the material at even the shortest pathlengths calculated. At wavelengths in the visible range (380-780nm) K decreases to 0.22 giving a low absorptance of 0.006-0.05, also within this range, K increases to 8.58 leading to an absorptance of up to 0.863 for L=23cm. Towards the infrared range of the spectrum, K increases further giving an absorptance of 0.909 for L =23cm.

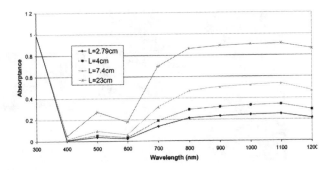

Figure 15 *Absorptance at specific wavelengths using calculated pathlengths.*

As a comparison, the value of K varies from approximately 4m⁻¹ for 'water white' glass (which appears white when viewed on the edge) to approximately 32m⁻¹ for poor (greenish cast of edge) glass[15]. Diamant is therefore found to be an appropriate encapsulation material for QDSCs.

7   A PROTOTYPE DESIGN FOR QDSC

To undertake an experimental evaluation of the modelled system, a number of QDSC systems have been fabricated incorporating QDs into plastic. For comparison, samples with fluorescent dyes and the plastic carrier material alone were also fabricated. BP Single Crystal[21] silicon solar cells are attached to the QDSC plates with silicone oil[22-25], a refractive index matching liquid. The performance of the QDSC was characterised using (i) a spectrometer to measure the absorption, fluorescence, transmission, and reflection and (ii) by attaching a PV cell and measuring the spectral response.

A QDSC of dimensions of 11.5x11.5x0.3cm has been fabricated to accommodate the dimensions of the PV cell (11.5x0.3cm). All other sides of the device have reflective coatings of reflectivity =1. The system is illustrated in figure 16a & b. prior to casting and in figure 17a & b after casting.

A discolouration could be seen in the cast samples. Thermocouples have been attached to the system to determine the effects of temperature rise on the PV cell efficiency and also the effects of an increase in temperature of the material on the overall efficiency of the PV cell. The thermal stability of the QDs is an important factor for the realization of the QDSC. After characterisation under the solar simulator, the system is placed in a UV chamber for varying time periods.

Practical applications need additional information about outdoor behaviour of the QDSCs including their response on sunny days as well as cloudy days and the evolution of this response with illumination variations during the day. The reported capability of these concentrators for collecting diffuse light may allow the possibility of increasing in the effective operating time for building integrated PV devices[26].

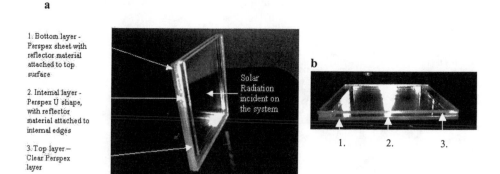

a

1. Bottom layer - Perspex sheet with reflector material attached to top surface

2. Internal layer - Perspex U shape, with reflector material attached to internal edges

3. Top layer - Clear Perspex layer

Solar Radiation incident on the system

b

1.   2.   3.

Figure 16 a & b   *System prototype prior to casting.*

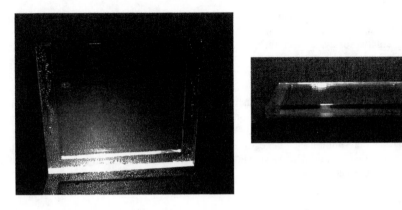

Figure 17 a & b   *The system of 0.1% volume fraction of CdSe/CdS QDs after casting.
A slight discolouration is visible in the cast samples.*

1. CONCLUSION

Buildings play a significant role in the global energy balance. Typically they account for some 20-30% of the total primary energy requirements of industrialized countries[27] and 40% in the EU. Therefore, the place to erect PV systems to supply such loads is the building itself. Applying PVs to buildings is expected to be the largest application for grid connected PVs[28]. Manufacturers are developing specialised products to capitalise on this opportunity. R&D continues to focus on improving efficiency, reducing material and fabrication costs, and enhancing reliability and flexibility, and continues to produce innovations in thin-film PV and the single and polycrystalline technologies, which currently dominate the market. One such novel device is the Quantum Dot Solar Concentrator. Fluorescent Solar Concentrators (FSCs) have advantages over geometric concentrators in that tracking is unnecessary and both direct and diffuse radiation can be collected. However, the development of the FSC has been limited by the performance of fluorescent dyes. By using quantum dots (QDs) the unrealised concept of a FSC can be turned into a practical concentrator[3]. The size-dependence of the absorption threshold, red-shift dependence on the spread of dot sizes and their luminescent efficiency, make QDs ideal replacements for organic dyes in a new type of fluorescent concentrator[29]. Apart from the effect of the solar cells' efficiency, the efficiency of the total system is mainly determined by the properties of the dots and the matrix. The most important features are the quantum yield and the overlap of the absorption and emission bands of the dots, in addition to high transparency in the matrix and a good quality in the finish of the surfaces. Also important are the matrix properties, for example absorption coefficient and extinction coefficient of the matrix as shown in the presented results.

A three-dimensional raytrace technique, providing a model to optimise the design of a QDSC has been developed. The developed model includes reflection, refraction and absorption of solar radiation allowing the prediction of the optical efficiency of a QDSC system. Using the developed model, a parametric analysis was performed and the optical properties of a selected system optimised. The shift in the emission spectra after a number of intersections with a QD has been shown assuming; (1) isotropic energy input; (2) a laser input[17]; (3) discretised solar irradiance spectrum for air mass 1.5[18]. The shift in the luminescence spectra has been clearly illustrated. If the QDs can retain their high quantum efficiency and be incorporated into a suitable transparent media[3] with a low extinction coefficient and achieving a short pathlength, QDSC, which perform in the upper range of efficiency predicted for dye concentrators[4] may be possible.

Acknowledgements

This project is supported by the UK Engineering and Physical Sciences Research Council, B.P. Solar and Saint-Gobain. The collaboration is gratefully acknowledged of Professor Keith Barnham, Dr Amanda Chatten and Dr Ned Ekins-Daukes of Blackett Laboratory, Physics Department, Imperial College of Science, Technology and Medicine, London, Professor Paul O'Brien and Dr Azad Malik, Department of Chemistry, University of Manchester, Manchester.

References

1   Schmela, M., 1999. BIPV, *Photon Magazine* Apr-May.
2   Eames, P. C., Zacharopoulos, A., McLarnon, D., Hyde, T. J., Norton, B., Development and experimental characterisation of low cost façade integrated concentrator photovoltaics, *20/20 Vision, CIBSE/ ASHRAE Conference*, Dublin 2000.
3   Barnham, K. W. J., Marques, J. L., Hassard, J. and O'Brien, P. 2000. Quantum-dot concentrator and thermodynamic model for the global redshift, *Applied Physics Letters*, Vol. 76, No.9, pp. 1197-1199.
4   Goetzberger, A., Stahl, W. and Wittwer, V. 1985. Physical limitations of the concentration of direct and diffuse radiation, *Proceedings of the 6th European Photovoltaic Solar Energy Conference*, pp. 209-215.
5   Wittwer, V., Stahl, W., Goetzberger, A., 1984. Fluorescent Planar Concentrators, *Solar Energy Materials*, Vol. 11, pp. 187 – 197.
6   Goetzberger, A. 1978. Fluorescent Solar Energy Collectors: Operating Conditions with Diffuse Light, *Applied Physics*, Vol. 16, pp. 399 - 404.
7   Goetzberger, A. and Greubel, W. 1977. Solar Energy Conversion with fluorescent Concentrators, *Applied Physics*, Vol. 14, pp. 123 - 139.
8   Alivisatos, A.P., 1998. Electrical Studies of Semiconductor-Nanocrystal Colloids, *MRS Bulletin*, pp. 18 - 23.
9.   Micic, O. I., Cheong, H. M., Fu, H., Zunger, A., Sprague, J. R., Mascarenhas, A., Nozik, A. J. 1997. Size-Dependent Spectroscopy of InP Quantum Dots, *Journal of Physical Chemistry B*, Vol. 101, pp. 4904 - 4912.
10   Blanton, S. A., Hines, M. A., Guyot-Sionnest, P., 1996. Photoluminescence wandering in single CdSe nanocrystals, *Applied Physics Letters*, Vol. 69, No. 25, pp. 3905 - 3907.
11   Batchelder, J. & Zewail, A., 1979. Luminescent Solar Concentrator. Theory of Operation and Techniques for Performance Evaluation, *Applied Optics*, Vol. 18, pp. 3090.
12   Levitt & Weber, 1977. Materials for luminescent greenhouse solar collectors, *Applied Optics,* Vol. 16, No. 10, pp. 2684 – 2689.
13   Goetzberger, A., Wittwer, V. 1979. Fluorescent Planar Collector-Concentrators for Solar Energy Conversion, *Advanced Solid State Physics*, Vol. 19, pp. 427 – 451.
14   Batchelder, J. & Zewail, A., 1981. Luminescent Solar Concentrator. Experimental and Theoretical Analysis of their Possible Efficiencies, *Applied Optics,* Vol. 21, pp. 3733.

15 Duffie, J. A. and Beckman, W. A., 1991. *Solar engineering of thermal processes*, John Wiley and Sons, New York.

16 Fu, H., Zunger, A., 1996. InP quantum dots: Electronic structure, surface effects and the redshifted emission, *Physical Review B*, Vol. 56, No. 3, 1496-1508.

17 Chatten et al., 2001. *Private communication*

18 ISO9845-1, 1995. *Solar energy-Reference solar spectral irradiance at the ground at different receiving conditions.*

19 Soti, R., Farkas, E., Hilbert, M., Farkas, Zs., Ketskemety, I., 1996. Photon transport in luminescent solar concentrators, *Journal of Luminescence*, Vol. 68 pp. 105 - 114.

20 Saint Gobain, 2001. *Private Communication.*

21 BP Solar, UK, 2001. Private Communication.

22 Mansour, A. F., 1997. Optical Efficiency and Optical Properties of Luminescent Solar Concentrators, *Polymer Testing, Vol. 17, pp. 333 - 343.*

23 Mansour, A. F., 1998. Outdoor Testing of Luminescent Solar Concentrators in a Liquid Polymer and Bulk Plate of PMMA, *Polymer Testing*, Vol. 17, pp. 153 - 162.

24 Bakr, N. A., Mansour, A. F., Hammam, M., 1999. Optical and Thermal Spectroscopic Studies of Luminescent Dye Doped Poly(methyl methacrylate) as Solar Concentrator, *Journal of Applied Polymer Science*, Vol. 74, No. 14, pp. 3316 – 3323.

25 Salem, A. I., Mansour, A. F., El-Sayed, N. M. & Bassyouni, A. H., 2000. Outdoor testing and solar simulation for oxazine 750 laser dye luminescent solar concentrator, *Renewable Energy*, Vol. 20, pp. 95 – 107.

26 Sidrach de Cardona, M., Carrascosa, M., Meseguer, F., Cusso, F., Jaque, F., 1985. Outdoor evaluation of luminescent solar concentrator prototypes, *Applied Optics*, Vol. 24, No. 13, pp. 2028-2032.

27 Sick, F. & Erge, T., 1996. *Photovoltaics in Buildings – A Handbook for Architects and Engineers*, International Energy Agency, Paris.

28 Hill, R., 1996. Major Markets for PV Over the Next 10 Years. *Proceedings of the International Conference – Photovoltaics: Clean Energy Today*, ETSU, London.

29 Barnham, K. W. J., Ballard, I, Bushnell, D. B., Chatten, A. J., Connolly, J. G., Ekins-Daukes, N. J., Kluftinger, B. G., Rohr, C., Mazzer, M., Hill, G., Roberts, J. S., Malik, M. A., O'Brien, P. O., 2001. Future Applications of Low Dimensional Structures in Photovoltaics, *Photovoltaics for the 21st Century.*

MICROSYSTEMS FOR THE CONTROL OF CABLE VIBRATION

Jan G. Korvink[1], F. Braun[2] and M. Schlaich[3]

[1]Institute for Microsystem Technology, Albert Ludwig University, Freiburg, Germany
[2]Diploma student, ILEK, University of Stuttgart, Germany
[3]Schlaich, Bergermann and Partners, Stuttgart, Germany
Email: korvink@imtek.de, m.schlaich@sbp.de

1   ABSTRACT

In this article we consider the suspension cables of bridges. Vibrations of bridge cables induced by wind, rain and traffic are the cause of fatigue and comfort problems in these structures. Conventional measures to eliminate or at least to reduce vibration phenomena are large macro systems such as hydraulic dampers, coupling devices, or surface spirals which are attached to the cables to change their dynamic behavior. These systems, however, are only a compromise as they are not only costly, but also increase the wind resistance of the cables and constitute a visual obstruction. Future systems to tackle the problem could involve the application of microsystems such as micro turbulence generators, micro flaps and nano structured surface coatings. We believe that such devices, which are hardly visible to the naked eye, can lead to the same effect as macrosystems, when they are applied to the cables in sufficiently large quantities[1].

The paper summarizes mechanisms of cable excitation. We briefly first review the history of cable damping devices in bridge design. This is followed by a look at current cable control mechanisms. Then, looking to the future, the potential of different types of microsystems to control vortex shedding is considered. The possible effects of micro vortex generators, which affect the boundary layer along the cable surface, have been studied in some more detail and initial findings will be presented.

2   HISTORY

The study of cable vibrations has a long history, which we briefly outline here. It appears that the first experiments on cable vibrations were performed by Mersenne in 1636, followed by the work of Noble and Pigot in 1667 describing cable modes and the positions of nodes. In the period 1732-1755 the Swiss scientist Daniel Bernoulli placed the theory of transverse cable vibrations on a solid footing, by expressing solutions of cable modes by Bessel functions, and demonstrating the independence of cable modes. In 1760, Lagrange derived the general equations of motion of a discrete system, and in 1788 described a procedure to discretize a massless string continuum. In 1820, Poisson derived the general partial differential equation of the string, and in 1843 Duhamel produced analytical solutions of the partial differential equation. In 1851 Rohrs and Stokes produced approximate solutions for the transverse vibration of a stretched string with small sag, and then, in 1868, Routh found an exact solution for this case. He extended the analysis to longitudinal vibrations, and to antisymmetric mode shapes. The next important result comes in 1941, when Rannie and von Karman described the symmetric and antisymmetric vibrations of a stretched cable in 3D, extended a year later by Klšppel and Lie to include prestrain. The first treatment of cables with large sag was done by Pugsley in 1949, and a first treatment of suspension bridges by Bleich et al. in 1950, with an analytical solution published in 1953 by Saxon and Cahn. From this point the publication density grows considerably. We note three recent advances only. In 1978 Irvine treated the skew cable, extended in 1984 by Triantafyllou to a general solution for skew, extensible cable with dynamic tension changes. In 1995 Kovacs described the influence of longitudinal vibrations on the swing susceptibility of stays. A summary of these theories and the appropriate references can be found in[2].

3   MECHANISMS OF CABLE EXCITATION

Cables are caused to vibrate by a number of environmental influences, which we now briefly consider.

3.1 Dynamic Vortex Shedding

Von Karman vortices are shed downstream from a cable that is subject to a wind stream, and these cause a fluctuating pressure perpendicular to the flow direction to be exerted on the cable (see e.g. the illustrations in Figure 2). If the vortex shedding frequency is close to the resonance frequency of the cable, these two effects can synchronize. It should be noted that the vortices are not uniformly distributed along the cable length. The position about the cable circumference at which the vortices are shed depends on the Reynolds number of the flow. For sub-critical flows, i.e. , Re $\leq$3 5 $\times 10^5$ the vortices shed at an angle of 80° from the direction of the inflow. In the super-critical domain where 5×10^5 $\leq$Re $\leq 3 \times 10^6$, we observe angles of 134° – 140°. For the trans-critical zone, Re $> 3 \times 10^6$ the shedding angle lies in the range 100° -120°.

3.2 Galloping

An irregularity in the cross section al shape of a cable can lead to a self-excited galloping behavior. This can, for example, arise due to ice or rainwater that changes the ideal profile of the cable. The cable experiences a combination of transverse force and torque, and its deflection response dynamically lock in at a characteristic frequency. The movement mimics the gallop of a horse. This phenomenon appears at relatively high wind speeds, where the fluid damping is also large, so that it does not represent a serious problem.

3.3 Wake Galloping

Any structure downstream from a vortex shedding feature will experience its vortices as a source of fluctuating pressure distributions, which in turn causes oscillations and galloping to set in.

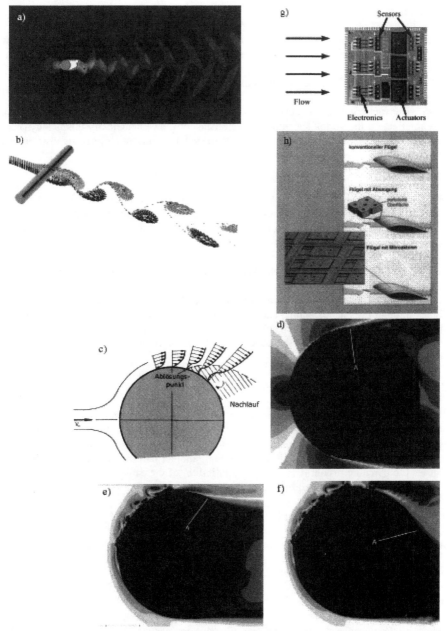

Figure 2 *Vortices around flow structures. a) and b) Simulation of vortex shedding from a round flow obstruction. c), d), e) and f) Simulation of vortex generation to control flow separation. g) and h) Microsystems for vortex shedding control. Picture credits: a),b) Phoenix CFD c)* [3]*, d),e),f)* [2] *g)* [4] *h)* [5]*.*

3.4 Buffeting

Nearby structures such as walls, large masts and of course buildings can induce flow irregularities that impinge on the bridge cables an in turn, cause fluctuations.

3.5 Rain and Wind Induced Vibrations

Rain water running down a cable causes a flow shape change of the cable, which in turn has an effect on the vortex shedding characteristics of the cable. Clearly this is a dynamic process, for the liquid will constantly adapt its shape to the current air pressure distribution.

3.6 Drag Crisis

At a Reynolds number of around the drag coefficient drops dramatically from a value of ca. to . If the wind speed fluctuates around this point, the cable experiences large force fluctuations in the direction of wind flow. For a cable diameter of , this phenomena is observed for wind speeds in the range .

3.7 Excitation of Anchors by deck movement or live load-induced vibrations

Varying traffic loads (trains and motorized transport) also cause fluctuating forces on the cables through the anchor points, and typically the excitation amplitudes drop with rising traffic speed. In addition, the wind produces forces on other bridge components, for example self-excited flutter on the deck, which in turn is transferred to the cables through the anchor points.

3.8 Parameter-Induced Vibrations

If the bridge structure has a resonant mode whose natural frequency is a multiple of the suspension cableÕs natural frequency, these two effects can strongly couple. It is called parameter resonance because a parameter that determines the resonant frequency (here the cable tension) is a function of a vibration mode of another part of the structure. This effect can lead to dynamic instabilities, i.e., the equilibrium of the cable becomes instable. Note that this effect is not described by linear vibration analysis.

3.9 Parameter Double Resonance

Axial resonances in the bridge deck cause strong cable vibrations. In particular, two adjacent high-order modes can be simultaneously excited because not only the cable force, but also its anchor position is fluctuating.

3.10   Ice Induced Impulse

When ice forms on a cable, the ice can dislodge in response to cable vibrations. This causes a momentum transfer, which is transferred to the rest of the structure. Power cable masts are particularly susceptible to damage due to the rather high torsional forces that can arise. It is notable that the seeding of ice formation is a function of surface roughness; ice formation on smooth surfaces is rather unlikely.

4   CONTROL OF CABLE VIBRATION

Cable vibration control measures are principally subdivided into passive methods, which remain in place in an unchanged form, and active measures, which can adapt to the level of excitation. The active measures are typically smaller in size and have a much wider range of applicability and hence a higher damping effect than passive systems. Many passive measures have a further disadvantage, namely, that they are present independently of the level of excitations, and that they for example always result in a higher level of wind resistance. In Figure 1 we show the currently available and future control possibilities, where the terms are now defined: Environment signifies the influence on cable aerodynamics due to the presence of nearby bodies, for example a plate; Shape refers to the cross-section of the cable and Surface refers to the aerodynamic properties of the surface of the cable; The term Continuum refers to either the entire cable or at least its sheathing. With Boundary Condition we refer to either the anchor points of the cable or to cable discontinuities such as stabilizers, dampers or weights.

Active control measures can be divided into macroscopic and microscopic systems. Active Macrosystems may contain microsystem components, such as sensors and control electronics, but are characterized by macro components such as jacks, dampers and motors that are placed at discrete positions in the structure. By contrast, Active Microsystems are to be seen as integrated units comprising micro components such as sensors, control and actuators, spread over a large area to yield a macroscopic effect. Thus the work done per unit is rather small, yet through massive parallelism over a large area they can compete with their macrosystem counterparts. We will discuss these systems in the next section.

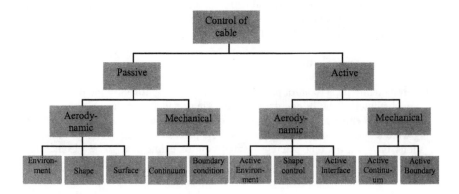

Figure 1 *Control mechanisms for cable vibrations.*

4.1 Passive Aerodynamic Control

Many passive measures exist to control the aerodynamics of cables. For example, by structuring the surface of the cable sheathing, much as is done for golf balls, it is possible to gain a tremendous reduction in wind resistance. Another method is to add small ribs along the length of the cable. Both of these measures help to diminish the influence of rainwater runoff on wind excitation of the cable. Other measures protrude further into the flow stream, such as elliptic plates, scruton spirals and splitter plates, whereby the latter is strongly direction dependent.

4.2 Passive Mechanical Control

Passive mechanical measures to control cable vibrations employ one or more of the following methods:

- Increasing the structural damping
- Enlarging the structural inertia
- Modifying the cable eigenfrequency
- Changing the system (introducing additional compliant nodes)

The methods directly or indirectly modify the modal shapes, the maximal amplitudes of vibration and the resonant frequencies.

Cross-ties generally modify the eigenshapes and eigenfrequencies of a cable through the introduction of additional, elastically-supported nodes. In addition, they introduce a significant amount of extra structural damping, since the ties also dissipate energy into the flow stream. Interference galloping is suppressed by tieing together parallel running cables. Damped couplers introduce additional damping, and so do chain dampers, whereby a chain is loosely suspended from the length of the cable at a number of points, dissipating energy through friction among the chain links. The Stockbridge damper is a short section of cable with a mass at each end that is tied at its midpoint to a cable. It has both a damping and spring-like action. Fluid dampers are cylinders wrapped around the cable. The cylinders have internal flow restrictions, which cause the contained liquid to dissipate energy during flow. Neoprene inserts at cable anchors are mainly intended to reduce bending loads, but also contribute to damping. Internal friction dampers consist of two parts, a fixed tube, and a movable tube that is connected to the cable. Friction between the tubes, which is caused by cable vibrations, dissipate a portion of the energy. External hydraulic dampers are also in use, and are otherwise unwanted visual obstructions.

4.3 Active Aerodynamic Control

Movable aerodynamic flaps (control panels) can be used to significantly increase the wind speed at which flutter of the bridge deck sets in. These panels, similar in shape to wing flaps, are placed on either side of the bridge deck, and can be moved in response to the impinging wind strength. They do not appear to be a practical solution for cables, though. Investigations have shown that placing a small control cylinder at some distance to a cableÕs surface and at an angle of or to the direction of wind incidence strongly influences vortex shedding. Again, this appears to be a technically inviable solution.

4.4 Active Mechanical Control

Active dampers have been developed for the automobile industry, whereby a magneto-rheological fluid is electrically activated to increase stiffness and dissipation. Because of the small time constant, one could imagine connecting such dampers with cables to achieve load dependent control of vibrations. A further possibility is to actively move the attachment point of a cable on the bridge deck, thereby modifying the mode frequencies of the cables. It should be noted that circular modes, whereby the cable moves circularly about its longitudinal axis, cannot be influenced by such a setup.

5   MICROSYSTEMS FOR CABLE VIBRATION

5.1 Autonomous structures

Bridges and other structures are designed to be passive, and in such a way that they can withstand a wide range of loads. In addition, they are designed to remain permanently in place, regardless of the loading conditions.

In contrast, autonomous structures are adaptive, intelligent and independently active. They consist of distributed, decentrally organized and intelligent components that communicate among one another and that can make independent decisions. To achieve real autonomy, they should also be able to learn from experience, and supply their own energy. Table 1 summarizes the various concepts.

Table 1: Classification of Systems.

System	Sensor	Actuator	Control	Processor	Energy Source & Store
Autonomous	●	●	●	●	●
Intelligent	●	●	●	●	
Adaptive	●	●	●		
Active	●	●			
Sensorial	●				
Passive					

5.2 Adaptronics

Adaptronics describes the co-integration of sensorial and actuatorial functions in one structure or material. If both sensors and actuators are present, they communicate via a controller with each other. For cable vibrations caused by vortex shedding, the interplay of these capabilities can be viewed as happening on three levels:

- Intelligent surface (smart surface - microsystem). Suppression of surface forces by controlling the boundary layer through momentum transfer, sucking away or changing the surface geometry. *The system controls the environment.*
- Intelligent continuum (smart continuum - microsystem or macrosystem). Control of the cable continuum by distributed muscles through piezoelectric or shape-memory alloys. *The system controls the internal stiffness.*
- Intelligent system (smart system - macrosystem). Control of damping through a single piezoelectric or hydraulic actuator. *The system changes the external boundary conditions.*

5.3 Microsystems

Microsystems are greatly miniaturized units usually made up of sensors, actuators and circuitry. Miniaturization is strongly driven by advances in microelectronic lithography, and indeed many microsystems today are produced by modified microelectronic manufacturing processes such as the CMOS (Complimentary Metal Oxide) low power electronics process. The modifications involve additional micro structuring post-processing steps, and include deposition steps of materials that are not tolerated in the usual CMOS process. But many more manufacturing techniques exist, and range from the miniaturization of traditional mechanical engineering processing (milling, turning, drilling, injection moulding) all the way to special lithography techniques using x-ray sources. A key feature of most micro processing steps is the way cost and effort scales with the number of devices produced. The processes are highly parallel, meaning that many devices are created at the same time, making micro processing ideal for large volume production.

Microsystems have tremendous technical advantages. Since it is usual to co-integrate sensors, actuators and control circuitry, we get some useful by-products. Each microsystem can act autonomously, and can interact with the decisions of other microsystems, resulting in a sophistication of response that can be made more than the sum of the individual actions. The close proximity of sensors, actuators and control units can result in faster and more precise responses than is possible using macrosystems with the same functionality. Finally, microsystems are predestined to be applied over large surfaces, and with their feature size being very small, they promise "invisibility", yet macroscopic effect. One is tempted to talk of "intelligent" surfaces. We will now take a closer look at how microsystem technology can have an effect on cable vibrations.

5.4 Force Actuators

Many macroscopic actuation principles are available at the micro scale, such as shape memory alloys, piezoelectric actuators, pneumatics, hydraulics and thermomechanics. Yet the specific force produced by an individual micro actuator is probably never large enough to counter the forces arising in bridge cables. We expect their strength to come from influencing the flow around the structural members. By reducing skin friction, and thereby reducing the force acting on the body, and by controlling the frequency of vortex shedding, which in turn changes the resonant coupling of wind forces with structural response, we believe that microsystems could have a tremendous effect on the way cable-stayed structures behave. Naturally, force actuators can be inverted and used as power generators, much as automatic winders in watches work. This raises the possibility of making systems equipped with such devices self-sufficient and eliminating the need for complex interconnection and power supply.

5.5 Flow Actuators

Recent research results, mainly in the USA, of micro actuators used to disturb the boundary layer on aircraft wings (up to supersonic frequencies) have shown that such devices have the potential to greatly reduce wing drag and hence both improve fuel consumption and increase the stability of the aircraft in turbulent air. The microsystems contain micro actuators that only affect the immediate boundary layer. These are either small flaps, bulging membranes, or zero mass flow rate membrane pumps that generate vortices which enter the flow stream. By carefully controlling the point of flow separation on an individual cable, one not only reduces drag, but reduces (and possibly eliminates) the oscillating cable forces caused by vortex shedding.

5.6 Sensors

Using exactly the same manufacturing technology, it is possible to create sensors that measure the flow response around a micro actuator. These sensors are based on traditional hot-wire anemometry, where the transport through the air of heat generated in a hot wire is measured and used to characterize the airflow speed. In addition, the technology can be used to create micro-accelerometers that make a point-wise measurement of the acceleration response of the cable. The tremendously short time constant of these sensors (they are use to make crash decisions in cars) make them virtually instantaneous when typical wind-induced structural responses are considered. Other currently available micro sensors include the effects of strain, temperature, surface fluid shear and air pressure, each with their uses. We foresee a scenario where actuators equipped with these sensors can autonomously, and interactively with other similar systems, produce an orchestrated response that intelligently counteracts the effects of wind on the underlying structure.

5.7 Intelligent Surfaces

Active systems certainly produce questions of complexity, reliability and of course energy supply. Micro structuring technology opens up an avenue of mimicking biological systems in reducing skin drag, with surface microstructure producing much the same effect as the dimples on golf balls or the skin structure of a shark. Such an effect is passive, in that it does not require any power supply.

6   SUMMARY

It may well be that an intelligent surface for cables requires a combination of active and passive systems. One can certainly imagine adaptive microstructures, with surface texture adapting in real time to impinging wind loads.

Acknowledgements

The authors wish to thank Dr. Evgenii B. Rudnyi for critically reading the manuscript.

References

1   M. Schlaich, J. G. Korvink, *Autonomous Bridges*, Proceedings of fib/IABSE Bridge Engineering Conference 2000, Sharm El Sheikh, Egypt, 2000
2   F. Braun, *Kontrolle von Seilschwingungen mit Mikrosystemen*, Diploma thesis, University of Stuttgart (2001) (Translation: *Control of Cable Vibrations with Microsystems*)
3   Ruscheweyh, *Dynamische Windwirkung an Bauwerken - Band I, Grundlagen Anwendungen*. Bauverlag GmbH, Wiesbaden und Berlin, 1982
4   Y. C. Tai, C. M. Ho, Micro-Electro-Mechanical-Systems (MEMS) and Fluid Flows, Ann. Rev. Fluid Mech., 1998, 579-612
5   Spektrum der Wissenschaft - Spezial 4: *SchlŸsseltechnologien*. Spektrum-der-Wissenschaft-Verlagsgesellschaft, Heidelberg, 1995

CARBON NANOTUBES AND THEIR APPLICATION IN THE CONSTRUCTION INDUSTRY

J.M. Makar and J.J. Beaudoin

Institute for Research in Construction, National Research Council Canada
1200 Montreal Road, Ottawa, Ontario K1A 0R6 Canada

1   INTRODUCTION

Carbon nanotubes (CNT) are the subject of one of the most important areas of research in nanotechnology. Their unique properties and potential for valuable commercial applications ranging from electronics to chemical process control have meant that an enormous amount of effort has been undertaken on the investigation of nanotubes in the last five years. Despite this high level of research activity, very little attention has been paid to potential applications in the construction industry. This paper seeks to bridge the gap between CNT and construction materials research. It describes carbon nanotubes, including their structure, how they are produced and their properties. Potential applications, both in general and specifically for the construction industry, are presented. The paper concludes by giving the results of initial research at the National Research Council's Institute for Research in Construction on carbon nanotube/cement composites.

2   TUBE STRUCTURE

Carbon nanotubes can be visualized as a modified form of graphite. Graphite is formed from many layers of carbon atoms that are bonded in a hexagonal pattern in flat sheets, with weak bonds between the sheets and strong bonds within them. A CNT can be thought of as a sheet or sheets of graphite that have been rolled up into a tube structure. CNT can be single walled nanotubes (SWNT), as if a single sheet had been rolled up, or multiwalled (MWNT), similar in appearance to a number of sheets rolled together. Figure 1 shows a schematic of a single walled nanotube.

CNT can be produced with different types of chirality (the orientation of the hexagons formed by the carbon atoms with respect to the tube axis). Two tubes with the same diameter may therefore have different structures despite being formed solely of carbon atoms. CNT are described by their chiral vector as (m,n) tubes, where m and n are integer numbers (Figure 2). A (m,0) tube {e.g. (10,0)} is described as an armchair tube since the hexagons in the tube run straight along its length in a manner similar to the arms of an armchair, while a tube with m=n {e.g. (5,5)} is called a zig-zag tube, since the hexagons zig-zag down the length of the tube. The nanotube shown in Figure 1 is also a zig-zag

tube. Tubes may also have chirality between these two extremes, such as the (5,4) tube
that would be formed by rolling up a tube with the vector shown at the bottom of Figure 2.

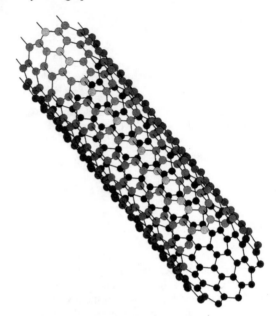

Figure 1 *Schematic of a Single Walled Nanotube*

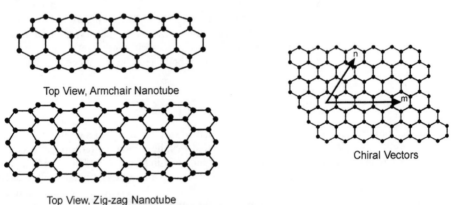

Top View, Armchair Nanotube

Top View, Zig-zag Nanotube

Chiral Vectors

Figure 2   *Nanotube Chiral Vectors and Top Views of Schematic Tubes*   *(The
                  black dots are carbon atoms)*

3   SYNTHESIS

Three different approaches have been taken to produce CNT.[1]   The first method to be
discovered uses an electric arc-discharge, where a high voltage electrical current is passed
through the air (or an inert or reactive gas) into a carbon electrode. The other methods are

laser ablation, using a high intensity laser beam directed at a carbon target, and chemical vapour deposition, which uses a carbon based gas such as methane at high temperatures and sometimes pressures. In all three cases free carbon atoms are obtained and given the energy necessary to form carbon nanotubes rather than graphite or amorphous carbon. Although MWNT can be produced without the presence of a catalyst, SWNT generally require the presence of a transition metal catalyst to form. The catalyst is typically present in the form of nanoparticles. Individual carbon atoms are believed to enter the particle and diffuse through it to its surface, where they join to the growing tube or tubes.[2] Depending on the shape and size of the particle and the surrounding environmental conditions, SWNT or MWNT will form. The final length of CNT also depends on growth conditions. Further details on the synthesis of CNT can be found in Dai.[1]

4 PROPERTIES

CNT have a number of unique properties. Their electronic behaviour varies depending on their chirality.[5] Armchair tubes are metallic, while other tubes are semiconductors. In addition, the degree of conductivity or semiconductivity can be controlled by doping. As an example, recent research[6] has shown that the presence of oxygen can dramatically affect the conductivity of CNT. Changes in size[7] or mechanical deformation[8] have also been shown to affect their electronic properties. CNT are also high quality field emitters. Placing a carbon nanotube in a strong electric field will cause electrons to be emitted at high efficiency without damaging the tubes.[9] A summary of the electronic behaviour of carbon nanotubes can be found in Louie.[5]

Mechanically, CNT appear to be the strongest material yet to be discovered. Experimental results have shown that they have moduli of elasticity that exceed 1 TPa in value.[10] Measurements of ultimate strength and strain have been more difficult to make, but measurements of the yield strength of SWNT of 63 GPa have been reported[11] as well as yield strains are on the order of 6%.[12] CNT are also highly flexible, being capable of bending in circles or forming knots. Like macroscopic tubes, they can buckle or flatten under appropriate loadings.[13] Yakobson and Avouris[14] summarize CNT mechanical behaviour.

Less work has been done on the thermal behaviour of CNT, but theoretical work[15] as well as experimental measurements performed on CNT suspended in liquids[16] suggest that their thermal conductivity may also be remarkably high, approaching the theoretical limit for carbon materials. If these estimates are accurate, CNT are the most thermally conductive material at room temperatures yet discovered. An additional feature is that the thermal conductivity of CNT is believed to be much higher along the tubes than across them, creating the potential for materials that have anisotropic heat conduction properties. However, direct measurements on CNT have yet to confirm these predictions.[17]

5 APPLICATIONS

Key areas of existing and potential CNT applications include electronics, sensors, structural materials, fillers and storage materials. The most highly developed commercial application for this material is the use of MWNT as a filler material in plastic composites and paints[18], sometimes as an improved substitute for carbon black. This market has been identified as having a multibillion dollar value.

The ability of CNT to be both metallic and semi-conductor with a change in structure, rather than a change in composition, creates significant possibilities for nanoscale electronics. Different types of CNT transistors and logic gates have already been demonstrated[19] and integrated circuits are under development. Ajayan and Zhou[9] describe the potential for electronic applications in more detail. A related application, and likely one of the first to go to market, is as emitters in field effect displays. The high efficiency and reliability of CNT field emitters is expected to lead to much cheaper displays with superior performance.[18]

The mechanical behaviour of CNT has created great interest in their use as structural materials. Potential uses of CNT ropes are still speculative and are discussed in more detail below as possible construction industry applications. Carbon nanotube composites appear to be more practical in their potential for shorter term development as the length of tube that is required is much closer to that which can currently be synthesized. CNTs are also close to ideal reinforcing fibers due to their strength and ultra high aspect ratios. As a result, considerable attention has been paid to research in this application, with a particular emphasis on polymer composites. Unfortunately, composite behaviour to date has not met performance expectations. There are two problems, which are common to all CNT composite applications. First, there is the question of dispersing the CNT into the matrix material. Dispersion is much more complex than simply mixing a powder of nanotubes into the liquid matrix material. Carbon nanotubes tend to adhere together after purification due to Van der Waal's forces, making an even distribution of individual tubes particularly difficult to achieve. This challenge has been met by a variety of methods. While initial work[20] required functionalizing the tubes, later research has only required the use of surfactants in combination with sonication (i.e. using a pzeio-electric system to deliver acoustic energy to nanotubes through the liquid medium).[21] The sonic energy breaks up the nanotube bundles and disperses the tubes, while the surfactant helps to ensure they remain dispersed. The dispersed tubes are generally mixed into the matrix under continued sonication. The second problem is achieving suitable CNT-matrix bonding. Typical CNT polymer composites experience fibre pullout under low loads and do not, as a result, achieve high strengths. Research in overcoming this problem continues.

Attempts have also been made to develop CNT/metal[22] and CNT/ceramic composites.[23] Similar problems with performance as those in polymer composites have been encountered, along with additional complications due to the higher temperatures needed to sinter the matrix materials. However, CNT/aluminum composites do not seem to experience the carbide formation seen in carbon fibre composites.[22] A very recent paper has described a successful technique for producing a strengthened alumina ceramic.[24] In this work, an alumina/ethanol slurry is added to CNT dispersed in ethanol. The resulting powder was then sieved and ball milled before spark sintering. The initial stages produce evenly distributed CNT, while the sintering method ensured a fully dense material while maintaining nanometric grain sizes in the alumina and avoiding damaging the CNT. This approach appears very promising for the development of future ceramic composites.

There are a number of other significant areas of application, including as sensors, for the storage and control of catalysts and as atomic force microscope tips. A less successful development has been in the area of hydrogen storage. While initial work was promising with very high storage by weight, more recent research has suggests that CNT are unlikely to be effective storage devices.

The extent of applications for CNT will depend on improvements in synthesis methods. Reducing costs is the most crucial factor, but the ability to regularly produce tubes with high lengths, tubes with specific chirality, individual tubes or ropes that are highly commensurate, and tubes with specific electronic properties will all be necessary for different commercial applications. Currently it appears that field emission displays are likely to be the first wide spread use of CNT beyond the current application of MWNT as a filler material. Other applications will depend both on further research on the applications itself and substantial improvements in CNT synthesis methods.

6 APPLICATIONS OF CARBON NANOTUBES IN THE CONSTRUCTION INDUSTRY

While many of the CNT applications developed for other industries will also find uses in construction, there are at least three broad areas of research that will lead to products intended specifically for the construction industry. These areas include CNT composites made with existing construction materials, CNT ropes for use as structural components and CNT heat transfer systems. The first applications are likely to be in CNT composite materials. As noted earlier, CNT are excellent reinforcing materials because of their extremely high strength, toughness and aspect ratios. Polymer, cement and glass are all potential candidates for CNT matrix materials. Current work on polymer/CNT composites has been discussed above, while cement/CNT composites will be covered in more detail in the following sections. Glass reinforced with CNT or other nanofibres are of interest due to the possibility that nanofibres or tubes may be able to provide reinforcement without interfering with light transmittance. However, little work has been done on this aspect of the optical behaviour of nanofibres and the success of this approach remains to be seen.

The production of longer CNTs that can be formed into ropes would create obvious possibilities for applications such as suspension bridges, where CNT strengths and moduli of elasticity would allow for the design of significantly longer spans than existing technology makes possible. Similarly, the use of CNT ropes can be envisioned in improved pre- or post- tensioned concrete structures. Carbon nanotubes have also been discussed as materials for the construction of very large, space based structures, including space elevators.[25] These cable systems have the theoretical capability of reaching from the earth's surface to far beyond geosynchronous orbit. Elevator cars running simultaneously up and down the cable would allow cargo and goods to be transferred from space to the earth and vice-versa with minimal energy requirements. While it remains to be seen whether such structures can be engineered and constructed, CNT appear to be the only material capable of bearing the immense structural loads that would be required.

The thermal conductivity of carbon nanotubes presents other applications. Improved thermally resistant composite materials may be possible, since a sufficient density of carbon nanotubes could allow heat to be conducted rapidly away from the contact surface to heat sinks. In the longer term, it may also be possible to develop insulating materials and heat pipes, taking advantage of the differences in thermal conductivity across the tubes and along their lengths. One application would be in the heating of buildings through the replacement of existing liquid based systems for heating floors.

7 CEMENT SYSTEMS

Concrete is the single most important construction material. Composed of fine and coarse aggregates held together by a hydrated cement binder, it forms a part of most construction projects, ranging from house foundations through high-rise tower components to dams and highway bridges. Ordinary portland cement (OPC) is the most common form of cement binder used in concrete, although there are a large number of specialty forms of cement[26].

OPC is typically formed by grinding amorphous masses of cement clinker and gypsum into powder. The primary constituents of the clinker are a series of oxides, including tricalcium silicate (C_3S in cement chemistry notation), dicalcium silicate (C_2S), tricalcium aluminate (C_3A) and tetracalcium aluminoferrite (C_4AF). C_3S and C_2S are the most important constituents in OPC. These oxides, when mixed with water, undergo hydration reactions to form the solid cement binder. Direct observations of cement grains show that many of them have dimensions on the order of 5 to 30 μm. However, smaller particles are also present in OPC (Figure 3). The impact of these finer particles, if any, on the cement hydration process is not yet understood.

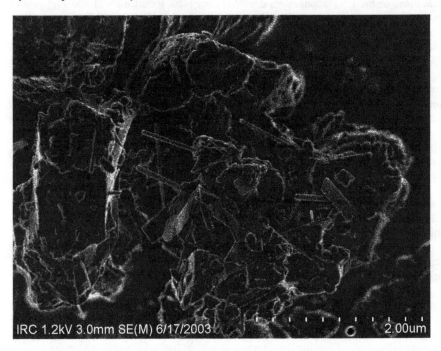

IRC 1.2kV 3.0mm SE(M) 6/17/2003   2.00um

Figure 3 *Small and intermediate size cement grains*

Hydrated cement is a brittle material that is much stronger in compression than in tension. Indirect measurements of tensile strength give values below 4 MPa, while the compressive strengths of the same concretes are more than an order of magnitude higher in value[27]. Various forms of reinforcement, typically in the form of rods or fibers, are added to concrete to compensate for its weakness in tension[28]. Pre- or post- tensioning of concrete beams and other structures is also used to provide improved mechanical performance of structures such as tanks, beams, nuclear reactor containment facilities, and water pipes. Other methods for improving the mechanical performance of concrete are needed to improve structure performance and develop new concrete applications.

The strength of concrete is dependent on a number of factors, which include, amongst others, the ratio of water to cement in the original mix of materials, degree and size of porosities present in the cement, the presence of micro-cracking in the binder and the quality of binding of the aggregate to the cement[29]. Cement itself has a complex, nanoscale structure[30]. Some of the properties that affect the strength of cement are expected to act at the nanoscale. However, this same nanostructure and the chemical processes that produce it open possibilities for using nanotechnology to modify cement and concrete behaviour.

Improved mechanical performance is one of the benefits expected to be obtained through the application of nanotechnology to cement systems. One approach to developing better performance is the addition of the nanoscale reinforcing materials, which might range from small spheres that would only act to interrupt cracking to nano-fibres or rods, which would act in a manner more similar to larger scale reinforcing systems. The work presented in the remaining part of the paper follows this route. Other approaches are also possible.

8   CEMENT AND CARBON NANOTUBES

CNT are expected to have several distinct advantages as a reinforcing material for cements as compared to more traditional fibers. First, they have significantly greater strengths than other fibres, which should improve overall mechanical behaviour. Second, CNT have much higher aspect ratios, requiring significantly higher energies for crack propagation around a tube as compared to across it than would be the case for a lower aspect ratio fibre. Thirdly, the smaller diameters of CNT means both that they can be more widely distributed in the cement matrix with reduced fibre spacing and that their interaction with the matrix may be different from that of the larger fibres. CNT, with their diameters being close in size to the thickness of the calcium silicate hydrate (C-S-H) layers hydrated cement, could show very different behaviour, including different bonding mechanisms. Finally, carbon nanotubes can be functionalized to chemically react with cement components, providing routes for other forms of interaction and cement system property control.

As with other CNT composites, the major issues to overcome in preparing high quality CNT/cement composites including distributing the CNT within the cement and obtaining suitable bonding between the two materials. One route is to disperse the CNT in a surfactant mixed with water or another solvent, as has commonly been used for polymer composites. There is an obvious correspondence with the use of superplasticizers to improve the workability of concrete. Initial research at the National Research Council Canada has shown that small amounts of CNT can be dispersed by sonication in water containing 5% superplasticizer (Eucon 37, Euclid Admixtures Canada, Inc.). However, it is not clear whether sufficient CNT can be dispersed in this medium to produce the 2-10% mass concentration of CNT that appears necessary for good mechanical performance in other ceramic composites. Further work is necessary to resolve this question.

A second route is to follow the approach taken in developing alumina/CNT composites. In this case the CNT are dispersed in ethanol under sonication. Initial results from this method are presented in this paper. An amount equal to 0.007 g. of commercially supplied CNT (1.4 nm single walled carbon nanotubes produced by Carbon Nanotechnologies Inc. and supplied in the purified form described as "Buckypearls") was sonicated at low power for 2 hours in ethanol. Ordinary portland cement powder (0.43 g) was then added to the liquid to form cement/CNT/ethanol slurry. The slurry was further sonicated for an additional 5 hours. The ethanol was then allowed to evaporate, leaving a fragile cement/CNT "paper", composed of carbon nanotubes bonded to cement grains and each other in a porous structure. It is worth noting that the sonication and evaporation process appears to affect the surface morphology of some cement particles, making them rougher in appearance. An additional effect of the sonication appears to be segregation of the particles, with finer cement particles being trapped in the CNT paper near its surface, while the larger particles settled to the bottom of the paper. Part of the cement/CNT paper was then ground using a mortar and pestle in order to break up the material and allow examination of individual grains of cement.

Both the paper and the ground powder were examined using a Hitachi S4800I cold field emission gun scanning electron electron microscope (SEM) at medium to high magnifications (25,000x to 200,000x). Figure 4 shows two images from the cement/CNT paper from near its outer surface. The scale bar in the images divides the scale into tenths of the indicated value, while the thickness of the lines that form the scale bar is a hundredth of the indicated value. In this region there is a porous structure and it is apparent that the sonication process has either not fully dispersed the carbon nanotubes, or that they have clustered together again during the evaporation process. The cement particles are attached both to each other and to multiple CNT. The central particle is coated with tube bundles of different diameters. It should be noted that individual CNT will be difficult or impossible to see in these images. While this microscope has the capability of resolving structures as small as 1.0 nm in size, it is more likely that the very fine structure, such as indicated by the arrow, are bundles of small numbers of CNT.

Figures 5 and 6 show grains of cement powder that have been well coated with CNT bundles after the paper has been broken apart by hand grinding. Figure 5 shows an image of a larger cement grain, while Figure 5 shows an image from agglomerations of smaller grains. In both cases large numbers of CNT can be seen on the grain surfaces. In figure 6 nanotubes can also be seen splitting off from larger bundles, along with loops and other structures. More nanotubes appear to be lying on the surface of the grain in Figure 5, while those on the smaller grains are more likely to be only anchored at one end.

Figures 3, 4 and 6 show different grain shapes, although all the cement grains in all three figures have a layered structure. Both the presence of layers in the unprocessed grains of Figure 3 and additional tests have shown that this layered structure is not due to the use of ethanol as a solvent. Instead, the sonication process used to distribute the carbon nanotubes appears to be breaking up some of the agglomerated cement grains. Shards from these grains are then caught at the top of the carbon nanotube paper as shown in Figure 4, while the coarser grains sink to the bottom of the paper. Additional experiments have shown that quality of the nanotube dispersion controls the size of the particles trapped at the top of the paper with smaller tube bundles trapping smaller cement particles. Grinding the nanotube paper frees up the larger particles but also allows clumps of smaller particles held together in part by nanotubes as shown in Figure 6 to form.

9   CONCLUSION

Carbon nanotubes are one of the most important materials under investigation for nanotechnology applications. Their unique properties, ranging from ultra high strength through unusual electronic behaviour and high thermal conductivity to an ability to store nanoparticles inside the tubes themselves has suggested potential applications in many different fields of scientific and engineering endeavour. As was the case with silicon transistor technology, these applications will grow in time as the capacity for industrial production and manipulation of CNT is created and as understanding of the physics of CNT continues to improve.

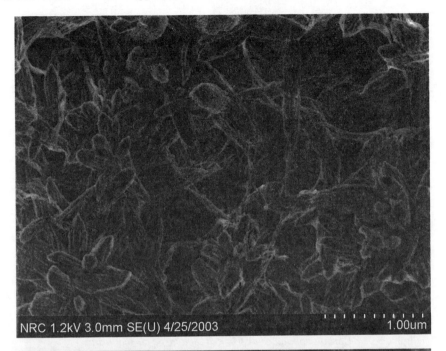

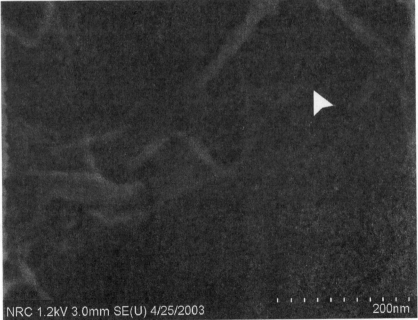

Figure 4   *Unhydrated CNT/cement "paper", showing sonicated cement grains*

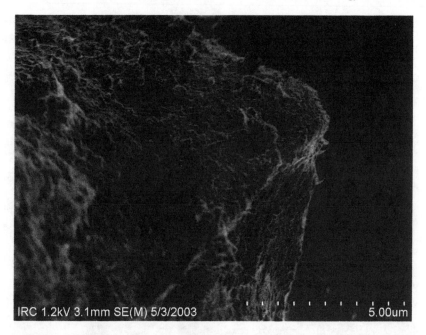

Figure 5   *Carbon nanotubes distributed on a large cement grain*

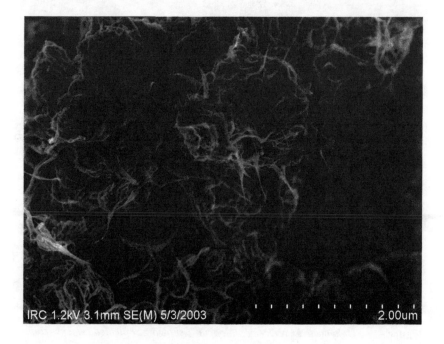

Figure 6   *Carbon nanotubes distributed on small cement grains*

Construction applications of CNT range from composite materials through high strength structural components to heat transfer technology. Cement and concrete CNT composites have particularly strong potential, since CNT act as a near ideal reinforcing material with diameters similar in scale to the layers in calcium-silicate-hydrate. Research has shown that it is possible to distribute CNT bundles across cement grains using an ethanol/sonication technique. Work is in progress to investigate the interaction between CNT and hydrated cement, and the mechanical performance of cement/CNT composites.

References

1. H. Dai, *Topics in Applied Physics*, 2001, **80**, 29.
2. J.-C. Charlier and S. Iijima, *Topics in Applied Physics*, 2001, **80**, 55.
3. J. S. Suh and J. S. Lee, *Applied Physics Letters*, 1999, **75**, 2047.
4. R. R. Schlittler, J. W. Seo, J. K. Gimzewski, C. Durkan, M. S. M. Saifullah, M. E. Welland, *Science*, 2001 **292**, 1136.
5. S. G. Louie, *Topics in Applied Physics*, 2001, **80**, 113.
6. P. G. Collins, K. Bradley, M. Ishigami, A. Zettl, *Science*, 2000, **287**, 1801.
7. M. Ouyang, *Science*, 2001, **292**, 702
8. P. J. de Pablo, C. Goméz-Navarro, M. T. Martinez, A. M. Benito, W. K. Maser, J. Colchero, J. Goméz-Herrero, and A. M. Basó, *Applied Physics Letters*, 2002 **80**, .
9. P. M. Ajayan and O. Z. Zhou, *Topics in Applied Physics*, 2001, **80**, 391.
10. J.-P. Salvetat, J.-M. Bonard, N. H. Thomson, A. J. Kulik, L. Forró, W. Benoit, and L. Zuppiroli, *Applied. Physics A*, 1999, **69**, 255.
11. M.-F. Yu, O. Lourie, M. J. Dyer, K. Moloni, T. F. Kelly, and R. S. Ruoff, *Science*, 2000, **287**, 637.
12. D. A. Walters, L. M. Ericson, M. J. Casavant, J. Liu, D. T. Colbert, K. A. Smith, R. E. Smalley, *Applied Physics Letters*, 1999, **74**, 3803.
13. O. Lourie, O., D. M. Cox and H.D. Wagner, *Physical Review Letters*, 1998, **81**, 1638.
14. B. I. Yakobson and Ph. Avouris, *Topics in Applied Physics*, 2001, **80**, 287.
15. S. Berber, Y.-K. Kwon, and D. Tomanek, *Physical Review Letters*, 2000, **84**, 4613.
16. S.U.S. Choi, Z. G. Chang, W. Yu, F. E. Lockwood, and E. A. Grulke, *Applied Physics Letters*, 2001, **79**, 2252.
17. J. Hone, *Topics in Applied Physics*, 2001, **80**, 273.
18. R. H. Baughman, A. A. Zakhidov and W. A. de Heer, *Science*, 2002, **297**, 787.
19. A. Javey, Q. Wang, A. Ural, Y. Li, H. Dai, *Nano Letters*, 2002, **2**, 929.
20. T. Tiamo, M. Roylance and J. Gassner, *Proceedings of the 32nd International SAMPE Technical Conference*, 2000, Boston, pp. 192.
21. H. J. Barraza, F. Pompeo, E. A. O'Rear, D. E. Resasco, *Nano Letters*, 2002, **2**, 797.
22. T. Kuzumaki, O. Ujiie, H. Ichinose, K. Ito, *Journal of Materials Research*, 2002, **13**, 2445.
23. R. Z. Ma J. Wu, B. Q. Wei, J. Liang, and D. H. Wu, *Journal of Materials Science*, 1998, **33**, 5243.
24. G.-D. Zhan, J. D. Kuntz, J. Wan and A. K. Mukherjee, *Nature Materials*, 2003, **2**, 38.
25. D. V. Smitherman, *Space Elevators: An Advanced Earth-Space Infrastructure for the New Millenium*, NASA, 2000.
26. A. M. Neville, *Properties of Concrete* (4th Edition), 1995, 2.
27. C. Perry and J.E. Gillott, *Cement and Concrete Research*, 1977, **7**, 553.
28. C. D. Johnston, *Advances in Concrete Technology* (2nd Edition), 1994, 603.
29. A. M. Neville, *Properties of Concrete* (4th Edition), 1995, 269.
30. R.F. Feldman and P.J. Sereda, Materials and Structures, 1968, 509.

NANO-SCIENCE AND -TECHNOLOGY FOR ASPHALT PAVEMENTS

M.N. Partl[1] , R. Gubler[1], M. Hugener[1]

[1] Road Engineering/Sealing Components, EMPA Swiss Federal Laboratories for Materials Testing and Research, Überlandstrasse 129, CH-8600 Dübendorf, Switzerland

1   INTRODUCTION

In spite of the fact that bituminous materials, such as asphalt, are mainly used on a large scale and in huge quantities for road constructions, the mechanical behavior of these materials depends to a great extend on structural elements and phenomena which are effective on a micro- and nano-scale. Although this is well known by researchers, material producers and engineers for many years, the aspect of nano-science and -technology has hardly found any special attention, so far. New efforts and possibilities of materials engineering on a nano-scale in other fields may well lead to a new leap forward in improving mechanical and physical properties as well as durability of this important group of composite construction materials.

This paper intends to stimulate the application and development of nano-scientific and -technological concepts for bituminous materials and asphalt pavements. Due to the wide range of possibilities, however, it does not claim to present a complete overview on the whole field. This would clearly go beyond the scope of this paper. It rather gives a short outline on material related problems and research topics where nano-science and -technology could produce a major contribution to the improvement of bituminous materials and where research needs to be done. This includes research with respect to the effect and engineering of nano-particles, such as mineral or organic filler and fibers, as well as the study of nano-structural questions related to binder/aggregate bonding and coating mechanisms. It includes also chemo-physical mechanisms during blending of bitumen with polymers and nano-structural modeling of materials behavior (e.g. damage and healing mechanisms).

2 OVERVIEW ON ASPHALT PAVEMENT MATERIALS

2.1   Components of Asphalt Pavement Materials

Asphalt is a particulate composite typically made of bitumen or polymer-modified bituminous binder, rigid cubical particles (mineral filler down to a nano-scale range, sand and stone mineral aggregates), additives (e.g. bonding and stripping agents like hydrated lime) and air (Table 1). In some cases other particles such as fibers (e.g. cellulose fibers, steel fibers for tank training areas), crumb rubber, glass, slag (glass-like or mineral-like)[1] etc. are added in order to fulfill special functional requirements in terms of visibility, friction etc. or to improve mechanical and climatic long-term performance, e.g. by reducing the risk of rutting, low temperature and fatigue cracking. Some components such as tar and asbestos fibers are no longer applied or under very restricted use (like volatile solvents for binders) in many countries for health and environmental reasons.

Table 1   *Selection of components used in asphalt pavement materials*

Binders	Additives	Particles	Air voids
Hot Process • Bitumen • Polymer Bitumen (Elastomers, Plastomers, Natural Latex, Reclaimed Synthetic Rubber,..) • Extenders (Sulfur, Lignin) • Synthetic Binders for Colored Asphalt •	• Bonding/Anti-Stripping Agents (Hydrated Lime, Amine Derivatives,..) • Hydraulic Cement • Natural Asphalt (Gilsonite, Trinidad Lake Asphalt) • Rejuvenators • Emulsifiers • Oxidants • Antioxidants • Polymer Additives • Reactive Nano-Ceramics • Self-Deicing Agents (CaCl Flakes,..) • ···	**Micro/Nano Particles** • Filler (Crusher Fines, Baghouse Fines, Fly Ash, Carbon Black, Silica Fume, ..) • Fibers (Cellulose, Mineral, Polymer,..) • Pigments • Crumb Rubber •	• Closed Pores • Inter-Connected Pores
Cold/Warm Process • Bitumen-Emulsions (Cold Water) • Foam Bitumen (Cold Water with Hot Bitumen) •		**Macro Particles** • Sand, Stones • Crushed Recycl. Glass • Slag • Recycled Tire Particles • Fibers (Steel, Fiberglass, Polyester, Polyprop.,..) • Reclaimed Asphalt Pavement (RAP) •	

2.1.1 Bituminous Binders.   Bituminous binders are commonly considered to be composed of saturates, aromatics, asphaltenes and resins. They are very complex mixes of macromolecules of different particle sizes. In case of polymer bitumen for pavements, bitumen is blended with about 5% macromolecules of a higher molecular weight. Typically, SBS (Styrene-Butadiene-Styrene block copolymer), SIS (Styrene-isoprene block copolymer), EVA (Poly-Ethylene-Vinyl-Acetate), EMA (Poly-Ethylene-Methacrylate) and others are used.

The differences in molecular weight and hence in the size of the polymers is depicted in Figure 1, where a gelpermation chromatogram of a SBS polymer bitumen sample in the original state and after overheating is shown. It is easily seen that overheating reduces the polymer peaks significantly, which means that the long and heavy polymer chains are reduced to smaller ones, raising the valley of the curve between the polymer and bitumen peak to a significantly higher level. This demonstrates that overheating considerably degrades the large polymer molecules and changes the mechanical behavior and durability of pavements significantly.

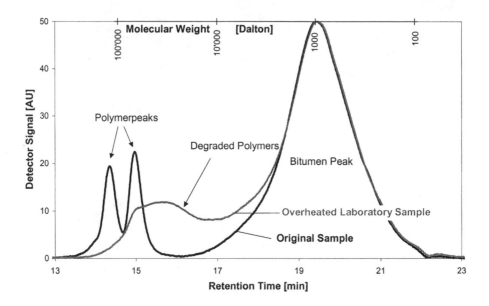

Figure 1 *Chromatogram of a SBS polymer-modified bituminous binder in the original state and after overheating*

However, the size of the polymer molecules is only one aspect, which influences the behavior of a polymer bitumen blend. Even more important is the micro- and nano-structural interaction between polymer and bitumen.

On the one hand polymers may play the role of a discontinuous nano-fiber reinforcement of the binder. In this case, the polymer molecules are embedded in the bitumen like dispersed nano-fibers that are only "glued" together by the bitumen with no strong physical connection in between. Due to this, the tension load is basically transferred to the nano-fibers by irreversible shear flow of the bitumen and the polymer bitumen basically behaves like a plastomer. An example of this type of polymer bitumen is APP polymer modified bitumen (APP = atactic polypropylene), which is used for bituminous waterproofing membranes.

On the other hand, the polymer may produce a linked network of polymeric macromolecules like the SBS modified polymer bitumen in Figure 2. In this case, the bitumen is trapped within a net of elastic nano-fibers, which are linked together with knots and therefore lead to an elastomeric behavior of the binder. These knots consist of a group of polystyrene clusters. Each cluster belongs to an individual SBS block-polymer and is attached to the elastic butadiene molecule in between. The polystyrene cluster is composed of up to 300 styrene molecules and the elastic poly-butadiene part in the middle is composed of up to 1300 butadiene molecules.

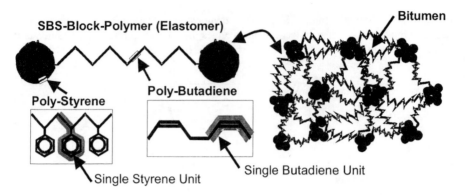

Figure 2 *Simplified nano-mechanical model of SBS Poly-Styrene-Butadiene-Styrene modified bitumen*

2.1.2 Special Binders. Asphalt pavements are typically constructed by hot mix compaction, which is energy consuming and requires special care during the production process (e.g. safety and temperature control issues, rain). This is particularly true for asphalt pavements with polymer bitumen, because this type of binder has found to be very sensitive to overheating.

In order to allow cold or warm asphalt mixtures and to apply bituminous sealcoats and unbound layer stabilization (e.g. subbase), cationic or anionic bitumen emulsions are used (Table 1). The particle size of these emulsions may go down to the nano-scale from 20□m to about 200nm.

A special approach is the so-called Carbonyte Process™ where the carbon bonds between the asphalt molecules are modified by reactive nano-ceramics for pavement protection systems and fast curing sealcoats[2]. This process is reported to be a result of the space shuttle thermal shield program and therefore appears a perfect example on how high-tech nano-technology may lead to innovative solutions in asphalt paving technology.

Recently, the use of foam bitumen has become increasingly important. Foam bitumen is produced by injecting a fine mist of water at ambient temperature into 170°C to 180°C hot bitumen in an expansion chamber[3]. The minimum particle size is around 7.5□m. Foam bitumen is applied for surface dressings, unbound layer stabilization, cold mix recycling with reclaimed asphalt pavement (RAP) materials sometimes contaminated with substances such as tar.

2.2   Structural and Micromechanical Aspects

2.2.1 Asphalt Pavement Concepts. From a structural point of view, asphalt works either through:

- *Corn-to-corn contact* by interaction and interlock of a coarse aggregate structure (Figure 3a, b), where the load is basically carried by the stone skeleton. The simplified structural model in Figure 3 implies, that failure mainly occurs through wedge effects within the load carrying stone skeleton producing local failure plains, which are oriented parallel to the vertical loading direction. Hence, it is very important for this type of pavements that a good lateral confinement within the pavement layer is achieved. Typical representatives of this type of pavements are stone mastic asphalt SMA with up to 7.5 Mass% binder and ca. 4 Vol% air voids as well as porous asphalt with up to 5 Mass% binder and over 20 Vol% air voids (for water drainage reasons).

* A *dense graded aggregate* concept (Figure 3c), which is based on the dense packing of the aggregates in order to minimize the binder content and to maximize the volume filled by mineral aggregates for stiffness and bearing capacity purposes. Here, the load is basically transferred by the stones through internal friction as well as the glue and shear effect of the binder. During loading, local hydrostatic pressure cells are produced in the room where the incompressible binder is trapped between the aggregates. Due to this pressure, the binder is squeezed out of these cells by local shear flow, thus producing deviatoric actions. Hence, the failure follows a shear pattern with cohesive shear failure plains. Typical representatives for this type of pavements are asphalt concrete with up to 6.5 Mass% binder and ca 4 Vol% air voids.

* A *dense mastic* concept (Figure 3d) where the aggregates "swim" within the bituminous matrix and the load is primarily carried by the mastic and to a minor extend by the stones. Dense mastic has a comparatively high binder content (up to 8.5 Mass%) and virtually no air voids. Due to the fact that the behavior of this type of materials approaches a frozen liquid, failure occurs again in a shear pattern

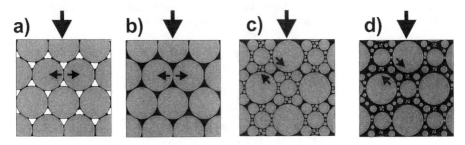

Figure 3 *Idealized structural models of asphalt pavement types with failure lines: a) porous asphalt (gap graded), b) stone mastic (gap graded), c) asphalt concrete (dense graded), d) mastic asphalt = gussasphalt (dense graded)*

2.2.2 Special Considerations on Porous Asphalt. In case of porous asphalt with its high air void content (Figure 3a), the coating of the stones and the binder joints between the aggregates are extremely important. Coating of the stones with a thick binder film is necessary not only to reduce the effect of oxidative aging and deterioration of the binder due to permanent airflow in the pores but also to provide sufficient long-term protection of the stones and the binder-stone adhesion interface from water and chemical agents. From the simplified model in Figure 3a follows that the binder joints do not take significant share of the load. However, they fulfill the important task of keeping the stones in place and taking the movements in the joints. Note, that in close vicinity of the contact points of the stones the binder films become thinner and thinner, which means that in these locations, even extremely small joint movements can produce locally high strains. The requirement to produce porous asphalt with a thick binder film and a high air void content has the disadvantage that there is a danger of binder drainage in these pavements during construction and later during summer under traffic conditions. To prevent this, binder contains organic or inorganic filler, fibers and/or polymer modified binder where the bitumen is trapped within a network of polymeric macromolecules (Figure 2).

3 FIELDS OF NANO-ACTIVITIES

3.1   General Overview

Generally, it can be assumed that nano-science and -technology have a great potential to advance asphalt pavement technology in the field of

- materials design (development and optimization)
- materials manufacturing (nano-technological production) and pavement construction
- materials properties (multifunctionality and sustainability)
- materials testing and pavement monitoring with sensors
- chemo-physical modeling of the material behavior down to a nano-scale

A schematic overview on focus areas where nano-science and -technology could improve asphalt pavement technology is given in Figure 4. In addition, Table 2 summarizes the general goals and topics for improvements in asphalt pavement technology by research on a nano-scale. It is clear that it will be very difficult if not impossible to satisfy all these partly contradictory requirements. On the other hand, it is also clear that the direction of research will certainly have to concentrate on optimizing the materials accordingly in order meet as many of these requirements as possible.

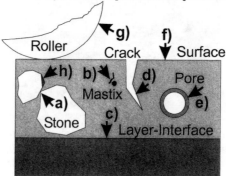

Figure 4   *Schema to visualize focus areas for nano-science and technology with respect to asphalt pavement structures:*
   a) Bond between stones (shear and tension)
   b) Mastic (stiffening, cohesion, durability, compaction improvers,..)
   c) Bond between layers (tack coats)
   d) Self-repair (healing) and rejuvenating agents
   e) Oxidation of binder films and binder inhomogeneities
   f) Surface properties (friction, optical properties, water repellent, abrasion resistant, self-cleaning), sealcoats for surface protection
   g) Anti-adhesion surface for rollers during compaction
   h) Bond, adhesion between stone and mastic

Table 2 *General goals and research topics for improvements in asphalt pavement technology by research on a nano-scale*

Planning	• Beneficial for user and environment • Affordable • Available in large quantities at any place
Construction	• Low energy consumption (cold/hot production process) • Fast and easy to handle (compaction improving agents, homogeneous material) • All-weather construction during summer and winter (heat, rain, frost, snow) and error tolerant • Instant opening to traffic possible • Non-toxic, harmless to environment and human health
Use	• Safety (non-hazardous, driving safety) • Mechanical resistance (crack, rutting, wearing) • Durable (no maintenance) • Multifunctional (load carrying function, noise reduction, visibility function, self-deicing, self-cleaning, traffic diagnosis and monitoring function, energy production function)
Repair/Replacement	• Easy to repair • Smart material (distress sensoring and self-healing structures) • Sustainable (reversible and repeatable transformation of resources)

3.2 Materials Design

3.2.1 General Considerations. Material design based on nano-scale considerations is a typical nano-scientific task, which requires fundamental research in terms of both experimental materials development and theoretical modeling. This research should deal with the optimization of existing asphalt pavement concepts as well as the development of innovative types of bituminous pavement materials.

However, due to the fact that road pavements are mass products, which require mega-quantities of asphalt material spread over many thousands of kilometers all over the continents, it should always be kept in mind that nano-particles for pavement materials must be non-hazardous low-cost products, which are easy to handle and available in high quantities at almost any place in the world (Table 2). They should also fulfill ecological requirements such as low energy consumption and environmental compatibility, i.e. without producing linear landfills. Furthermore, of course, they should improve long-term performance and functional properties of the materials in a significant way without reducing the clear advantages of existing asphalt pavement materials, which are generally recyclable and easy to repair. This is a very positive factor in terms of sustainability, i.e. the use of materials by reversible and repeatable transformation and not by irreversible consumption of resources. Hence, nano-technological concepts should never interfere with this important benefit.

3.2.2 Nano-Filler. Mineralic filler particles are particles smaller than 0.09mm. With simple methods, like sedimentation or washing method, the grain distribution of filler between 1□m up to 0,09mm can be determined on a routine basis. Other methods, like laser granulometry, allow detecting particle sizes distributions down to only a few nanometers, but are not commonly used in asphalt technology.

It is well known that the influence of filler particles on the binder increases with decreasing size of the filler particles. However, in the Swiss Standards SN7670135 only rough instructions are given on the dosage such as the acceptable content of filler < 0.02 mm (25...60 Mass-%) and < 0.005 mm (10...25 Mass-%). This demonstrates that the influence of the particle size distribution of fillers is given only little attention, in common practice, so far, in spite of the fact that there are considerable differences in the particle size distribution of different fillers. This is shown in Figure 5 taken from an ongoing research[4] at EMPA.

The investigation of filler on a phenomenological basis has been a focus of researchers for many years[5]. However, the influence and mechanisms of fine filler particles on the material behavior of asphalt are still not well understood and further research on a micro- and nano-scale needs to be conducted. This research should also deal with the following questions: Can bitumen be improved by adding organic or inorganic nano-filler and how should this nano-filler be produced, processed and tested? What are the best methods to determine the particle size distribution below 1000 nm?

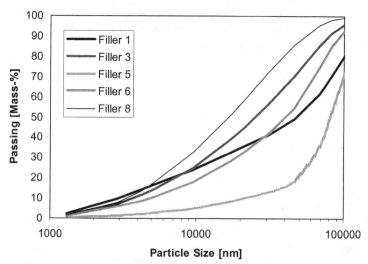

Figure 5   *Particle size distribution of five different mineral fillers obtained by laser granulometry (only part of the gradation obtained by this method is shown)*

3.2.3 Experience with Tar.   Tar is a highly cancerogenous material and is therefore no longer used in modern asphalt pavement technologies. However, from old pavements with tar bitumen, we learn that those pavements showed in many cases extremely good performance and were in some cases even superior to pavements produced exclusively with bituminous binders. One difference between bitumen and tar is the fact, that tar contains up to 5 Mass% fine particles, which are not soluble in toluene. These particles can behave like special filler. For tar a much smoother transition between binder and filler can be observed than for bitumen where a gap in the very fine particle size exists.

3.2.4 Natural Asphalt. In Switzerland, natural asphalt, like Trinidad lake asphalt often combined with cellulose fibers, is used for pavements in order to increase rut resistance or to produce thick binder films in case of porous asphalt. The effect is again of micro- and nano-technological nature but not well understood

3.2.5 Polymer Bitumen. The distribution of polymers within the polymer bitumen binder is of significant importance. This distribution is characterized with fluorescence microscopy in the □m-range. It is observed that some polymer bitumen binders show a homogeneous polymer distribution that raises the question if this is also the case in the nm-range.

SBS block copolymers are composed of three blocks (polystyrene, polybutadiene and polystyrene) and have proven superior to other polymers for bitumen modification. The reason for this observation is not well understood to this day. The resistance to high temperatures SBS, on the other hand, is less advantageous. An improved understanding of the distribution and interaction of SBS with bitumen on a nano-scale range could possibly be beneficial to develop even better performing and temperature resistant polymer bitumen binders.

3.3 Materials Manufacturing and Pavement Construction

3.3.1 Energy Consumption during Construction. Energy consumption during hot mix asphalt paving is one major concern nowadays for ecological and economical reasons, in particular in heavily populated urban areas. It may also become a quality issue when the hot mix cannot be delivered just in time to the construction site due to congestions and traffic problems.

Here, emulsions for "cold" production processes should be promoted and further developed by research on a micro- and nano-scale, i.e. by introducing innovative technologies like the use of reactive nano-ceramics. Such efforts could also lead to new approaches, which allow all-weather installation of asphalt pavements or instant opening to traffic.

Nano-scientific research could also help to reduce heat energy consumption during paving and to increase the placing efficiency and workability by developing special micro-bubbles or other particles, which produce a compaction improving effect (similar to self-compaction for concrete) or reduce the viscosity of the binder temporarily in order to increase the depth effect of the compactors thus allowing for higher layer thickness per work cycle.

3.3.2 Anti-adhesive Surfaces for Rollers and Mixers. In addition to construction related improvements of pavement materials, nano-science and -technology could also help to develop special bitumen-repellent nano-coatings for the inner surfaces of mixers and for drums of steel rollers in order to prevent bituminous material from sticking to the working tools and to avoid the use of water during the compaction process of hot mixes. These special surfaces should be not only bitumen-repellent but also mechanically hard and wear-resistant. The benefit would be twofold, as it would increase both the efficiency in the construction process and the quality of the hot mix asphalt pavement by excluding the possibility that roller water could penetrate into the mix causing damage.

3.4 Material Properties

With respect to materials properties (multifunctionality and sustainability) nano- science and -technology could contribute not only by improving the load carrying function and durability by introducing special nano-designed binders, additives or particles, as mentioned above, but also by adding nano-components to improve multifunctionality of pavements with respect to noise reduction, increase of safety (visibility of pavements surface during the night and in tunnels, environmentally friendly self-deicing and self-cleaning concepts, pavement surfaces with display function for remote controlled active markers), traffic sensitive smart pavements with sensor properties for distress diagnosis and traffic monitoring purposes. Another issue in this context is the development of new pavement materials with self-healing properties.

3.5   Material Testing

3.5.1   General   Considerations.   Materials   testing   and   pavement   monitoring addresses the experimental tools which have to be adopted or developed in order to improve the understanding of the materials properties on a nano-scale. This is true in particular with respect to the investigation of adhesion properties in the binder/stone interface and of nano-mechanical effects of filler as well as the study and modeling of the properties of binder films.

3.5.2   Measuring Binder Properties on a Micro- and Nano-Scale. There is an ongoing discussion if the mechanical binder properties such as viscosity are size dependent. Measurements with sliding plate viscometers suggest that very thin layers have a remarkably higher viscosity than thick layers. Measurements on mixes of binders and fillers support that idea[5]. Viscosity in such systems is sharply dropping under a significant loading. Such a drop is hard to explain by classical models (e.g. heating up by the test) but could be explained by a model on a micro- and nano-scale. One could imagine that a rearrangement of the particles in such mixes takes place by producing larger gaps (by making others smaller). This would lead to a substantial local decrease in viscosity. Since the contribution of these larger gaps to the total deformation is important, a lower apparent viscosity would result.

However, validating this model is difficult since the measurement of the size in very small gaps is hard to achieve with known viscometers (and even the preparation of samples with a uniform gap). A completely new test set-up designed on a micro-scale (most probably based on piezoelectric principles) could overcome such problems. In addition, working with very small sample would allow to observe aging phenomena directly, not disturbed by any diffusion effects.

On the one hand, such a research could lead to new fields. On the other hand, it is clearly of practical relevance due to the fact that the thickness of binder films between the mineral particles in a pavement is mostly in the order of a micrometer or even below.

Such devices are certainly not necessarily restricted to pure binders. In addition, one could think of measuring properties of mixes of fillers and binders. The behavior of theses materials deviates fundamentally from pure binders, and can be seen as the first step from the pure binder towards the composite material asphalt.

3.5.3   Single Plate Approach. Probably the simplest possible device consists of a piezoelectric crystal plate on which a defined binder layer has to be applied. Using the crystal as actuator and measuring the response would lead to conclusion on the mechanical properties of the binder.

The advantages of this approach read as follows:

- Comparatively simple sample preparation
- Binder in full contact with the environment, reaction to influences can be studied very quickly.
- Ideal to investigate aging, water sensitivity and the influence of chemical agents

The disadvantages are:

- Requires a complicated numerical model
- Risk of lack in precision of the knowledge of the binder's shape
- Results are influenced by environment
- Sample may change during test.

Hence, the single plate approach can be expected to be very useful for studying environmental and other influences on the binder. It appears ideal for measuring the change in properties with time but is probably not appropriate for high precision measurements.

3.5.4   Single Plate Approach with Protective Measures.   A protective coverage applied after sample preparation could be used to allow only for the desirable environmental contact of the approach above. It has to be said, however, that this would lead to a series of other problems:

- Interaction of coverage and binder
- Contribution of the coverage to the mechanical behaviour of the test set-up
- Tailoring of the coverage, e.g. to allow oxidation but to avoid distillative losses

This approach could be useful for the study of very specific interactions with the environment and for the investigation of time dependence excluding undesirable influences for the environment. Again, this approach appears not appropriate for high precision measurements.

3.5.5   Dual Plate Approach.   The binder is squeezed between two plates. One might be used as actuator, the other as response unit. There is still the problem of measuring the gap accurately and the mechanical model might still be rather sophisticated. However, the stability of the set-up can be expected to be much better than in the cases discussed above. The advantages of this approach are

- Comparatively simple sample preparation may be achieved
- Reasonably accurate knowledge of the test geometry is possible
- Interaction with the environment is controlled

However, there are also some disadvantages

- Need for a complicated numerical model
- Most complicated devices
- Limited interaction with the environment

This approach looks promising for reasonably accurate measuring in realistic dimensions (film thickness), allowing the study of binders over time without much interaction with the environment.

3.5.6   Torsional Dynamic Resonance Rheometer. Recently EMPA has presented a new High-Frequency Torsional resonance Rheometer (HFTR) developed at ETH Zurich[6,7]. The rheometer basically consists of a rod which is embedded or dipped in the binder sample and vibrates in the kHz range. This rod interacts with the viscoelastic medium by changing both its resonance frequency and its damping characteristics.

The HFTR appears to be a promising tool for continuous characterization of the mechanical effects due to environmental contact of the surface film of the bitumen sample. However, as this approach is actually not working in a micro- and nano-scale range yet, there is still some additional development of instrumentation to be done, e.g. by reducing the geometry of the device down to a size which is appropriate for micro- and nano-scale investigations. In addition, this approach requires improved modeling and an extensive study of the binder film parameters (geometry, skin effect) influencing the measurement data.

3.6   Modeling of the Material Behaviour

The mechanical behaviour of asphalt is a result of the interaction between its different components. Micro- and nano-mechanical modeling may lead to an improved understanding of the mechanisms of interaction between the stone skeleton and the bituminous mastic. Advanced modeling may be very useful to explain the influence of different aggregate grading curves, anisotropy effects, local stress and strain redistribution during loading as well as crack-formation and -propagation phenomena. However, in order to understand local stone/mastic adhesion phenomena and the effect of thin binder films or binder inhomogeneities (e.g. from oxidation) as well as the influence of nano-particles like filler, fibres and macromolecules, nano-mechanical models have to be developed. These models should also focus on damage and aging mechanisms.

The nano-mechanical effects of polymers within the binder are still not sufficiently understood. It appears, that, nowadays, the pragmatic try and error approach is still the most common way of research, whereas the nano-mechanical asphalt design is only in a very early stage of development.

4   CONCLUSIONS

Nano-science and -technology may lead to significant improvements and further development in asphalt pavement technology and help to find answers to many questions of the material behavior, which are not well understood, so far. In particular, progress appears possible in the field of materials design, testing, modeling and manufacturing as well as pavement construction and sensing. First examples are encouraging but mostly achieved more through empirical trial and error procedures than through systematic research on a nano-scale. In terms of testing on a micro- and nano- scale, new methods and devices need to be developed and implemented based on experience with existing testing devices, e.g. by down-scaling those devices to micro- or nano-size.

Some ideas and suggestions outlined in this paper may seem a bit futuristic and some nano-related questions may still remain unanswered for quite a period of time. Research on a micro-scale has already started and needs to be intensified significantly whereas research on a nano-scale is still at the very beginning of being established for this important type of building materials. However, regardless of the promising prospects of this fascinating field of research, it has to be kept in mind that in spite of finding solutions by focusing on a nano-scale, these solutions, in the end, will have to be applied for mega-quantities of asphalt pavement materials.

References

1   J.M. Reid, R.D. Evans, R. Holnsteiner, F. Berg, K.A. Pihl, O. Milvang-Jensen, O. Hjelmar, H. Rathmeyer, D. François, G. Raimbault, H.G. Johansson, K. Håkansson, U. Nilsson, M. Hugener, *ALT-MAT Contract No.: RO-97-SC.2238. European Commission.* Published on www.trl.co.uk/altmat/index.htm.(2001)

2   Steel Guard™ and Carbonyte Process ™ information by *California Pavement Maintenance Co* available on www.cpmamerica.com, Asphalt Advisor, October 2000

3   K.J. Jenkins, M.F.C. van de Ven, J. de Groot,. *Conference on Asphalt pavements for Southern Africa, CAPSA, Victoria Falls, Zimbabwe,*1999

4   R. Gubler, et al, *Research Project VSS1998/070* (unpublished, in progress)

5   R. Gubler, Y. Liu, D.A. Anderson, M.N. Partl, *Journal of the AAPT*, 1999, **68**, 284

6.   M.B. Sayir, A. Hochuli, M.N. Partl, *Eurobitume Workshop '99 – Performance Related Properties for Bituminous Binders*, 1999, Paper No. 106

7   L.D. Poulikakos, M.B. Sayir, M.N. Partl, *Proc. 6th Int RILEM Symp. on Performance Testing and Evaluation of Bituminous Materials, PTEBM '03*, 2003, Paper No 14

NATURAL ROOFING SLATE: THE USE OF INSTRUMENTED INDENTATION TECHNIQUE TO MEASURE CHANGES IN THE ELASTIC MODULUS AND HARDNESS DUE TO WEATHERING

Joan A Walsh[1] and Pavel Trtik[2]

[1]Advanced Concrete and Masonry Centre, University of Paisley, Paisley, U.K
[2]Institute for Building Materials, ETH, Zurich, Switzerland[*]
[*]formerly Scottish Centre for Nanotechnology in Construction Materials (NANOCOM), University of Paisley, UK

1   INTRODUCTION

The durability of natural roofing slates varies widely, some slates lasting over a hundred years while others last only 20 to 25 years. This research monitors the changes in the physical and chemical properties of the material due to weathering in order to predict its durability. The most important of these changes are increased water absorption and loss of strength (measured as the modulus of rupture). The water absorption procedure is non-destructive, making it possible to measure the rate of change in water absorption due to weathering by repeatedly testing the same sample. However, it is not possible to measure the rate of change in modulus of rupture in the same way since the test is destructive. Instead this research investigates the use of micro-indentation procedures to measure the change in elasticity and hardness of a slate due to weathering. Using samples of slate, weathered using the BS680 wetting and drying procedure, the elasticity and hardness were measured at the centre and close to the weathered edge of the test pieces. Comparing the results with those found in an unweathered sample, the elasticity at the centre of the test piece was found to be unaltered while that at the middle and the edge was progressively reduced. Although similar trends were observed in the hardness the results were too imprecise to make predictions. The effect of orientation of the test piece was also investigated. The loss of elasticity was found to penetrate to a greater depth along the grain of the slate.

2   BACKGROUND

Natural slate is a traditional roofing material used in the historic and conservation areas of cities and towns in Britain. Slates are low-grade metamorphic argillaceous mudstones which are found in orogenic belts which have undergone crustal shortening. During metamorphism, clay minerals present in the protolith recrystallise as phyllosilicates, chlorite and white mica. These, plus detrital quartz, are the principal minerals found in all slates. The phyllosilicates grow normal to the maximum stress, giving slate a well-developed cleavage. This is the property of a fine-grained rock that allows it to be split into thin sheets suitable as a roofing material. The durability of natural roofing slates varies widely, some lasting several hundred years while others fail within months of being exposed to the extremes of the British climate. Kessler & Sligh[1] examined the effect of weathering on 61 samples of old slates exposed for between 12 and 131 years and found that there was a loss of strength and an increase in water absorption with time. Comparing the data with those for fresh samples, it was possible to determine the average annual increase in water absorption and the average annual decrease in strength. This is substantiated by the examination of unpublished data gathered by the Building Research Establishment (BRE) on weathered slate, where there was shown to be an increase in water absorption with the age of the slate. The BRE have no equivalent data for loss of strength with age.

Traditionally the quality of slates has been assessed using various national standards. The ranges of tests and compliance requirements of the different standards vary considerably, often reflecting the qualities of the slate produced in the country of origin. These standards are currently being replaced within the European Union by a common standard, prEN12326.[2]

3   WATER ABSORBENCY AND LIFE EXPECTANCY

Most of the national standards have water absorption tests which follow similar testing procedures, yet there is wide disparity between the compliance requirements of a good quality slate (Table 1). In many cases it appears that the limits were set pragmatically in an attempt to exclude slate of perceived poor quality. The American ASTM standard[3] is the only standard which estimates the life expectancy of a slate. Based on initial water absorption and depth of softening when exposed to acid vapour, this standard classifies slate into one of three grades, to each of which is assigned a corresponding life expectancy value (Table 1). These estimates of life expectancy are frequently used by the slate industry and often inappropriately. It does however highlight the need for a procedure to estimate the life expectancy of a slate. The ASTM and all the other standards investigated concentrate on the initial properties of a slate.

Table 1 *Compliance requirements set by different standards and the corresponding life expectancy estimated by the American standard.*

Limits		Grade 1	Grade2	Grade3
		Increase in weight (%)		
PrEN12326	E.U.	0.6		
BS 680	British	0.3*		
NF P32	French	0.4	0.7	0.9
UNI 8626 & 8635	Italian	No equivalent test		
UNE 220 201 85	Spanish	0.3		
C119 -74	American	0.25	0.36	0.45
Life expectancy		75-100 years	40-75 years	20-40 years
All standards, except BS680, determine the increase in weight after 48 hours immersion in water. In the BS680 test, samples are refluxed for 48 hours which in general yields higher absorption values.				

However, recent research[4] has found poor correlation between the initial water absorbency values and the reputation of a slate for durability. For example a Welsh slate from Pen yr Orsedd quarry has a water absorption of 0.24%, which is higher than that of a Spanish slate at 0.15%, yet the Welsh slate is expected to last over 100 years whereas the Spanish slate is estimated to last approximately 30 years (Maol in 'Builder' 1995). Instead of concentrating on initial values, Walsh [4,5] assessed the effect of weathering by measuring the rate of increase in water absorbency in slates weathered experimentally, using the wetting and drying test prescribed in BS680.[6] It was found possible to model the results and, by comparison with those observed in naturally weathered slates, to estimate the life expectancy of the slate.[4] The next step was to supplement the water absorption data and investigate the changes in the mechanical properties due to weathering. However, the traditional tests being destructive, an alternative approach was needed.

4   FLEXURAL STRENGTH OF A SLATE; MODULUS OF RUPTURE R α

Failure of a slate is often caused by tensional stresses due to gravity, or bending stresses resulting from the uplifting effect of the wind. Flexural strength is the usual criterion for the strength of a slate, as it is considered a more appropriate measure than the more standard compression test used in the evaluation of other building materials. Results are reported as the modulus of rupture, a measure of the intrinsic strength of a material

$$R\alpha = \frac{3Pl}{2be^2} \tag{1}$$

$R\alpha$ = Modulus of rupture in newtons/mm^2
P = Failure load in newtons
l = Distance between support bars of the three point bending machine
b = Width of test piece in mm
e = Thickness of slate in mm

As slate is an anisotropic material, the strength depends on the direction of applied stress, i.e. the strength of a slate along the grain is less than that across the grain (Figure.1)

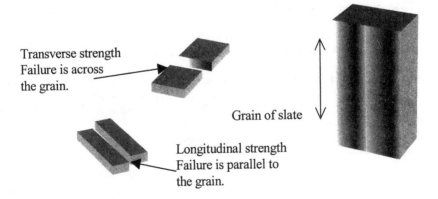

Transverse strength Failure is across the grain.

Grain of slate

Longitudinal strength Failure is parallel to the grain.

Figure 1 *Transverse strength; the slate fails across the grain, longitudinal strength failure is along the grain.*

Kessler & Sligh[1] found that there is no correlation between intrinsic strength of a slate and its reputation for durability. Hence not all national standards have a minimum strength requirement and the limits set by those that do vary considerably (Table 2). The British Standard has no minimum requirement, as the strength of indigenous slate is greater than that specified for other roofing materials[6]. This has resulted in the importation of weaker slates into Britain. For example, Spanish slate is reported (trade literature and press) to have a low intrinsic strength and when split too thinly for the British climate often fails. This has been rectified to some extent by producing a thicker slate for the British market. Although there is poor correlation between the modulus of rupture and the durability of a slate, there is a case for a minimum requirement to be specified. The Comité Européen de Normalisation[2] has effectively set a minimum apparent strength requirement by setting the minimum thickness as a function of the modulus of rupture.

Table 2 *Variation in the minimum requirement for the flexural strength of a slate as specified by different national standards.*

Standard		Grade 1	Grade 2	Grade 3
		MPa		
PrEN12326	E.U.	A function of the thickness		
BS 680	British	No requirement		
NF P32	French	70 dry 40 wet	50 dry 33 wet	33 dry 24 wet
UNI 8626 & 8635	Italian	A function of the thickness		
UNE 220 201 85	Spanish	28 dry, 23 wet		
C119 -74	American	62	62	62

4.1 Changes in the flexural strength due to weathering

This research has found that there is a loss of flexural strength in both experimental and natural weathering. This can be seen by comparing the strength of a used slate from the roof of a Scottish castle, known to have reached the end of its serviceable life, with that of a new Scottish slate of similar thickness (Figure 2). It can be seen that the flexural strength of the used slate is less than half that of the new. There is a corresponding loss in elasticity. Slates weathered experimentally using the BS680 wetting and drying procedures[4] also exhibited an overall reduction in the modulus of rupture (Figure 3). However, due to the large variation found in all natural materials, a large number of samples would need to be analysed in order to predict reliably the rate of change due to weathering.

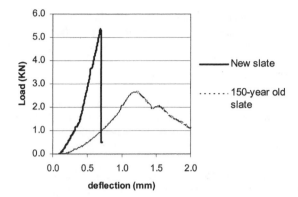

Figure 2 *The flexural strength of an used slate which has reached the end of its serviceable life is compared with that of a new Scottish slate of equivalent thickness.*

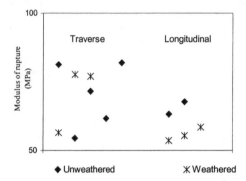

Figure 3 *The loss of flexural strength due to weathering is overshadowed by the large standard deviation of the test results.*

5   METHODOLOGY

Instead of testing for loss of strength, this research investigates changes due to experimental weathering in different physical properties such as elasticity and hardness.

5.1 Weathering Procedure

Two slates were used; a Welsh Blaenau Ffestiniog slate (W/F-8) and a Spanish slate (S/F-1). Test pieces (50mm x 50mm) were cut from each of the slates and weathered by subjecting them to repeated cycles of wetting and drying as prescribed in BS680[6]. The Welsh slate was weathered for 140 cycles while the Spanish slate was weathered for 47 cycles.

The water absorption was measured before weathering and at intervals throughout the weathering regime. Two methods were used; (1) that prescribed by prEN12326 in which the increase of water is measured after 48 hours immersion in water at ambient temperature; (2) the water absorption test as prescribed by BS680 in which the samples are refluxed for 48 hours. Greater changes are observed in slates which have been tested by the second method. This is due to refluxing which accelerates the weathering of the slate.

5.2 Testing Procedure

To assess the effect of orientation on the micro-mechanical properties, three sections were prepared from unweathered and weathered test pieces each as follows;
1. Horizontal section: parallel to the cleavage surface
2. Longitudinal section: normal to the cleavage surface and parallel to the grain of the slate
3. Transverse section: normal to both the cleavage surface and the grain of the slate.
The slate sections were then embedded in replicate cylindrical resin blocks.

The elastic modulus and hardness were measured using Nanoindenter XP, manufactured by MTS. Three sets of tests were performed on each sample, each set consisting of twelve indentations arranged in an array (4 by 3) set 100 µm apart. The depth of the indentations was set at 2000 nm. The test results provide values of indentation hardness and elastic modulus derived from unloading segment as described by the Oliver and Pharr method.[7] The mean and standard deviation values are shown in Tables 3 and 4. There is considerable variation in the results, in particular the hardness values which have an average coefficient of variation of 67%. The variation of the elastic modulus is considerable less i.e. the average coefficient of variation is 11%. Hence hardness results are not included in Table 4 and the subsequent discussion is limited to the elastic modulus results

Table 3 *Mean and standard deviation values of the elastic modulus and hardness*

Slate sample	Orientation	Elasticity modulus (GPa)	Standard deviation	Hardness (GPa)	Standard deviation
W/F-8	Horizontal	41.3			
	Transverse	84.5	7.0	2.78	0.49
	Parallel	64.8	7.2	2.07	0.71
S/SF-1	Horizontal	48.5			
	Transverse	79.5	7.9	1.93	0.30
	Parallel	73.0	9.7		

It can be seen that in both cases there is considerable variation in the micro-mechanical properties of the different orientations, e.g. the elastic modulus is greatest in the transverse sections and the least in the section parallel to cleavage.

The next stage was to determine whether it was possible to detect changes in the elasticity due to experimental weathering. A section close to the weathered surface was prepared from the sample, cutting across the grain. The elasticity was found to be reduced from 79.5 GPa observed in the unweathered slates to 62.7 GPa, a decrease of 21%. This showed that it is possible by this method to detect changes in elasticity due to experimental weathering. It was then decided in addition to assess the change in the micro-mechanical properties of a single test piece by comparing the elasticity and hardness at the centre with that at the weathered edge. Two sections (200 mm by 20 mm) were cut from weathered 500 mm by 500 mm test pieces, starting at the weathered edge and ending at the middle.

The first section was cut across the grain and the second was cut parallel to the grain as shown in Figure 4. Two sets of indentations were performed on each section following the procedure described above, the one at the weathered edge referred to as the "outside" and the second at the opposite end referred to as the "middle". The results are shown in Table 4.

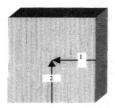

Figure 4 *Diagram showing the orientation of the two sections; 1 is a transverse section cutting across the grain of the slate, 2 is a section parallel (longitudinal) to the grain.*

Table 4 *Variation in the elastic modulus (GPa) due to weathering. *Greater degree of weathering is observed in samples which have been tested according to BS680 due to refluxing.*

	Unweathered		Middle		Outside	
	GPa	St Dev	GPa	St Dev	GPa	St Dev
Section 1	**Transverse section cut across grain**					
S/SF-1	79.5	7.9	78.3	7.8	76.7	10.1
					62.7*	6.1
W/F-8	84.9	7.0	78.5*	8.0	70.8*	8.3
Section 2	**Longitudinal section cut parallel to grain**					
S/SF-1	73.0	9.7	74.9	9.4	67.8	10.3
W/F-8	64.8	7.2	69.1*	11.8	62.4*	7.1

6   DISCUSSION

The aim of this preliminary study was to determine whether it was possible to detect significant changes in the elastic modulus due to weathering. Two slates, which had been experimentally weathered using the wetting and drying routine, were tested and the effect on the elastic modulus of the following discussed:

- Orientation of the unweathered test piece
- The degree of weathering
- Comparison of the two slates

As seen in Table 3, there is considerable variation in the elastic moduli of the unweathered samples depending on their orientation. This is due to the orientation of the individual minerals, in particular the platy phyllosilicates, which became aligned during metamorphism. The lowest values are found in the section parallel to the cleavage surface, i.e. the [001] plane of the phyllosilicates. There is also a difference in the elastic moduli of the transverse and parallel sections, the transverse section being the greater. This difference is more pronounced in W/F-8, which has a more developed grain, than in S/SF-1 i.e. the elasticity of the transverse section of the former is 30% greater than the corresponding parallel section as compared to 9% for the later. This is assumed to be related to the grain of the slate. However it is not possible at this stage to determine whether this is due to different orientations of the same minerals or alternatively it may be due to differences in composition where the proportion of the tested surface occupied by platy minerals varies depending on its orientation. As in all aspect of this research, more work is needed to relate individual indentations to the actual mineral indented.

As seen in Table 4 it is also possible to detect changes in the micro-mechanical properties of the slate due to experimental weathering. It is not possible to make direct comparison between the properties of the two slates as different weathering regimes were used throughout. However it is possible to make the following generalisation. Comparing the unweathered test pieces with the inside and outside sections of the weathered test pieces, it can be seen that there is overall loss of elasticity. The greater the weathering the greater the effect; this can be seen by comparing the two S/SF-1 outside sections. The elastic modulus of the refluxed sample is 63 GPa as compared to the sample tested using simple immersion, which has an elastic modulus of 77 GPa.

Although W/F-8 was weathered for almost three times as long as S/SF-1, the loss in elasticity is less for both the transverse and longitudinal sections e.g 16% to 21% in the case of the former, from this it can be inferred that the rate of change in the elasticity is greater for the S/SF-1 sample.

Comparing the middle of the weathered test pieces with the unweathered, there is a small increase in the elasticity of the longitudinal sections. This may be a spurious result, however it may be due to increased crystallinity due to the prolonged heating during the weathering regime. A similar effect of weathering was observed in related research using XRD analysis. XRD scans of weathered samples are more crystalline, as measured by the sharpness of individual peaks, than their unweathered counterparts (Walsh in preparation).

7   CONCLUSIONS

It has been established that it is possible to detect changes in the micro-mechanical properties of the slates due to experimental weathering. The next stage is to select different slates, which have been subjected to the same weathering regime, and to measure the changes in the effects of weathering. Further refinement will be got by relating individual indentations to the mineralogy of the sample. It will then be possible to quantify the changes in the micro-mechanical properties and relate these changes to the vulnerability of the slate and thereby make prediction about the durability of the material.

References

1   D. W. Kessler and W.H. Sligh 'Physical properties and weathering characteristics of slate' in *Bureau of Standards Journal of Research* 1932, **9**, pp.377-411
2   Comité Européen de Normalisation 'Slate and Stone Products for discontinuous roofing and cladding'. *Part 1 Product Specification. prEN12326-1, Part 2 Methods of test prEN12326-2*, 1999.
3   American National Standards Institute, 'Standard specification for roofing slate C406 –58', 1971.
4   J. A. Walsh 'Predicting the service-life of natural roofing slates in a Scottish environment' *Ninth international conference on durability of building materials and components*, Brisbane, Paper **216**, 2002 pp.1-10.

5   J. A. Walsh, 'Methods of evaluating slate and their application to the Scottish slate quarries', PhD Thesis, University of Glasgow, 1999.

6   British Standards Institution, 'Specification for roofing slates. Metric Units', *British Standard BS 680: part 2,* 1971.

7   W. C. Oliver and G. M. Pharr, 'An improved technique for determining hardness and elastic modulus using load and displacement sensing indentation experiments', *Journal of Materials Research Society,* 1992, **7**, No 6, pp. 1564-1583.

USE OF INSTRUMENTED INDENTATIONS FOR QUALITY CONTROL OF BUILDING MATERIALS

K. Trtík , O. Vlasák

Czech Technical University in Prague, Prague, The Czech Republic
Advanced Concrete and Masonry Centre, University of Paisley, Scotland

1   INTRODUCTION

Instrumented indentation technique is one of the most sensitive methods that give the possibility to test the properties of materials. The high sensitivity brings the problem, that the testing of materials is fundamentally different to classical way of testing. The results of nanoindentation offer information about material characteristics on very local scale of the test specimen. This could be one of the reasons why this method is hitherto being used mainly for laboratory research, while the use in industry, especially in the case of building materials, is hitherto seldom.

One of the fields in which the instrumented indentations could found very practical industrial application is the area of the quality control of building materials. Since only very small samples are necessary for instrumented indentations, the technique shows a potential for assessment of quality of building materials on samples taken from existing structures, i.e. with need of only minimal damage imposed on the structure itself by sampling technique. This could be particularly true in the case of quality control of the fibre reinforcement used in the fibreconcrete and other fibre reinforced cement composites. In order to assess the appropriateness of the technique for quality control the authors decided to undertake pilot investigation. The aim of the investigation has been: (i) to assess the micromechanical properties of fibre reinforcement by instrumented indentation testing, (ii) correlated these results with those obtained from 'classical' test methods. Also, one of the questions posed by the authors has been: Is it possible to discern two materials with different but relatively very close mechanical properties by instrumented indentation? In order to meet these aims the authors decided to carry out measurements on two testing series of two kind of steel fibres by two test methods – (i) instrumented indentation and (ii) 'classical' tensile test. Both techniques were applied to steel fibres, which are commonly utilised for reinforcing concrete structures and the material characteristics (i.e tensile strengths) proclaimed by the respective manufacturers were not to be very different from each other. The aim of this article is to provide the report about these tests.

Figure 1   *Steel fibres embedded in cementitious mortar*

Bekaert and Trefil Arbed steel fibres embedded in cementitous mortar were tested in the investigation. The composition of the cement mortar was the same in both samples and was likely not to have any significant influence on the test results. Figure 1 shows the samples of the cement mortar with embedded Bekaert fibres in the left sample and Trefil Arbed fibres in the right-hand side sample. The size of the samples produced has been approximately 10 by 10 by 5 millimetres.

The fibres of not that different mechanical properties advertised by the producers were tested in two different ways. First by instrumented indentation techniques and these results were than compared with classical tensile tests on the segments cut out of the 'infinite' wire from which the fibres are produced.

2   INSTRUMENTED INDENTATION

Small samples of fibre reinforced mortars were produced and indentation test surfaces of these samples were produced by polishing. The operating principles of the instrumented indentation technique are documented elsewhere[1]. For the instrumented indentation tests the authors used Nanoindenter XP manufactured by MTS with continuous stiffness measurement option. Total number of indents was 24 for Trefil Arbed fibres and 16 for Bekaert fibres. Total number of valid results was 19 and 14, respectively. The final indentation depth was set to be two microns. A standard Berkovich indenter has been used throughout the investigation and distance between indents was hundred microns (see Figures 2 and 3).

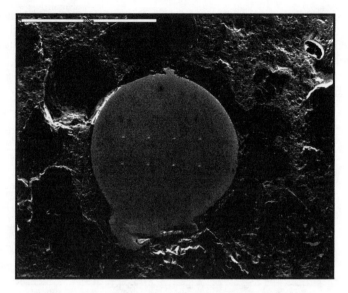

Figure 2 *Set of indents in a Bekaert fibre, scale bar corresponds to 500µm*

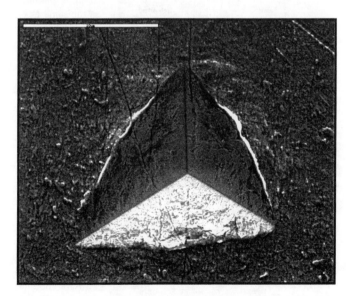

Figure 3 *Individual indent in a Bekaert fibre, scale bar corresponds to 10µm*

Several material characteristics can be derived from instrumented indentation technique. Apart form others these include indentation hardness and indentation modulus. Both hardness and modulus can be derived for (i) the unloading part of the load vs.

displacement diagram and (b) from continuous stiffness measurement during the loading part of the experiment. For the evaluation of the characteristics derived from the continuous stiffness measurement the authors decided to average the results over defined range from 100nm to 2000nm.

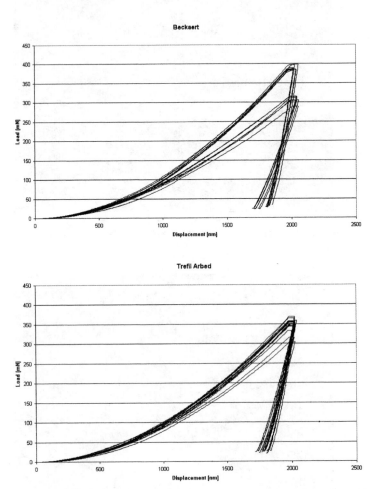

Figures 4 and 5   *Load vs displacement diagrams for Bekaert and TrefilArbed fibres*

Figures 4 and 5 show the load versus displacement diagrams of the instrumented indentation tests for Bekaert and TrefilArbed fibres, respectively. The summary of the micromechanical characteristics derived from these test is given in the Table 1.

Table 1 *Hardness and elastic moduli derived from instrumented indentation*

Fibre No.	Modulus Average Over Defined Range	Modulus From Unload	H Average Over Defined Range	Hardness From Unload
Tre fil Arbed	GPa	GPa	GPa	GPa
1	192.9	138.4	4.28	4.02
2	179.4	106.4	4.38	3.69
3	210.0	166.5	4.34	4.15
average	**195.0**	**138.7**	**4.33**	**3.97**
Bekaert	GPa	GPa	GPa	GPa
1	206.2	232.2	3.61	4.40
2	225.0	198.4	4.49	4.02
average	**215.6**	**215.3**	**4.05**	**4.21**
TA vs. B	90.4	64.4	06.9	94.1

3   TENSILE TEST

The authors intended to compare the characteristics obtained from instrumented indentation with comparable characteristics measured by 'classical' test methods on macroscale. For that the authors obtained from both respective producers 'infinite' wires, from which the steel fibre reinforcement is produced by cutting. The tensile tests were carried out on the segments taken from the 'infinite' wires. The results are summarised in Table 2.

Table 2 *Results of the tensile tests*

Segment No.	tensile strength [MPa]		modulus [GPa]	
	Bekaert	TrefilArbed	Bekaert	TrefilArbed
1	1389	1240	210.9	212.8
2	1385	1258	210.0	211.8
3	1351	1245	211.6	212.4
average	1375	1248	210.8	212.3
Standard deviation	17.05	7.59	0.65	0.41
variation coefficient [%]	1.24	0.61	0.31	0.19

4   COMPARISON OF RESULTS FROM INDENTATION AND TENSILE TESTS

In the Table 2, the test results obtained from nanoindentation and tensile test are compared. Concerning the comparison between the hardness from instrumented indentation and strength obtained from the tensile test, the authors were pleased to find that the ratio between the respective values for Bekaert and TrefilArbed fibres has been very similar, that is 94.1 vs. 90.8 per cent and that the indentation hardness vs. tensile strength ratio was very close to value of 3 in both cases.

The situation has been more complicated concerning the results of elastic moduli for the respective fibres. In this case three characteristics were compared - the elastic modulus from unloading, the elastic modulus derived from continuous stiffness measurement and averaged over range between 100nm and 2000nm, and third, modulus derived from tensile test. Concerning the values for Beakart fibres it is apparent from the Table No. 3 that all the values obtained are very close indeed. However, significantly lower values of elastic modulus from instrumented indentation were recorded for TrefilArbed fibres in the case of the indentation modulus derived from unloading part of the load-displacement diagram.

Table 3 *Summarised comparison of the instrumented indentation and tensile test results*

		Bekaert	Trefil Arbed	Ratio
Hardness From Unload	GPa	4.21	3.96	94.1
Tensile Strength	GPa	1.37	1.24	90.8
Hardness/Strength		3.07	3.18	

		Bekaert	Trefil Arbed
Modulus Over Defined Range	GPa	215.6	194.9
Modulus From Unload	GPa	215.4	138.7
Modulus From Tensile Test	GPa	210.8	212.3

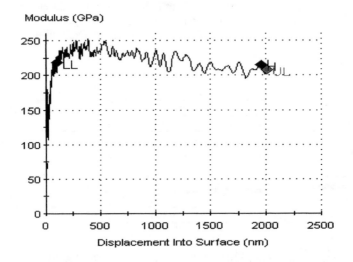

Figure 6 *Modulus derived from continuous stiffness measurements vs. displacement into surface for typical indentation test performed on Bekaert fibres*

The reason for this discrepancy can be found, when the closer attention is paid to the diagrams of indentation modulus versus 'displacement into surface' for both types of fibres. In Figure 6, typical modulus vs. displacement diagrams for Bekaert fibre is shown while in Figure 7 presents the respective diagram for Trefil Arbed fibre. Both curves exhibit the initial increase of elastic modulus on the first approximately 100nm which is likely to be due to surface roughness (i.e. closing the asperities and getting full contact between the indenter and the surface). However, while the Bekaert modulus remains approximately constant over the displacement, the modulus of Trefil Arbed fibres shows significant decrease with increase in the displacement.

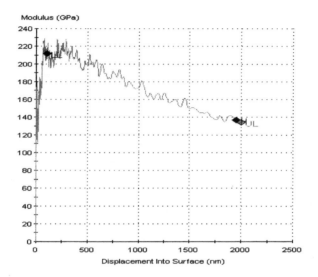

Figure 7 *Modulus derived from continuous stiffness measurements vs. displacement into surface for typical indentation test performed on Trefil Arbed fibres*

5   CONCLUSIONS

Despite the above mentioned problems with elastic modulus of one of the type of fibres, the first conclusion is, that the scatter of the result on individual fibres nanomechanical testing has been quite low which indicates the ability to use this technique for quality control of steel fibre reinforcement. The authors draw this conclusion despite being aware that steel is - in comparison to other building materials – a material fairly homogeneous for which the scatter of results was expected to be low.

The authors would like to highlight the pilot character project and there is naturally an option, which is even more appropriate for use of instrumented indentation technique than testing of fibres of 500μm in diameter. By utilising the sub-micrometre positioning accuracy of instrumented indentation facilities the technique could be even more appropriate for quality control of microfibrous reinforcement of cementitious materials.

Acknowledgement

The authors would like to express sincere thanks to Dr Pavel Trtik and Prof Peter J. M. Bartos for assistance with the instrumented indentation part of the experiment.

The authors would like to acknowledge Ministry of Education of the Czech Republic for financial support of the testing programme via research grant MSM 210000004 "Experimental Investigation of Building Construction Materials and Technologies".

References

1   W. C. Oliver and G. M. Pharr, 'An improved technique for determining hardness and elastic modulus using load and displacement sensing indentation experiments', *Journal of Materials Research Society,* 1992, **7**, No 6, pp. 1564-1583

Subject Index